Handbook of Cartography

Handbook of Cartography

Edited by **Marina De Lima**

R **C**ALLISTO **R**EFERENCE

New York

Published by Callisto Reference,
106 Park Avenue, Suite 200,
New York, NY 10016, USA
www.callistoreference.com

Handbook of Cartography
Edited by Marina De Lima

International Standard Book Number: 978-1-63239-377-7 (Hardback)

Printed in the United States of America.

Contents

Preface

Every book is a source of knowledge and this one is no exception. The idea that led to the conceptualization of this book was the fact that the world is advancing rapidly; which makes it crucial to document the progress in every field. I am aware that a lot of data is already available, yet, there is a lot more to learn. Hence, I accepted the responsibility of editing this book and contributing my knowledge to the community.

The space where natural systems and social systems interact is called terrestrial space. Cartography is a means to understand these intricate systems, their interaction and evolution. Hence, cartography has acquired a significant place in the modern scenario. This book provides an overview of the activities which state the significance of cartography in the perception and organization of the territory into distinct areas. Studying the past or understanding the present, cartography has always been a tool to explore all themes of knowledge.

While editing this book, I had multiple visions for it. Then I finally narrowed down to make every chapter a sole standing text explaining a particular topic, so that they can be used independently. However, the umbrella subject sinews them into a common theme. This makes the book a unique platform of knowledge.

I would like to give the major credit of this book to the experts from every corner of the world, who took the time to share their expertise with us. Also, I owe the completion of this book to the never-ending support of my family, who supported me throughout the project.

Editor

Mathematical Analysis in Cartography by Means of Computer Algebra System

Shao-Feng Bian and Hou-Pu Li

Additional information is available at the end of the chapter

1. Introduction

Theory of map projections is a branch of cartography studying the ways of projecting the curved surface of the earth and other heavenly bodies into the plane, and it is often called mathematical cartography. There are many fussy symbolic problems to be dealt with in map projections, such as power series expansions of elliptical functions, high order differential of transcendental functions, elliptical integrals and the operation of complex numbers. Many famous cartographers such as Adams (1921), Snyder (1987), Yang (1989, 2000) have made great efforts to solve these problems. Due to historical condition limitation, there were no advanced computer algebra systems at that time, so they had to dispose of these problems by hand, which had often required a paper and a pen. Some derivations and computations were however long and labor intensive such that one gave up midway. Briefly reviewing the existing methods, one will find that these problems were not perfectly and ideally solved yet. Formulas derived by hand often have quite complex and prolix forms, and their orders could not be very high. The most serious problem is that some higher terms of the formulas are erroneous because of the adopted approximate disposal.

With the advent of computers, the paper and pen approach is slowly being replaced by software developed to undertake symbolic derivations tasks. Specially, where symbolic rather than numerical solutions are desired, this software normally comes in handy. Software which performs symbolic computations is called computer algebra system. Nowadays, computer algebra systems like Maple, Mathcad, and Mathematica are widely used by scientists and engineers in different fields(Awang, 2005; Bian, 2006). By means of computer algebra system Mathematica, we have already solved many complicated mathematical problems in special fields of cartography in the past few years. Our research results indicate that the derivation efficiency can be significantly improved and formulas impossible to be obtained by hand can be easily derived with the help of Mathematica, which renovates the traditional analysis methods and enriches the mathematical theory basis of cartography to a certain extent.

The main contents and research results presented in this chapter are organized as follows. In Section II, we discuss the direct transformations from geodetic latitude to three kinds of auxiliary latitudes often used in cartography, and the direct transformations from these auxiliary latitudes to geodetic latitude are studied in Section III. In Section IV, the direct expansions of transformations between meridian arc, isometric latitude, and authalic functions are derived. In Section V, we discuss the non-iterative expressions of the forward and inverse Gauss projections by complex numbers. Finally in Section VI, we make a brief summary. It is assumed that the readers are somewhat conversant with Mathematica and its syntax.

2. The forward expansions of the rectifying, conformal and authalic latitudes

Cartographers prefer to adopt sphere as a basis of the map projection for convenience since calculation on the ellipsoid are significantly more complex than on the sphere. Formulas for the spherical form of a given map projection may be adapted for use with the ellipsoid by substitution of one of various "auxiliary latitudes" in place of the geodetic latitude. In using them, the ellipsoidal earth is, in effect, transformed to a sphere under certain restraints such as conformality or equal area, and the sphere is then projected onto a plane (Snyder, 1987). If the proper auxiliary latitudes are chosen, the sphere may have either true areas, true distances in certain directions, or conformality, relative to the ellipsoid. Spherical map projection formulas may then be used for the ellipsoid solely with the substitution of the appropriate auxiliary latitudes.

The rectifying, conformal and authalic latitudes are often used as auxiliary ones in cartography. The direct transformations form geodetic latitude to these auxiliary ones are expressed as transcendental functions or non-integrable ones. Adams (1921), Yang (1989, 2000) had derived forward expansions of these auxiliary latitudes form geodetic one through complicated formulation. Due to historical condition limitation, the derivation processes were done by hand and orders of these expansions could not be very high. Due to these reasons, the forward expansions for these auxiliary latitudes are reformulated by means of Mathematica. Readers will see that new expansions are expressed in a power series of the eccentricity of the reference ellipsoid e and extended up to tenth-order terms of e. The expansion processes become much easier under the system Mathematica.

2.1. The forward expansion of the rectifying latitude

The meridian arc from the equator $B = 0$ to B is

$$X = a(1 - e^2) \int_0^B (1 - e^2 \sin^2 B)^{-3/2} dB \tag{1}$$

where X is the meridian arc; B is the geodetic latitude; a is the semi-major axis of the reference ellipsoid;

(1) is an elliptic integral of the second kind and there is no analytical solution. Expanding the integrand by binomial theorem and itntegrating it item by item yield:

$$X = a(1 - e^2)(K_0 B + K_2 \sin 2B + K_4 \sin 4B + K_6 \sin 6B + K_8 \sin 8B + K_{10} \sin 10B) \qquad (2)$$

where

$$
\left\{
\begin{aligned}
K_0 &= 1 + \frac{3}{4}e^2 + \frac{45}{64}e^4 + \frac{175}{256}e^6 + \frac{11025}{16384}e^8 + \frac{43659}{65536}e^{10} \\
K_2 &= -\frac{3}{8}e^2 - \frac{15}{32}e^4 - \frac{525}{1024}e^6 - \frac{2205}{4096}e^8 - \frac{72765}{131072}e^{10} \\
K_4 &= \frac{15}{256}e^4 + \frac{105}{1024}e^6 + \frac{2205}{16384}e^8 + \frac{10395}{65536}e^{10} \\
K_6 &= -\frac{35}{3072}e^6 - \frac{105}{4096}e^8 - \frac{10395}{262144}e^{10} \\
K_8 &= \frac{315}{131072}e^8 + \frac{3465}{524288}e^{10} \\
K_{10} &= -\frac{693}{1310720}e^{10}
\end{aligned}
\right.
\qquad (3)
$$

The rectifying latitude ψ is defined as

$$\psi = \frac{X}{X(\frac{\pi}{2})} \cdot \frac{\pi}{2} \qquad (4)$$

Inserting (2) into (4) yields

$$\psi = B + \alpha_2 \sin 2B + \alpha_4 \sin 4B + \alpha_6 \sin 6B + \alpha_8 \sin 8B + \alpha_{10} \sin 10B \qquad (5)$$

where

$$
\left\{
\begin{aligned}
\alpha_2 &= K_2 / K_0 \\
\alpha_4 &= K_4 / K_0 \\
\alpha_6 &= K_6 / K_0 \\
\alpha_8 &= K_8 / K_0 \\
\alpha_{10} &= K_{10} / K_0
\end{aligned}
\right.
\qquad (6)
$$

Yang (1989, 2000) gave an expansion similar to (5) but expanded ψ up to $\sin 8B$. For simplicity and computing efficiency, it is better to expand (6) into a power series of the eccentricity. This process is easily done by means of Mathematica. As a result, (6) becomes:

$$
\begin{cases}
\alpha_2 = -\dfrac{3}{8}e^2 - \dfrac{3}{16}e^4 - \dfrac{111}{1024}e^6 - \dfrac{141}{2048}e^8 - \dfrac{1533}{32768}e^{10} \\[2mm]
\alpha_4 = \dfrac{15}{256}e^4 + \dfrac{15}{256}e^6 + \dfrac{405}{8192}e^8 + \dfrac{165}{4096}e^{10} \\[2mm]
\alpha_6 = -\dfrac{35}{3072}e^6 - \dfrac{35}{2048}e^8 - \dfrac{4935}{262144}e^{10} \\[2mm]
\alpha_8 = \dfrac{315}{131072}e^8 + \dfrac{315}{65536}e^{10} \\[2mm]
\alpha_{10} = -\dfrac{693}{1310720}e^{10}
\end{cases}
\tag{7}
$$

2.2 The forward expansion of the conformal latitude

Omitting the derivation process, the explicit expression for the isometric latitude q is

$$
q = \int_0^B \frac{1-e^2}{(1-e^2\sin^2 B)\cos B}dB = \ln\left[\tan\left(\frac{\pi}{4}+\frac{B}{2}\right)\left(\frac{1-e\sin B}{1+e\sin B}\right)^{e/2}\right]
\tag{8}
$$
$$
= \operatorname{arctan} h(\sin B) - e \cdot \operatorname{arctan} h(e\sin B)
$$

For the sphere, putting $e = 0$ and rewriting the geodetic latitude as the conformal one φ, (8) becomes

$$
q = \ln\left[\tan\left(\frac{\pi}{4}+\frac{\varphi}{2}\right)\right] = \operatorname{arctan} h(\sin\varphi)
\tag{9}
$$

Comparing (9) with (8) leads to

$$
\tan\left(\frac{\pi}{4}+\frac{\varphi}{2}\right) = \tan\left(\frac{\pi}{4}+\frac{B}{2}\right)\left(\frac{1-e\sin B}{1+e\sin B}\right)^{e/2}
\tag{10}
$$

Therefore, it holds

$$
\varphi = 2\operatorname{arctan}\left[\tan\left(\frac{\pi}{4}+\frac{B}{2}\right)\left(\frac{1-e\sin B}{1+e\sin B}\right)^{e/2}\right] - \frac{\pi}{2}
\tag{11}
$$

Since the eccentricity is small, the conformal latitude is close to the geodetic one. Though (11) is an analytical solution of φ, (11) is usually expanded into a power series of the eccentricity

$$
\varphi(B,e) = \varphi(B,0) + \frac{\partial\varphi}{\partial e}\bigg|_{e=0} e + \frac{1}{2!}\frac{\partial^2\varphi}{\partial e^2}\bigg|_{e=0} e^2 + \frac{1}{3!}\frac{\partial^3\varphi}{\partial e^3}\bigg|_{e=0} e^3 + \cdots \frac{1}{9!}\frac{\partial^9\varphi}{\partial e^9}\bigg|_{e=0}
$$
$$
e^9 + \frac{1}{10!}\frac{\partial^{10}\varphi}{\partial e^{10}}\bigg|_{e=0} e^{10} + \cdots
\tag{12}
$$

as the conventional usage in mathematical cartography.

Through the tedious expansion process, Yang (1989, 2000) gave a power series of the eccentricity e for the conformal latitude φ as

$$\varphi = B + \beta_2 \sin 2B + \beta_4 \sin 4B + \beta_6 \sin 6B + \beta_8 \sin 8B \tag{13}$$

Due to that (11) is a very complicated transcendental function, the coefficients β_2, β_4, β_6, β_8 in (13) derived by hand are only expanded to eighth-order terms of e. They are not accurate as expected and there are some mistakes in the eighth-order terms of e.

In fact, Mathematica works perfectly in solving derivatives of any complicated functions. By means of Mathematica, the new derived forward expansion expanded to tenth-order terms of e reads

$$\varphi = B + \beta_2 \sin 2B + \beta_4 \sin 4B + \beta_6 \sin 6B + \beta_8 \sin 8B + \beta_{10} \sin 10B \tag{14}$$

The derived coefficients in (13) and (14) are listed in Table 1 for comparison.

Coefficients derived by Yang(1989, 2000)	Coefficients derived by the author
$\beta_2 = -\dfrac{1}{2}e^2 - \dfrac{5}{24}e^4 - \dfrac{3}{32}e^6 - \dfrac{1399}{53760}e^8$ $\beta_4 = \dfrac{5}{48}e^4 + \dfrac{7}{80}e^6 + \dfrac{689}{17920}e^8$ $\beta_6 = -\dfrac{13}{480}e^6 - \dfrac{1363}{53760}e^8$ $\beta_8 = \dfrac{677}{17520}e^8$	$\beta_2 = -\dfrac{1}{2}e^2 - \dfrac{5}{24}e^4 - \dfrac{3}{32}e^6 - \dfrac{281}{5760}e^8 - \dfrac{7}{240}e^{10}$ $\beta_4 = \dfrac{5}{48}e^4 + \dfrac{7}{80}e^6 + \dfrac{697}{11520}e^8 + \dfrac{93}{2240}e^{10}$ $\beta_6 = -\dfrac{13}{480}e^6 - \dfrac{461}{13440}e^8 - \dfrac{1693}{53760}e^{10}$ $\beta_8 = \dfrac{1237}{161280}e^8 + \dfrac{131}{10080}e^{10}$ $\beta_{10} = -\dfrac{367}{161280}e^{10}$

Table 1. The comparison of coefficients of the forward expansion of conformal latitude derived by Yang (1989, 2000) and the author

Table 1 shows that the eighth order terms of e in coefficients given by Yang(1989, 2000) are erroneous.

2.3. The forward expansion of the authalic latitude

From the knowledge of mapping projection theory, the area of a section of a lune with a width of a unit interval of longitude F is

$$F = a^2(1-e^2)\int_0^B \frac{\cos B}{(1-e^2 \sin^2 B)^2} dB = a^2(1-e^2)\left[\frac{\sin B}{2(1-e^2 \sin^2 B)} + \frac{1}{4e}\ln\frac{1+e\sin B}{1-e\sin B}\right] \tag{15}$$

where F is also called authalic latitude function.

Denote

$$A = \frac{1}{2(1-e^2)} + \frac{1}{4e}\ln\frac{1+e}{1-e} \tag{16}$$

Suppose that there is an imaginary sphere with a radius

$$R = a\sqrt{(1-e^2)A} \tag{17}$$

and whose area from the spherical equator $\vartheta = 0$ to spherical latitude ϑ with a width of a unit interval of longitude is equal to the ellipsoidal area F, it holds

$$R^2\sin\vartheta = a^2(1-e^2)A\sin\vartheta = F \tag{18}$$

Therefore, it yields

$$\vartheta = \arcsin\left[\frac{1}{A}\left(\frac{\sin B}{2(1-e^2\sin^2 B)} + \frac{1}{4e}\ln\frac{1+e\sin B}{1-e\sin B}\right)\right] \tag{19}$$

where ϑ is authalic latitude. Yang(1989, 2000) gave its series expansion as

$$\vartheta = B + \gamma_2\sin 2B + \gamma_4\sin 4B + \gamma_6\sin 6B + \gamma_8\sin 8B \tag{20}$$

(19) is a complicated transcendental function. It is almost impossible to derive its eighth-order derivate by hand. There are some mistakes in the high order terms of coefficients γ_2, $\gamma_4, \gamma_6, \gamma_8$. The new derived forward expansion expanded to tenth-order terms of e by means of Mathematica reads

$$\vartheta = B + \gamma_2\sin 2B + \gamma_4\sin 4B + \gamma_6\sin 6B + \gamma_8\sin 8B + \gamma_{10}\sin 10B \tag{21}$$

The derived coefficients in (20) and (21) are listed in Table 2 for comparison.

Coefficients derived by Yang(1989, 2000)	Coefficients derived by the author
$\gamma_2 = -\frac{1}{3}e^2 - \frac{31}{180}e^4 - \frac{59}{560}e^6 - \frac{126853}{518400}e^8$	$\gamma_2 = -\frac{1}{3}e^2 - \frac{31}{180}e^4 - \frac{59}{560}e^6 - \frac{42811}{604800}e^8 - \frac{605399}{11975040}e^{10}$
$\gamma_4 = \frac{17}{360}e^4 + \frac{61}{1260}e^6 + \frac{3622447}{94089600}e^8$	$\gamma_4 = \frac{17}{360}e^4 + \frac{61}{1260}e^6 + \frac{76969}{1814400}e^8 + \frac{215431}{5987520}e^{10}$
$\gamma_6 = -\frac{383}{43560}e^6 - \frac{6688039}{658627200}e^8$	$\gamma_6 = -\frac{383}{45360}e^6 - \frac{3347}{259200}e^8 - \frac{1751791}{119750400}e^{10}$
$\gamma_8 = -\frac{27787}{23522400}e^8$	$\gamma_8 = \frac{6007}{3628800}e^8 + \frac{201293}{59875200}e^{10}$
	$\gamma_{10} = -\frac{5839}{17107200}e^{10}$

Table 2. The comparison of coefficients of the forward expansion of conformal latitude derived by Yang (1989, 2000) and the author

Table 2 shows that the eighth-orders terms of e in coefficients given by Yang(1989, 2000) are erroneous.

2.4. Accuracies of the forward expansions

In order to validate the exactness and reliability of the forward expansions of rectifying, conformal and authalic latitudes derived by the author, one has examined their accuracies choosing the CGCS2000 (China Geodetic Coordinate System 2000) reference ellipsoid with parameters $a = 6378137\,\mathrm{m}$, $f = 1 / 298.257222101$ (Chen, 2008; Yang, 2009), where f is the flattening. The accuracies of the forward expansions derived by Yang (1989, 2000) are also examined for comparison. The results show that the accuracy of the forward expansion of rectifying latitude derived by Yang (1989, 2000) is higher than 10^{-5}", while the accuracy of the forward expansion (5) derived by the author is higher than 10^{-7}". The accuracies of the forward expansion of conformal and authalic latitudes derived by Yang (1989, 2000) are higher than 10^{-4}", while the accuracies of the forward expansions derived by the author are higher than 10^{-8}". The accuracies of forward expansions derived by the author are improved by 2~4 orders of magnitude compared to forward expansions derived by Yang (1989, 2000).

3. The inverse expansions of rectifying, conformal and authalic latitudes

The inverse expansions of these auxiliary latitudes are much more difficult to derive than their forward ones. In this case, the differential equations are usually expressed as implicit functions of the geodetic latitude. There are neither any analytical solutions nor obvious expansions. For the inverse cases, to find geodetic latitude from auxiliary ones, one usually adopts iterative methods based on the forward expansions or an approximate series form. Yang (1989, 2000) had given the direct expansions of the inverse transformation by means of Lagrange series method, but their coefficients are expressed as polynomials of coefficients of the forward expansions, which are not convenient for practical use. Adams (1921) expressed the coefficients of inverse expansions as a power series of the eccentricity e by hand, but expanded them up to eighth-order terms of e at most. Due to these reasons, new inverse expansions are derived using the power series method by means of Mathematica. Their coefficients are uniformly expressed as a power series of the eccentricity and extended up to tenth-order terms of e.

3.1. The inverse expansions using the power series method

The processes to derive the inverse expansions using the power series method are as follows:

1. To obtain their various order derivatives in terms of the chain rule of implicit differentation;
2. To compute the coefficients of their power series expansions;
3. To integrate these series item by item and yield the final inverse expansions.

One can image that these procedures are quite complicated. Mathematica output shows that the expression of the sixth order derivative is up to 40 pages long! Therefore, it is unimaginable to derive the so long expression by hand. These procedures, however, will become much easier and be significantly simplified by means of Mathematica. As a result, the more simple and accurate expansions yield.

3.1.1. The inverse expansion of the rectifying Latitude

Differentiation to the both sides of (1) yields

$$\frac{dX}{dB} = \frac{a(1-e^2)}{(1-e^2\sin^2 B)^{3/2}} \tag{22}$$

From (4) and (2), one knows

$$X = a(1-e^2)K_0\psi \tag{23}$$

Inserting (23) into (22) yields

$$\frac{dB}{K_0 d\psi} = (1-e^2\sin^2 B)^{3/2} \tag{24}$$

To expand (24) into a power series of $\sin\psi$, we introduce the following new variable

$$t = \sin\psi \tag{25}$$

therefore

$$\frac{d\psi}{dt} = \frac{1}{\cos\psi} \tag{26}$$

and then denote

$$f(t) = \frac{dB}{K_0 d\psi} = (1-e^2\sin^2 B)^{3/2} \tag{27}$$

Making use of the chain rule of implicit differentiation

$$f'_t = \frac{df}{dB}\frac{dB}{d\psi}\frac{d\psi}{dt} + \frac{df}{d\psi}\frac{d\psi}{dt}, \quad f''_t = \frac{df'}{dB}\frac{dB}{d\psi}\frac{d\psi}{dt} + \frac{df'}{d\psi}\frac{d\psi}{dt}, \quad \cdots\cdots \tag{28}$$

It is easy to expand (27) into a power series of $\sin\psi$

$$f(t) = f(0) + f'_t(0)t + \frac{1}{2!}f''_t(0)t^2 + \frac{1}{3!}f'''_t(0)t^3 + \cdots + \frac{1}{10!}f_t^{(10)}(0)t^{10} + \cdots \tag{29}$$

Omitting the detailed procedure, one arrives at

$$\frac{dB}{K_0 d\psi} = 1 + A_2 \sin^2 \psi + A_4 \sin^4 \psi + A_6 \sin^6 \psi + A_8 \sin^8 \psi + A_{10} \sin^{10} \psi \tag{30}$$

where

$$\begin{cases} A_2 = -\dfrac{3}{2}e^2 - \dfrac{27}{8}e^4 - \dfrac{729}{128}e^6 - \dfrac{4329}{512}e^8 - \dfrac{381645}{32768}e^{10} \\[2mm] A_4 = \dfrac{21}{8}e^4 + \dfrac{621}{64}e^6 + \dfrac{11987}{512}e^8 + \dfrac{757215}{16384}e^{10} \\[2mm] A_6 = -\dfrac{151}{32}e^6 - \dfrac{775}{32}e^8 - \dfrac{621445}{8192}e^{10} \\[2mm] A_8 = \dfrac{1097}{128}e^8 + \dfrac{57607}{1024}e^{10} \\[2mm] A_{10} = -\dfrac{8011}{512}e^{10} \end{cases} \tag{31}$$

Multiplying K_0 and integrating at the both sides of (30) give the inverse expansion of rectifying latitude as

$$B = \psi + a_2 \sin 2\psi + a_4 \sin 4\psi + a_6 \sin 6\psi + a_8 \sin 8\psi + a_{10} \sin 10\psi \tag{32}$$

where

$$\begin{cases} a_2 = \dfrac{3}{8}e^2 + \dfrac{3}{16}e^4 + \dfrac{213}{2048}e^6 + \dfrac{255}{4096}e^8 + \dfrac{20861}{524288}e^{10} \\[2mm] a_4 = \dfrac{21}{256}e^4 + \dfrac{21}{256}e^6 + \dfrac{533}{8192}e^8 + \dfrac{197}{4096}e^{10} \\[2mm] a_6 = \dfrac{151}{6144}e^6 + \dfrac{151}{4096}e^8 + \dfrac{5019}{131072}e^{10} \\[2mm] a_8 = \dfrac{1097}{131072}e^8 + \dfrac{1097}{65536}e^{10} \\[2mm] a_{10} = \dfrac{8011}{2621440}e^{10} \end{cases} \tag{33}$$

3.1.2. The inverse expansion of the conformal latitude

Differentiating the both sides of (10) yields

$$\frac{d\varphi}{\cos\varphi} = \frac{1 - e^2}{(1 - e^2 \sin^2 B)\cos B} dB \tag{34}$$

Therefore, it holds

$$\frac{dB}{d\varphi} = \frac{(1 - e^2 \sin^2 B)\cos B}{(1 - e^2)\cos\varphi} \tag{35}$$

For the same reason, we introduce the following new variable

$$t = \sin\varphi \tag{36}$$

and then denote

$$f(t) = \frac{dB}{d\varphi} = \frac{(1 - e^2 \sin^2 B)\cos B}{(1 - e^2)\cos\varphi} \tag{37}$$

Using the same procedure as described in the former section, one arrives at

$$\frac{dB}{d\varphi} = \frac{1}{1 - e^2} + B_2 \sin^2\varphi + B_4 \sin^4\varphi + B_6 \sin^6\varphi + B_8 \sin^8\varphi + B_{10}\sin^{10}\varphi \tag{38}$$

where

$$\begin{cases} B_2 = -2e^2 - \dfrac{11}{2}e^4 - \dfrac{21}{2}e^6 - 17e^8 - 25e^{10} \\[2mm] B_4 = \dfrac{14}{3}e^4 + \dfrac{62}{3}e^6 + \dfrac{1369}{24}e^8 + \dfrac{3005}{24}e^{10} \\[2mm] B_6 = -\dfrac{56}{5}e^6 - \dfrac{614}{9}e^8 - \dfrac{4909}{20}e^{10} \\[2mm] B_8 = \dfrac{8558}{315}e^8 + \dfrac{7367}{35}e^{10} \\[2mm] B_{10} = -\dfrac{4174}{63}e^{10} \end{cases} \tag{39}$$

Integrating the both sides of (38) gives the inverse expansion of conformal latitude as

$$B = \varphi + b_2 \sin 2\varphi + b_4 \sin 4\varphi + b_6 \sin 6\varphi + b_8 \sin 8\varphi + b_{10}\sin 10\varphi \tag{40}$$

where

$$\begin{cases} b_2 = \dfrac{1}{2}e^2 + \dfrac{5}{24}e^4 + \dfrac{1}{12}e^6 + \dfrac{13}{360}e^8 + \dfrac{3}{160}e^{10} \\[2mm] b_4 = \dfrac{7}{48}e^4 + \dfrac{29}{240}e^6 + \dfrac{811}{11520}e^8 + \dfrac{81}{2240}e^{10} \\[2mm] b_6 = \dfrac{7}{120}e^6 + \dfrac{81}{1120}e^8 + \dfrac{3029}{53760}e^{10} \\[2mm] b_8 = \dfrac{4279}{161280}e^8 + \dfrac{883}{20160}e^{10} \\[2mm] b_{10} = \dfrac{2087}{161280}e^{10} \end{cases} \tag{41}$$

3.1.3 The inverse expansion of the authalic latitude

Inserting (18) into (15) yields

$$A \sin \vartheta = \int_0^B \frac{\cos B}{(1 - e^2 \sin^2 B)^2} dB \qquad (42)$$

Differentiating the both sides of (42) yields

$$\frac{dB}{d\vartheta} = \frac{A(1 - e^2 \sin^2 B)^2 \cos \vartheta}{\cos B} \qquad (43)$$

For the same reason, we introduce the folllowing new variable

$$t = \sin \vartheta \qquad (44)$$

and then denote

$$f(t) = \frac{dB}{d\vartheta} = \frac{A(1 - e^2 \sin^2 B)^2 \cos \vartheta}{\cos B} \qquad (45)$$

(45) can be expanded into a power series of $\sin \vartheta$. Using the chain rule of implicit function differentiation, one similarly arrives at

$$f(t) = \frac{dB}{d\vartheta} = A + C_2 \sin^2 \vartheta + C_4 \sin^4 \vartheta + C_6 \sin^6 \vartheta + C_8 \sin^8 \vartheta + C_{10} \sin^{10} \vartheta \qquad (46)$$

where

$$
\begin{cases}
C_2 = -\dfrac{4}{3}e^2 - \dfrac{41}{15}e^4 - \dfrac{4108}{945}e^6 - \dfrac{58427}{9450}e^8 - \dfrac{28547}{3465}e^{10} \\[2mm]
C_4 = \dfrac{92}{45}e^4 + \dfrac{6574}{945}e^6 + \dfrac{223469}{14175}e^8 + \dfrac{2768558}{93555}e^{10} \\[2mm]
C_6 = -\dfrac{3044}{945}e^6 - \dfrac{28901}{1890}e^8 - \dfrac{21018157}{467775}e^{10} \\[2mm]
C_8 = \dfrac{24236}{4725}e^8 + \dfrac{2086784}{66825}e^{10} \\[2mm]
C_{10} = -\dfrac{768272}{93555}e^{10}
\end{cases}
\qquad (47)
$$

To get the inverse expansion of the authalic latitude, one integrates (46) and arrives at

$$B = \vartheta + c_2 \sin 2\vartheta + c_4 \sin 4\vartheta + c_6 \sin 6\vartheta + c_8 \sin 8\vartheta + c_{10} \sin 10\vartheta \qquad (48)$$

where

$$
\begin{cases}
c_2 = \dfrac{1}{3}e^2 + \dfrac{31}{180}e^4 + \dfrac{517}{5040}e^6 + \dfrac{120389}{181400}e^8 + \dfrac{1362253}{29937600}e^{10} \\[2mm]
c_4 = \dfrac{23}{360}e^4 + \dfrac{251}{3780}e^6 + \dfrac{102287}{1814400}e^8 + \dfrac{450739}{997920}e^{10} \\[2mm]
c_6 = \dfrac{761}{45360}e^6 + \dfrac{47561}{1814400}e^8 + \dfrac{434501}{14968800}e^{10} \\[2mm]
c_8 = \dfrac{6059}{1209600}e^8 + \dfrac{625511}{59875200}e^{10} \\[2mm]
c_{10} = \dfrac{48017}{29937600}e^{10}
\end{cases}
\tag{49}
$$

3.2. The inverse expansions using the Hermite interpolation method

In mathematical analysis, interpolation with functional values and their derivative values is called Hermite interpolation. The processes to derive the inverse expansions using this method are as follows:

1. To suppose the inverse expansions are expressed in a series of the sines of the multiple arcs with coefficients to be determined;
2. To compute the functional values and their derivative values at specific points;
3. To solve linear equations according to interpolation constraints and obtain the coefficients.

The detailed derivation processes are given by Li (2008, 2010). Confined to the length of the chapter, they are omitted. Comparing the results derived by this method with those in 3.1, one will find that they are consistent with each other even though they are formulated in different ways. This fact substantiates the correctness of the derived formulas.

3.3. The inverse expansions using the Lagrange's theorem method

We wish to investigate the inversion of an equation such as

$$
y = x + f(x) \tag{50}
$$

with $|f(x)| \ll |x|$ and $y \approx x$. The Lagrange's theorem states that in a suitable domain the solution of (50) is

$$
x = y + \sum_{n=1}^{\infty} \frac{(-1)^n}{n!} \frac{d^{n-1}}{dy^{n-1}} [f(y)]^n \tag{51}
$$

The proof of this theorem is given by Whittaker (1902) and Peter (2008).

The processes to derive the inverse expansions using the Lagrange series method are as follows:

1. To apply the Lagrange theorem to a trigonometric series;
2. To write the inverse expansions of the rectifying, conformal and authalic latitude;
3. To express the coefficients of the inverse expansions as a power series of the eccentricity.

The detailed derivation processes are given by Li (2010). Confined to the length of the chapter, they are also omitted. Comparing the results derived by this method with those in 3.1 and 3.2, one will find that they are all consistent with each other even though they are also formulated in different ways. This fact substantiates the correctness of the derived formulas, too.

3.4. Accuracies of the inverse expansions

The accuracies of the inverse expansions derived by Yang (1989, 2000) and the author has been examined choosing the CGCS2000 reference ellipsoid.

The results show that the accuracy of the inverse expansion of rectifying latitude is higher than $10^{-5}''$, while the accuracy of the inverse expansion (32) derived by the author is higher than $10^{-7}''$. The accuracies of the inverse expansion of conformal and authalic latitudes derived by Yang (1989, 2000) are higher than $10^{-4}''$, while the accuracies of the inverse expansions derived by the author are higher than $10^{-8}''$. The accuracies of inverse expansions derived by the author are improved by 2~4 orders of magnitude compared to those derived by Yang (1989, 2000).

4. The direct expansions of transformations between meridian arc, isometric latitude and authalic latitude function

The meridian arc, isometric latitude and authalic latitude function are functions of rectifying, conformal and authalic latitudes correspondingly. The transformations between the three variables are indirectly realized by computing the geodetic latitude in the past literatures such as Yang (1989, 2000), Snyder (1987). The computation processes are tedious and time-consuming. In order to simplify the computation processes and improve the computation efficiency, the direct expansions of transformations between meridian arc, isometric latitude and authalic latitude function are comprehensively derived by means of Mathematica.

4.1. The direct expansions of transformations between meridian arc and isometric latitude

4.1.1. The direct expansion of the transformation from meridian arc to isometric latitude

Inserting the known meridian arc X into (23) yields the rectifying latitude ψ. Using the inverse expansion of the rectifying latitude (32) and the forward expansion of the conformal latitude (14), one obtains the conformal latitude φ. Inserting it into (9) yields the isometric latitude q. The whole formulas are as follows:

$$
\begin{cases}
\psi = \dfrac{X}{a(1-e^2)K_0} \\
B = \psi + a_2 \sin 2\psi + a_4 \sin 4\psi + a_6 \sin 6\psi + a_8 \sin 8\psi + a_{10} \sin 10\psi \\
\varphi = B + \beta_2 \sin 2B + \beta_4 \sin 4B + \beta_6 \sin 6B + \beta_8 \sin 8B + \beta_{10} \sin 10B \\
q = \operatorname{arctanh}(\sin \varphi)
\end{cases}
\tag{52}
$$

Since the coefficients a_{2i}, β_{2i} ($i = 1, 2, \cdots 5$) are expressed in a power series of the eccentricity, one could expand q as a power series of the eccentricity e at $e = 0$ in order to obtain the direct expansion of the transformation from X to q. It is hardly completed by hand, but could be easily realized by means of Mathematica. Omitting the main operations by means of Mathematica yields the direct expansion of the transformation from meridian arc to isometric latitude

$$
\begin{cases}
\psi = \dfrac{X}{a(1-e^2)K_0} \\
q = \operatorname{arctanh}(\sin \psi) + \xi_1 \sin \psi + \xi_3 \sin 3\psi + \xi_5 \sin 5\psi + \xi_7 \sin 7\psi + \xi_9 \sin 9\psi
\end{cases}
\tag{53}
$$

where

$$
\begin{cases}
\xi_1 = -\dfrac{1}{4}e^2 - \dfrac{1}{64}e^4 + \dfrac{1}{3072}e^6 + \dfrac{33}{16384}e^8 + \dfrac{2363}{1310720}e^{10} \\
\xi_3 = -\dfrac{1}{96}e^4 - \dfrac{13}{3072}e^6 - \dfrac{13}{8192}e^8 - \dfrac{1057}{1966080}e^{10} \\
\xi_5 = -\dfrac{11}{7680}e^6 - \dfrac{29}{24576}e^8 - \dfrac{2897}{3932160}e^{10} \\
\xi_7 = -\dfrac{25}{86016}e^8 - \dfrac{727}{1966080}e^{10} \\
\xi_9 = -\dfrac{53}{737280}e^{10}
\end{cases}
\tag{54}
$$

4.1.2. The direct expansion of the transformation from isometric latitude to meridian arc

The whole formulas for the transformation from isometric latitude to meridian arc are as follows:

$$
\begin{cases}
\varphi = \arcsin(\tanh q) \\
B = \varphi + b_2 \sin 2\varphi + b_4 \sin 4\varphi + b_6 \sin 6\varphi + b_8 \sin 8\varphi + b_{10} \sin 10\varphi \\
\psi = B + \alpha_2 \sin 2B + \alpha_4 \sin 4B + \alpha_6 \sin 6B + \alpha_8 \sin 8B + \alpha_{10} \sin 10B \\
X = a(1-e^2)K_0\psi
\end{cases}
\tag{55}
$$

Expanding X as a power series of the eccentricity e at $e = 0$ by means of Mathematica yields the direct expansion of the transformation from isometric latitude to meridian arc

$$
\begin{cases}
\varphi = \arcsin(\tanh q) \\
X = a\left(j_0\varphi + j_2\sin 2\varphi + j_4\sin 4\varphi + j_6\sin 6\varphi + j_8\sin 8\varphi + j_{10}\sin 10\varphi\right)
\end{cases}
\tag{56}
$$

where

$$
\begin{cases}
j_0 = 1 - \dfrac{1}{4}e^2 - \dfrac{3}{64}e^4 - \dfrac{5}{256}e^6 - \dfrac{175}{16384}e^8 - \dfrac{441}{65536}e^{10} \\[2mm]
j_2 = \dfrac{1}{8}e^2 - \dfrac{1}{96}e^4 - \dfrac{9}{1024}e^6 - \dfrac{901}{184320}e^8 - \dfrac{16381}{5898240}e^{10} \\[2mm]
j_4 = \dfrac{13}{768}e^4 + \dfrac{17}{5120}e^6 - \dfrac{311}{737280}e^8 - \dfrac{18931}{20643840}e^{10} \\[2mm]
j_6 = \dfrac{61}{15360}e^6 + \dfrac{899}{430080}e^8 + \dfrac{14977}{27525120}e^{10} \\[2mm]
j_8 = \dfrac{49561}{41287680}e^8 + \dfrac{175087}{165150720}e^{10} \\[2mm]
j_{10} = \dfrac{34729}{82575360}e^{10}
\end{cases}
\tag{57}
$$

4.2. The direct expansions of transformations between meridian arc and authalic latitude function

4.2.1. The direct expansion of the transformation from meridian arc to authalic latitude function

The whole formulas for the transformation from meridian arc to authalic latitude function are as follows:

$$
\begin{cases}
\psi = \dfrac{X}{a(1-e^2)K_0} \\[2mm]
B = \psi + a_2\sin 2\psi + a_4\sin 4\psi + a_6\sin 6\psi + a_8\sin 8\psi + a_{10}\sin 10\psi \\[1mm]
\vartheta = B + \gamma_2\sin 2B + \gamma_4\sin 4B + \gamma_6\sin 6B + \gamma_8\sin 8B + \gamma_{10}\sin 10B \\[1mm]
F = R^2\sin\vartheta
\end{cases}
\tag{58}
$$

Expanding F as a power series of the eccentricity e at $e = 0$ by means of Mathematica yields the direct expansion of the transformation from meridian arc to authalic latitude function

$$
\begin{cases}
\psi = \dfrac{X}{a(1-e^2)K_0} \\[2mm]
F = a^2\left(\varepsilon_1\sin\psi + \varepsilon_3\sin 3\psi + \varepsilon_5\sin 5\psi + \varepsilon_7\sin 7\psi + \varepsilon_9\sin 9\psi + \varepsilon_{11}\sin 11\psi\right)
\end{cases}
\tag{59}
$$

where

$$\begin{cases} \varepsilon_1 = 1 - \dfrac{5}{16}e^2 - \dfrac{17}{256}e^4 - \dfrac{121}{4096}e^6 - \dfrac{137}{8192}e^8 - \dfrac{1407}{131072}e^{10} \\[2mm] \varepsilon_3 = \dfrac{1}{48}e^2 + \dfrac{1}{384}e^4 - \dfrac{103}{196608}e^8 - \dfrac{1775}{3145728}e^{10} \\[2mm] \varepsilon_5 = \dfrac{3}{1280}e^4 + \dfrac{43}{30720}e^6 + \dfrac{17}{24576}e^8 + \dfrac{467}{1572864}e^{10} \\[2mm] \varepsilon_7 = \dfrac{37}{86016}e^6 + \dfrac{5}{10752}e^8 + \dfrac{563}{1572864}e^{10} \\[2mm] \varepsilon_9 = \dfrac{59}{589824}e^8 + \dfrac{1853}{11796480}e^{10} \\[2mm] \varepsilon_{11} = \dfrac{1543}{57671680}e^{10} \end{cases} \qquad (60)$$

4.2.2. The direct expansion of the transformation from authalic latitude function to meridian arc

The whole formulas for the transformation from authalic latitude function to meridian arc are as follows:

$$\begin{cases} \vartheta = \arcsin(\dfrac{F}{R^2}) \\[2mm] B = \vartheta + c_2 \sin 2\vartheta + c_4 \sin 4\vartheta + c_6 \sin 6\vartheta + c_8 \sin 8\vartheta + c_{10} \sin 10\vartheta \\[2mm] \psi = B + \alpha_2 \sin 2B + \alpha_4 \sin 4B + \alpha_6 \sin 6B + \alpha_8 \sin 8B + \alpha_{10} \sin 10B \\[2mm] X = a(1 - e^2)K_0\psi \end{cases} \qquad (61)$$

Expanding X as a power series of the eccentricity e at $e = 0$ by means of Mathematica yields the direct expansion of the transformation from authalic latitude function to meridian arc

$$\begin{cases} \vartheta = \arcsin(\dfrac{F}{R^2}) \\[2mm] X = a\left(k_0\vartheta + k_2 \sin 2\vartheta + k_4 \sin 4\vartheta + k_6 \sin 6\vartheta + k_8 \sin 8\vartheta + k_{10} \sin 10\vartheta\right) \end{cases} \qquad (62)$$

where

$$\begin{cases} k_0 = 1 - \dfrac{1}{4}e^2 - \dfrac{3}{64}e^4 - \dfrac{5}{256}e^6 - \dfrac{175}{16384}e^8 - \dfrac{441}{65536}e^{10} \\[2mm] k_2 = -\dfrac{1}{24}e^2 - \dfrac{7}{1440}e^4 - \dfrac{61}{107520}e^6 + \dfrac{2719}{8294400}e^8 + \dfrac{30578453}{61312204800}e^{10} \\[2mm] k_4 = -\dfrac{29}{11520}e^4 - \dfrac{1411}{967680}e^6 - \dfrac{180269}{232243200}e^8 - \dfrac{4110829}{10218700800}e^{10} \\[2mm] k_6 = -\dfrac{1003}{2903040}e^6 - \dfrac{341}{921600}e^8 - \dfrac{36598301}{122624409600}e^{10} \\[2mm] k_8 = -\dfrac{40457}{619315200}e^8 - \dfrac{3602683}{35035545600}e^{10} \\[2mm] k_{10} = -\dfrac{1800439}{122624409600}e^{10} \end{cases} \qquad (63)$$

4.3. The direct expansions of transformations between isometric latitude and authalic latitude function

4.3.1. The direct expansion of the transformation from isometric latitude to authalic latitude function

The whole formulas for the transformation from isometric latitude to authalic latitude function are as follows:

$$
\begin{cases}
\varphi = \arcsin(\tanh q) \\
B = \varphi + b_2 \sin 2\varphi + b_4 \sin 4\varphi + b_6 \sin 6\varphi + b_8 \sin 8\varphi + b_{10} \sin 10\varphi \\
\vartheta = B + \gamma_2 \sin 2B + \gamma_4 \sin 4B + \gamma_6 \sin 6B + \gamma_8 \sin 8B + \gamma_{10} \sin 10B \\
F = R^2 \sin \vartheta
\end{cases}
\tag{64}
$$

Expanding F as a power series of the eccentricity e at $e = 0$ by means of Mathematica yields the direct expansion of the transformation from isometric latitude to authalic latitude function

$$
\begin{cases}
\phi = \arcsin(\tanh q) \\
F = a^2 \left(\eta_1 \sin \phi + \eta_3 \sin 3\phi + \eta_5 \sin 5\phi + \eta_7 \sin 7\phi + \eta_9 \sin 9\phi + \eta_{11} \sin 11\phi \right)
\end{cases}
\tag{65}
$$

where

$$
\begin{cases}
\eta_1 = 1 - \dfrac{1}{4}e^2 - \dfrac{1}{12}e^4 - \dfrac{7}{192}e^6 - \dfrac{113}{5760}e^8 - \dfrac{7}{560}e^{10} \\[2mm]
\eta_3 = \dfrac{1}{12}e^2 - \dfrac{7}{960}e^6 - \dfrac{1}{192}e^8 - \dfrac{43}{13440}e^{10} \\[2mm]
\eta_5 = \dfrac{1}{60}e^4 + \dfrac{1}{192}e^6 + \dfrac{1}{20160}e^8 - \dfrac{13}{11520}e^{10} \\[2mm]
\eta_7 = \dfrac{31}{6720}e^6 + \dfrac{7}{2304}e^8 + \dfrac{41}{40320}e^{10} \\[2mm]
\eta_9 = \dfrac{41}{26880}e^8 + \dfrac{1}{640}e^{10} \\[2mm]
\eta_{11} = \dfrac{167}{295680}e^{10}
\end{cases}
\tag{66}
$$

4.3.2. The direct expansion of the transformation from authalic latitude function to isometric latitude

The whole formulas for the transformation from authalic latitude function to isometric latitude are as follows:

$$
\begin{cases}
\vartheta = \arcsin(\dfrac{F}{R^2}) \\
B = \vartheta + c_2 \sin 2\vartheta + c_4 \sin 4\vartheta + c_6 \sin 6\vartheta + c_8 \sin 8\vartheta + c_{10} \sin 10\vartheta \\
\varphi = B + \beta_2 \sin 2B + \beta_4 \sin 4B + \beta_6 \sin 6B + \beta_8 \sin 8B + \beta_{10} \sin 10B \\
q = \operatorname{arctanh}(\sin \varphi)
\end{cases}
\tag{67}
$$

Expanding q as a power series of the eccentricity at $e = 0$ by means of Mathematica yields the direct expansion of the transformation form authalic latitude function to isometric latitude

$$
\begin{cases}
\vartheta = \arcsin(\dfrac{F}{R^2}) \\
q = \operatorname{arctanh}(\sin \vartheta) + l_1 \sin \vartheta + l_3 \sin 3\vartheta + l_5 \sin 5\vartheta + l_7 \sin 7\vartheta + l_9 \sin 9\vartheta
\end{cases}
\tag{68}
$$

where

$$
\begin{cases}
l_1 = -\dfrac{1}{3}e^2 - \dfrac{1}{30}e^4 - \dfrac{11}{1890}e^6 - \dfrac{107}{32400}e^8 + \dfrac{1513}{1663200}e^{10} \\[2mm]
l_3 = -\dfrac{1}{90}e^4 - \dfrac{61}{11340}e^6 - \dfrac{2321}{907200}e^8 - \dfrac{1021}{831600}e^{10} \\[2mm]
l_5 = -\dfrac{1}{756}e^6 - \dfrac{5}{4032}e^8 - \dfrac{151}{166320}e^{10} \\[2mm]
l_7 = -\dfrac{71}{302400}e^8 - \dfrac{41}{123200}e^{10} \\[2mm]
l_9 = -\dfrac{61}{1197504}e^{10}
\end{cases}
\tag{69}
$$

4.4 Accuracies of the direct expansions

The accuracies of the indirect and direct expansions given by Yang(1989, 2000) derived by the author has been examined choosing the CGCS2000 reference ellipsoid.

The results show that the accuracy of the indirect expansion of the transformation from meridian arc to isometric latitude is higher than 10^{-3}", while the accuracy of the direct expansion (53) is higher than 10^{-7}". The accuracy of the indirect expansion of the transformation from isometric latitude to meridian arc is higher than 10^{-2} m, while the accuracy of the direct expansion (56) is higher than 10^{-7} m. The accuracy of the indirect expansion of the transformation from meridian arc to authalic latitude function is higher than $0.1\,\text{km}^2$, while the accuracy of the direct expansion (59) is higher than $10^{-5}\,\text{km}^2$. The accuracy of the indirect expansion of the transformation from authalic latitude function to meridian arc is higher than 10^{-2} m, while the direct expansion (62) is higher than 10^{-4} m. The accuracy of the indirect expansion of the transformation from isometric latitude to authalic latitude function is higher than $0.1\,\text{km}^2$, while the accuracy of the direct expansion (65) is higher than 10^{-7}km^2. The accuracy of the indirect expansion of the transformation from authalic latitude function to isometric latitude is higher than $10^{-2}''$, while the accuracy of the direct expansion (67) is higher than $10^{-6}''$. The accuracies of the direct expansions derived by the author are improved by 2~6 orders of magnitude compared to the indirect ones derived by Yang (1989, 2000).

5. The non-iterative expressions of the forward and inverse Gauss projections by complex numbers

Gauss projection plays a fundamental role in ellipsoidal geodesy, surveying, map projection and geographical information system (GIS). It has found its wide application in those areas.

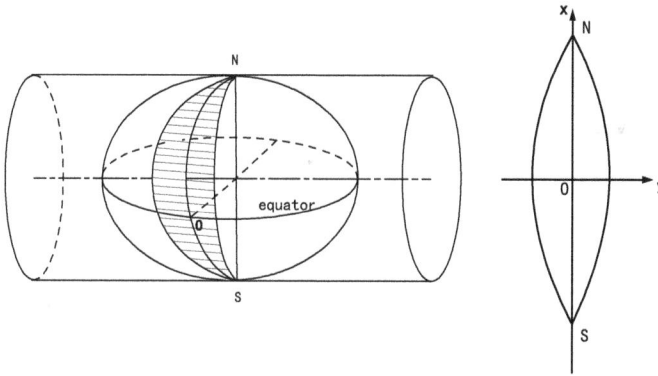

Figure 1. Gauss Projection, where x and y are the vertical and horizontal axes after the projection respectively, O is the origin of the projection coordinates.

As shown in Figure 1, Gauss projection has to meet the following three constraints:

① conformal mapping;
② the central meridian mapped as a straight line (usually chosen as a vertical axis of x) after projection;
③ scale being true along the central meridian.

Traditional expressions of the forward and inverse Gauss projections are real functions in a power series of longitude difference. Though real functions are easy to understand for most people, they make Gauss projection expressions very tedious. Mathematically speaking, there is natural relationship between the conformal mapping and analytical complex functions which automatically meet the differential equation of the conformal mapping, or the "Cauchy-Riemann Equations". Complex functions, a powerful mathematical method, play a very special and key role in the conformal mapping. Bowring (1990) and Klotz (1993) have discussed Gauss projection by complex numbers. But the expressions they derived require iterations, which makes the computation process very fussy. In terms of the direct expansions of transformations between meridian arc and isometric latitude given in section

IV, the non-iterative expressions of the forward and inverse Gauss projections by complex numbers are derived.

5.1. The non-iterative expressions of the forward Gauss projection by complex numbers

Let w be complex numbers consisting of isometric latitude q and longitude difference l before projection, z be complex numbers consisting of corresponding coordinates x, y after projection.

$$\begin{cases} w = q + il \\ z = x + iy \end{cases} \tag{70}$$

where $i = \sqrt{-1}$.

In terms of complex functions theory, analytical functions meet conformal mapping naturally. Therefore, to meet the conformal mapping constraint, the forward Gauss projection should be in the following form

$$z = x + iy = f(w) = f(q + il) \tag{71}$$

where f is an arbitrary analytical function in the complex numbers domain. According to the second constraint, when $l = 0$, imaginary part disappears and only real part exists, (71) becomes

$$x = f(q) \tag{72}$$

(72) shows that the central meridian is a straight line after the projection when $l = 0$.

Finally, from the third constraint, "scale is true along the central meridian", one knows that x in (72) should be nothing else but the meridian arc X, and (72) is essentially consist with the direct expansion of the transformation from isometric latitude to meridian arc (56). Substituting X in (56) with x gives the explicit form of (72)

$$\begin{cases} \varphi = \arcsin(\tanh q) \\ x = a\left(j_0\varphi + j_2 \sin 2\varphi + j_4 \sin 4\varphi + j_6 \sin 6\varphi + j_8 \sin 8\varphi + j_{10} \sin 10\varphi\right) \end{cases} \tag{73}$$

(73) defines the functional relationship between meridian arc and isometric latitude. If one extends the definition of q in a real number variable to a complex numbers variable, or substitutes q with $w = q + il$, the original real number conformal latitude φ will be automatically extended as a complex numbers variable. We denote the corresponding complex numbers latitude as Φ, and insert it into (73). Rewriting a real variable x at the left-hand of the second equation in (73) as a complex numbers variable $z = x + iy$, one arrives at

$$\begin{cases} \Phi = \arcsin(\tanh w) \\ z = x + iy = a\left(j_0\varphi + j_2\sin 2\varphi + j_4\sin 4\varphi + j_6\sin 6\varphi + j_8\sin 8\varphi + j_{10}\sin 10\varphi\right) \end{cases} \quad (74)$$

(74) is the solution of the forward Gauss projection by complex numbers. Its correctness can be explained as follows:

The two equations in (74) are all elementary complex functions. Because elementary functions in their basic interval are all analytical ones in the complex numbers domain, the mapping defined by (74) form $w = q + il$ to $z = x + iy$ meets the conformal mapping constraint. When $l = 0$, the imaginary part disappears and (74) restores to (73). Therefore, (74) meets the second and third constraints of Gauss projection when $l = 0$. Hence, it is clear that (74) is the solution of the forward Gauss projection indeed.

5.2. The non-iterative expressions of the inverse Gauss projection by complex numbers

In principle, the inverse Gauss projection can be iteratively solved in terms of the forward Gauss projection (74). In order to eliminate the iteration, one more practical approach is proposed based on the direct expansion of the transformation from meridian arc to isometric latitude (53).

In order to meet the conformal mapping constraint, the inverse Gauss projection should be in the following form

$$w = q + il = f^{-1}(z) = f^{-1}(x + iy) \quad (75)$$

where f^{-1} is the inverse function of f. According to the second constraint, when $l = 0$, imaginary part disappears and only real part exists, (75) becomes

$$q = f^{-1}(x) \quad (76)$$

Finally, from the third constraint, one knows that x in (76) should be the meridian arc X, and (76) is essentially consist with the direct expansion of the transformation from meridian arc to isometric latitude as (53) shows. Substituting X in (53) with x gives the explicit form of (76)

$$\begin{cases} \psi = \dfrac{x}{a(1 - e^2)K_0} \\ q = \operatorname{arctanh}(\sin\psi) + \xi_1\sin\psi + \xi_3\sin 3\psi + \xi_5\sin 5\psi + \xi_7\sin 7\psi + \xi_9\sin 9\psi \end{cases} \quad (77)$$

If one extends the definition of x in a real number variable to a complex numbers variable, or substitutes x with $z = x + iy$, the original real number rectifying latitude ψ will be automatically extended as a complex numbers variable. We denote the corresponding complex number latitude as Ψ, and insert it into (77). Rewriting a real variable q at the left-hand of the second equation in (77) as a complex numbers variable $w = q + il$, one arrives at

$$\begin{cases} \Psi = \dfrac{x+iy}{a(1-e^2)K_0} \\ w = q + il = \text{arctanh}(\sin\Psi) + \xi_1 \sin\Psi + \xi_3 \sin 3\Psi + \xi_5 \sin 5\Psi + \xi_7 \sin 7\Psi + \xi_9 \sin 9\Psi \end{cases} \qquad (78)$$

Therefore, the isometric latitude q and longitude l is known. Inserting q into (78) yields the conformal latitude

$$\varphi = \arcsin(\tanh q) \qquad (80)$$

Then one can compute the geodetic latitude through the inverse expansion of the conformal latitude (40).

(77) is the solution of the inverse Gauss projection by complex numbers. Its correctness can be explained as follows:

The two equations in (78) are all elementary complex functions, so the mapping defined by (78) form $z = x + iy$ to $w = q + il$ meets the conformal mapping constraint. When $l = 0$, the imaginary part disappears and (78) restores to (77). Therefore, (78) meets the second and third constraints of Gauss projection when $l = 0$. Hence, it is clear that (78) is the solution of the inverse Gauss projection indeed.

6. Conclusions

Some typical mathematical problems in map projections are solved by means of computer algebra system which has powerful function of symbolical operation. The main contents and research results presented in this chapter are as follows:

1. Forward expansions of rectifying, conformal and authalic latitudes are derived, and some mistakes once made in the high orders of traditional forward formulas are pointed out and corrected. Inverse expansions of rectifying, conformal and authalic latitudes are derived using power series expansion, Hermite interpolation and Language's theorem methods respectively. These expansions are expressed in a series of the sines of the multiple arcs. Their coefficients are expressed in a power series of the first eccentricity of the reference ellipsoid and extended up to its tenth-order terms. The accuracies of these expansions are analyzed through numerical examples. The results show that the accuracies of these expansions derived by means of computer algebra system are improved by 2~4 orders of magnitude compared to the formulas derived by hand.

2. Direct expansions of transformations between meridian arc, isometric latitude and authalic latitude function are derived. Their coefficients are expressed in a power series of the first eccentricity of the reference ellipsoid, and extended up to its tenth-order terms. Numerical examples show that the accuracies of these direct expansions are improved by 2~6 orders of magnitude compared to the traditional indirect formulas.

3. Gauss projection is discussed in terms of complex numbers theory. The non-iterative expressions of the forward and inverse Gauss projections by complex numbers are derived based on the direct expansions of transformations between meridian arc and isometric latitude, which enriches the theory of conformal projection. In USA, Universal

Transverse Mercator Projection (or UTM) is usually implemented. Mathematically speaking, there is no essential difference between UTM and Gauss projections. The only difference is that the scale factor of UTM is 0.9996 rather than 1. With a slight modification, the non-iterative expressions of the forward and inverse Gauss projections can be extended to UTM projection accordingly.

Author details

Shao-Feng Bian
Department of Navigation, Naval University of Engineering, Wuhan, China

Hou-Pu Li
Department of Navigation, Naval University of Engineering, Wuhan, China
Key Laboratory of Surveying and Mapping Technology on Island and Reef, State Bureau of Surveying, Mapping and Geoinformation, Qingdao, China

Acknowledgement

This work was financially supported by 973 Program (2012CB719902), National Natural Science Foundation of China (No. 41071295 and 40904018), and Key Laboratory of Surveying and Mapping Technology on Island and Reef, State Bureau of Surveying, Mapping and Geoinformation, China (No.2010B04).

7. References

[1] Adam O S (1921). Latitude Developments Connected with Geodesy and Cartography with Tables, Including a Table for Lambert Equal-Area Meridional Projection. Spec. Pub. No.67, U. S. Coast and Geodetic Survey.
[2] Snyder J P (1987). Map Projections-a Working Manual. U. S. Geological Survey Professional Paper 1395, Washington.
[3] Yang Qihe (1989). The Theories and Methods of Map Projection. PLA Press, Beijing. (in Chinese)
[4] Yang Qihe, Snyder J P, Tobler W R (2000). Map Projection Transformation: Principles and Applications. Taylor and Francis, London.
[5] Awange J L, Grafarend E W (2005). Solving Algebraic Computational Problems in Geodesy and Geoinformatics. Springer, Berlin.
[6] Bian S F, Chen Y B (2006). Solving an Inverse Problem of a Meridian Arc in Terms of Computer Algebra System. Journal of Surveying Engineering, 132(1): 153-155.
[7] Chen Junyong (2008). Chinese Modern Geodetic Datum-Chinese Geodetic Coordinate System 2000 (CGCS2000) and its Frame. Acta Geodaetica et Cartographica Sinica, 37(3): 269-271. (in Chinese)
[8] Yang Y X (2009). Chinese Geodetic Coordinate System 2000. Chinese Science Bulletin, 54(16): 2714-2721.

[9] Li Houpu, Bian Shaofeng (2008). Derivation of Inverse Expansions for Auxiliary Latitudes by Hermite Interpolation Method. Geomatics and Information Science of Wuhan University, 33(6): 623-626. (in Chinese)

[10] Li Houpu (2010). The Research of the Precise Computation Theory and Its Application Based on Computer Algebra System for Geodetic Coordinate System. Naval University of Engineering, Wuhan. (in Chinese)

[11] Whittaker C E (1902). Modern Analysis. Cambridge.

[12] Peter O (2008). The Mercator Projections. Edinburgh.

[13] Bowring B R (1990). The Transverse Mercator Projection-a Solution by Complex Numbers. Survey Review, 30(237): 325-342.

[14] Klotz J (1993). Eine Analytische Loesung der Gauss-Krüger-Abbildung. Zeitschrift für Versicherungswesen, 118(3): 106-115.

Chapter 2

Analysis of Pre-Geodetic Maps in Search of Construction Steps Details

Gabriele Bitelli, Stefano Cremonini and Giorgia Gatta

Additional information is available at the end of the chapter

1. Introduction

Cartographic Heritage consists in the whole amount of ancient cartographic documents (not only maps, but also atlas, planispheres, globes, …) the history has brought us, today perceived as a cultural value to be necessarily preserved due to its historical and geographical content as well as its artistic value. It is a great but often poorly known heritage, because usually ancient cartographic documents are kept in places closed to the public, and only occasionally they are proved valuable outside of specific research activities.

The recovery of ancient cartography is intended to save, and possibly to spread throughout a wide public, Cartographic Heritage, making use of the potential it offers. Unfortunately, ancient cartographic documents often suffer from preservation problems of their analogue support (an organic material, thus subject to wear), mostly due to aging. Therefore, the recovery of ancient cartography firstly consists in traditional restoration, intended to safeguard the analogue support against the damaging effects of time. Beside this, a recovery of the content of historical documents is also possible, carrying the cartographic document to a different support, usually a digital one. In such a way, regeneration of ancient cartography in a digital environment is an interesting way of Cultural Heritage preservation and valorisation.

Digital regeneration is not exhaust by the digitization step: modern digital techniques allow new chances of using the map information, which would be unachievable on an analogue support. In particular, georeferencing and analysis of map deformations help in metric analysis of ancient cartography. In fact, usually the metric precision of an ancient cartographic document can be very different from that of a present map, due to an amount of deformations and errors that can be very high with respect to our standards. For example, graphical deformations can be induced by the old type of cartographic transformation (if one exists, it can be different from the modern ones), whereas other deformations can be due to an alteration in the analogue support; other errors can be recorded in the cartographic document, for instance errors that were made by the cartographer during survey or draft steps, or errors inherent to the surveying instruments at that time. In order to compare an ancient cartographic document to a modern one (usually a modern map used as reference), a georeferencing process is performed in a digital environment, assigning cartographic coordinates to a number of still existing and recognizable Ground Control Points (GCPs). This way, the native metric content of the map can be reproduced in the digital image, and furthermore an analysis of the existing map deformations is allowed. Thus, it becomes possible to understand the characteristics of metric precision of the original product (e.g. the projection type) in respect to the present reference cartographic base, as well as to evaluate and represent the degree of deformation recorded in the ancient document. The historical map, now in digital form, can be easily exploited and compared with other cartographic databases, thanks also to current web services; change analysis and analytical procedures can be performed through GIS applications.

This way, regeneration of ancient maps in digital form appears to be useful for many users: not only the public and institutions who collect them, but also experts who exploit this kind of documents to derive information for their studies, ranging from urban development to geomorphological or environmental topics. Many institutions today are digitalizing their cartographic heritage, in order to preserve and catalogue it and give online access to it (Adcock et al., 2004). On the evidence of growing interest in the argument, the International Cartographic Association (ICA) instituted in 2007 the "Commission on Digital Technologies in Cartographic Heritage", whose aim is to encourage digital approaches to cartographic heritage.

The present research would demonstrate the usefulness of the digital regeneration of ancient cartography; it provides an example of studies that can be performed after digital regeneration of ancient cartography, with a non-conventional approach mainly focused on technical considerations about the map-making procedures.

2. Materials

In this study, a set of three maps, depicting the northern coast of the Adriatic Sea along the Po river delta (South of Venice, Italy) at the end of the 16th century, is analysed (Figure 1). The first two maps were both drafted in the year 1592, whereas the third one was drafted few years later (1599).

The maps represent a rare case where the authors of documents are known. The same cartographer, Ottavio Fabri, was author of the first map (hereinafter "F map") and co-author in the other two (hereinafter "P map" and "L map"), in which the main authors were Gerolamo Pontara and Bonaiuto Lorini, respectively. All of them were very famous land-surveyors in Renaissance Venice (*Savi ed Esecutori delle Acque della Serenissima Repubblica*).

The dimensions of these documents are very large, and their average scales range between about 1:12,000 and 1:13,000, not being constant throughout the entire maps. The original documents are kept today in the Venice National Archive (ASVe), and two of them (F and L maps) have been examined on site. The original documents are drawn on several watercolor papers stuck on canvas supports. For the aim of the present study, digital copies of the originals, made by ASVe, were used: high resolution (300-400 dpi) copies were derived from high quality scanning (F and L maps), and a lower quality copy was derived from digitization of photographic images (P map), the original paper support being afflicted by wear problems. No images may be reproduced, in any form or by any means, without permission of the Venice National Archive; it is necessary to apply to the Photo-reproduction Section of the Venice State Archive in order to obtain the release of copies. Granted for these reproductions: n. 81/2010.

A further fundamental tool of analysis was the book edited by the main author, Fabri, containing the description of the so called *squadra zoppa* (or *squadra mobile*, i.e. mobile square), a new topographical instrument probably invented and used by Fabri himself (Panepinto, 2009). That instrument was useful for performing every type of topographic measurement (i.e. heights, distances, depths) in urban and land surveying, and also for map drawing (reporting the measurements on the paper). Surprisingly, the handbook seems to be a powerful record of the author's whole technical experience originated by the surveying operations performed in the geographical areas depicted in the maps here studied (Figure 2). A copy of the book, today preserved at the *Dore* Library of the Engineering Faculty of the University of Bologna, has been examined on site (Fabri 1673).

3. The reasons of the choice and the previous studies

Even if the above mentioned existence of a textbook probably related to the chosen maps would not be taken into account, other strong motivations appear to exist to focus our analysis on those cartographic samples.

As these maps were made during the very short period between the years 1592 and 1599, i.e. the lapse of time immediately forerunning a series of very important works aiming at the Po river channel diversion, they stimulate a compelling geomorphological analysis (Cremonini 2007a; Cremonini & Samonati 2009) focused on the easternmost peripheral areas that today no longer exist, due to erosional dynamics of seashore evolution developed during the last four centuries (Cremonini 2007b; Cremonini 2010). A further problem arises, due to the fact that the maps depict in a quite different manner the same landforms, although they appear to have been drawn in the same years by the same author or co-author (Ottavio Fabri). For these reasons the maps have already been studied from various viewpoints and metrically analysed in a digital environment (Bitelli et al. 2009, 2010), to try to overcome the merely qualitative comparison between the available maps.

Although the modern digital techniques, in particular georeferencing of the cartographic samples coupled with a study of the map deformations, help in metric analysis of ancient cartography, in pre-geodetic cartography studies specific analytical tools need to be used, e.g. in the step-by-step solution here proposed.

(a) F map

(b) P map

(c) L map

Figure 1. The three analyzed maps proportionally scaled (red bar = 1 m): a) *Delta del fiume Po*, by Ottavio Fabri, 1592. *Savi ed esecutori delle acque, serie Po,* dis. 9bis. Size: about 3.5 x 2.5 m; b) *Delta del fiume Po*, by Gerolamo Pontara, Ottavio Fabri, 1592. *Savi ed esecutori delle acque, serie Po,* dis. 8. Size: about 2.9 x 1.6 m; c) *Delta del fiume Po*, by Bonaiuto Lorini, Ottavio Fabri, Gerolamo Pontara, Alessandro Betinzuoli, Bartolomeo Montini, 1599. *Savi ed esecutori delle acque, serie Po,* dis. 10. Size: about 2.3x1.4 m.

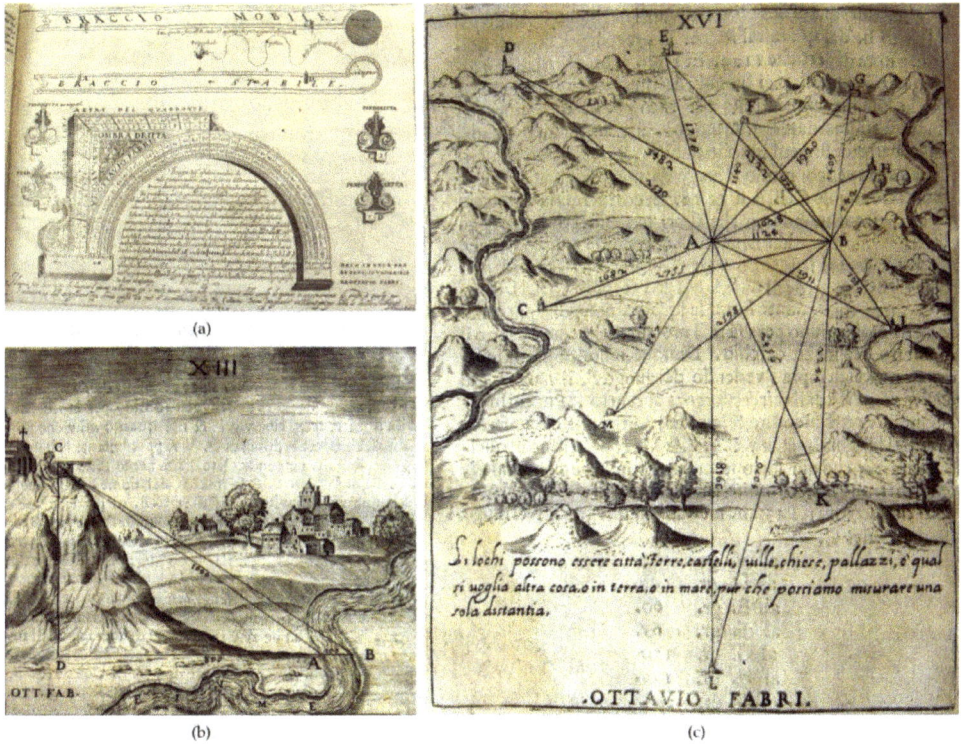

Figure 2. Images derived from O. Fabri, *L'uso della squadra mobile* (1673 edition, kept in the Engineering Faculty Library of the University of Bologna): (a) a picture of the *squadra mobile*; (b) measurement of river widths; (c) an exercise of forward intersection.

3.1. The quantitative approach

3.1.1. Georeferencing

Georeferencing is the technique of inserting a map into a reference system, usually a modern cartographic one. The process is performed by selecting in the ancient map a proper number (usually as large as possible) of peculiar points (GCPs, Ground Control Points), still existing today, and deriving their cartographic coordinates from the present cartography or a specifically designed survey. In the peculiar case of ancient cartography, the task can be very difficult or also impossible, because of the remarkable landscape evolution over time (Benavides & Koster, 2006).

When a sufficient number of points is available, a *one-to-one* correspondence between the two set of control points lying on two different plane surfaces (i.e. points on the digital image of the ancient map, expressed in image coordinates, and reference points, expressed in cartographic coordinates) is established through a "best-fit" process, that finishes with the calculation of the transformation parameters. The number of involved parameters can varies with the transformations, each transformation requiring a different number of control points.

In fact, many kinds of georeferencing methods exist, and they produce results that can be qualitatively and quantitatively different. In particular, the georeferencing algorithms can be grouped in two different classes: global and local transformations (Balletti 2006; Boutoura & Livieratos, 2006). In a global transformation (conformal, affine, projective, generic order polynomial) the unknown parameters are calculated for the whole area. On the other hand, in a local transformation (finite elements, morphing) the unknown parameters are calculated for a small area, defined by a small number of control points or close to each control point. The *best* georeferencing method probably do not exist, because the choice of a specific georeferencing method depends on the specific case (map characteristics, number of available GCPs, etc.) and on the purpose which the georeferenced images will be used for.

By means of a georeferencing process, the native metric content of the map being reproduced in the digital image, the historical map can be compared with the present cartography or other cartographies (also coeval ancient maps) in the same reference system. The process generates a new aspect of the ancient map, showing the typical deformation induced by its cartographic characteristics (and partly by the applied algorithm): in this way it is possible to understand the metric quality of the map representation (e.g. by means of the residuals errors associated to each single point, output by the geroreferencing process) and the projection features of the historical map, but also to perform many other kinds of analysis, e.g. studies related to change of the landscape.

In this specific case, after careful analysis of the three cartographic samples, a set of about 80 common GCPs, clearly identifiable also on the IGM (Italian Military Geographic Institute) topographic sheet, was recognized on the three ancient maps. It has to be stressed that in this phase a great deal of problems arose, concerning the basic characters of the points themselves (e.g. their planimetric precision, their graphic representation on the ancient maps, etc.), in addition to the difficulty in finding points that are still existing. North and East coordinates were attributed to each selected point according to the UTM-ED50 (fuse 33) grid. Different georeferencing methods were tested on the three map samples, and the most useful in order to compare them with the present landscape resulted polynomial transformations. In particular, the second order polynomial transformation resulted a good compromise between adaptation of the ancient maps to inland area details of the present landscape, on one hand, and constraint of deformations (indicated by the mean residual errors) throughout the maps, on the other (Bitelli et al. 2009).

Polynomial transformations coincide with a linear transformation (6-parameter affine) in the first order, and a non-linear one at higher degrees. A linear transformation corrects for scale, offset, rotation and reflection effects, whereas a non-linear transformation (for example, the 2nd order polynomial transformation) corrects for non-linear distortions: the final result depends very much on the number of control points and their spatial distribution in the image plane.

In Figure 3 an overlay of the three maps on present high resolution satellite images (in *Bing Maps*™ environment) is reported. The mean residual error (expressed as RMS, Root Mean

Squared error) was about 588 m in F and P maps, ranging between 18 and 1,320 m in F map and between 85 and 1,650 in P map; it was more constrained in L map (mean: 452 m; range: 29 ÷ 1,068 m). The residuals appear to be lower in the map centre, whereas they increase in size in the peripheral areas. This is the classical border effect due to the polynomial transformation associated to the lack of reference points in the area. But in this specific case it can be supposed that the effect can be due also to other reasons, such as the survey technique locally adopted, accidental or intentional drawing errors, etc.

(a) A red circle indicates the Po river delta area (Italy)

(b) F map   (c) P map   (d) L map

Figure 3. Overlay of the three ancient maps, georeferenced by means of a second order polynomial transformation, on present high resolution satellite images (in *Bing Maps*™ environment).

3.1.2. Study of map deformation

Comparing an ancient map with a present one allows evaluation and representation of the deformation degree of the former. Map deformations can be induced by physical alteration of the analogical support (a very frequent case for ancient maps) or by the old type of cartographic transformation (that frequently is unknown to us, and usually quite different from that used today and inducing less constrained deformations), or finally by surveying and drafting errors. Map deformations can be very high for ancient cartography; therefore, their assessment is essential in order to use the old samples for further studies (Livieratos, 2006). The assessment of map deformations consists in calculating some parameters from a comparison of the ancient map with a modern one as a reference (usually a present cartography to an appropriate scale, i.e. comparable with the scale of the ancient map), such as the rotation angle with respect to the present cartographic North, the scale variation throughout the map and the distortion of the present cartographic grid as resulting after its adaption onto the old map. The process is performed recognizing in the ancient map a proper number of still existing points, whose cartographic coordinates can be derived from the present cartography. A specifically designed software tool can result very useful for analysis of map deformations, allowing the calculation of all the aforementioned parameters and their subsequent drafting in an intuitive way (Jenny & Hurny, 2011).

The study of map deformation characterising the cartographic samples here analysed showed scale factors quite variable throughout the maps, being slightly more homogenous in F map than in P map. The average scale resulting from the calculation was 1:12,300 (1:14,300 ÷ 1:10,300) and 1:13,400 (1:16,600 ÷ 1:10,200) in F and P maps, respectively. Notwithstanding this, the former map showed two severe anomalous variation areas near the northern and southern delta lobe corners (Bitelli et al. 2009). In particular, the northern gross deformation affecting F map is supposed to derive from a shift of the drawing, intentionally made by the author for unclear reasons. L map presented a bit more constrained scale factor (mean 1:11,200, range 1:13,200 ÷ 1:11,200) and a smaller deformation than the other two, but it has to be taken into account that L map depicts a smaller area in respect to F and P maps (Figure 4). Moreover, the calculated rotation angles were 15.7°, 8.9° and 3.6° for F, P and L maps, respectively: they indicate an angular displacement of about 7° between the first and the second map, and 12° between the first and the third map.

In this specific case, where large areas, depicted in the maps, correspond to disappeared coastal (i.e. peripheral) belts, points suitable to be used as GCPs cannot be found in the present landscape, and the insertion of GCPs all around the deltaic area becomes obviously impossible. The metric analysis that was possible to perform on the set of maps highlighted gross deformations in all maps (especially in F and P maps). Notwithstanding some differences in the results showed by the maps, this analysis alone resulted insufficient to state which map has to be considered as the most faithful to the real asset of the ancient landscape.

Thus, a question still remains open: why does exist a family of so severe deformations?

4. A second approach: a search for survey and draft step details

Due to their representation scale and their aims, the maps could be regarded as true precursors of the modern "technical maps". But it is likely that these wide-sized maps were simply due to a scenic use of the products (e.g. for a good view during technical meetings held in the presence of political authorities) rather than to a real technical need. Unfortunately, up to now we possess neither the original field drafts nor the related field notebook that could allow us to do an enormous quality jump in the detailed inspection of the map generation process. Thus, any other analytical attempt must be taken into account to try to find out information about survey and draft phases.

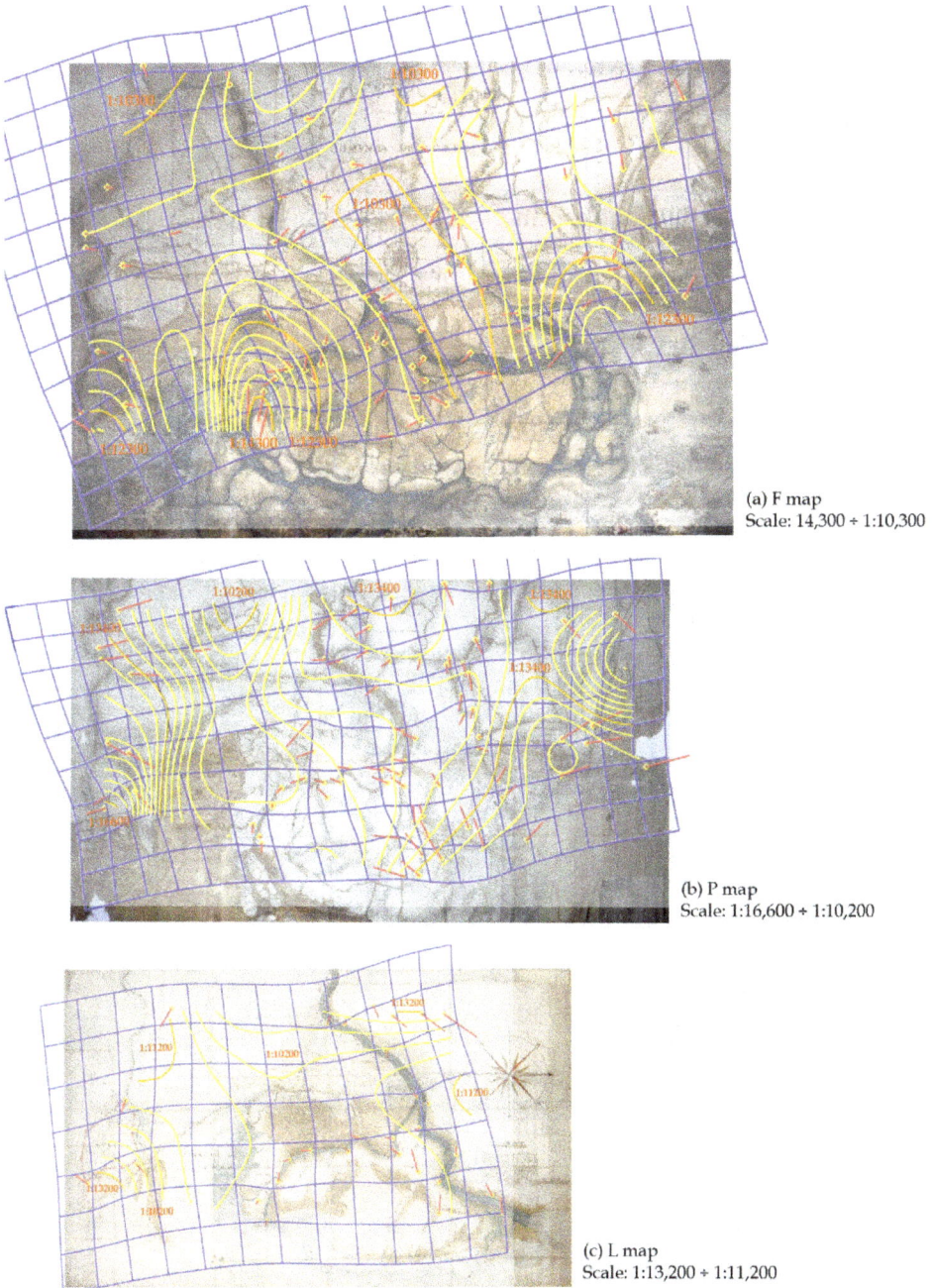

(a) F map
Scale: 14,300 ÷ 1:10,300

(b) P map
Scale: 1:16,600 ÷ 1:10,200

(c) L map
Scale: 1:13,200 ÷ 1:11,200

Figure 4. Graphical results from a deformation analysis performed on the three maps (see Jenny & Hurny 2011 for software description): in blue the UTM-ED50 grid (mesh size = 2 km), in red the residual vectors on the GCPs, in yellow the scale isolines, in orange some values of the calculated map scale.

In order to achieve this aim, late Renaissance land-surveying techniques (in particular the use of the *squadra zoppa* probably invented by Fabri himself) were studied and linked to technical signs preserved in the maps. Various signs, preserved either in the palimpsest of the maps or in their final drawing, were taken into account, and nine classes of evidence have been recognized (Bitelli et al. 2010): i) written information; ii) technical grids; iii) topographic measurements; iv) sighting tracks; v) "lost landmarks"; vi) preparing/correcting/updating drawings; vii) additional iconography; viii) unresolved questions; ix) restoration problems. The meaning of some signs is partly explained in the legend, whereas the meaning of others can be inferred by Fabri's methodological textbook. They will be briefly analysed in the next paragraphs.

4.1. Written information

In their cartouche, F and L maps state that the documents were made by merging some previous maps or other cartographic drawings and partly by means of direct topographic measurements (Figure 5). This kind of capital information helps us to understand that pre-geodetic maps can be composite products (at least in the case of large scale maps), therefore they can be very difficult to be analyzed and their usual georeferencing could be partially inappropriate. L map cartouche states that different surveying techniques were adopted according to the different interest level of the various topographic domains (see Paragraph 4.3). Unfortunately, due to the low resolution of the digital image, on P map we cannot read the text content; moreover it seems to be an unfinished product regarding the cartouche decorations. Besides, the fundamental message is that the maps can be considered a sort of patchwork probably generated by merging some local maps, previously surveyed by the authors themselves or other colleagues, or partially *ad hoc* surveyed.

(a) F map

(b) P map

(c) L map

Figure 5. The cartouche in the three maps.

4.2. Technical grids

A small-sized square mesh grid, drawn in charcoal, is distinctly visible on all the three maps (Figure 6a). The grids are characterized by different cell sizes: 10 cm on paper (about 1 km in world) in F map, 7.7 cm (about 1 km) in P map, 4 cm (about 400 m) in L map. In the first case, the cell size is exactly as long as half of the scale bar. This kind of grid covers the whole map and is drawn parallel to the North branches of the compass-card and to the map edges. Hence, the grid was probably helpful in assisting the transposition of the field-bearings onto the map support. We cannot state if the numbering of the grid squares (characterising the P map) had been used for this or other purposes, such as a mere scale reduction for copying.

Another very large sized grid exists on F map alone, and it is characterized by cells about five times larger than those of the previous one (Figure 6b). The grid orientation does not fit with the magnetic North, but coincides with a topographic direction clearly identifiable with the most rectilinear street reach located along a wide artificial canal (*Rettinella*) between the towns of *Loreo* and *Tornova*. It can be supposed that the canal was the starting baseline for the survey of the central part of the mapped lands, according to the forward intersection scheme showed by Fabri in his textbook (Figure 2c), assuming the still existing *Loreo* and *Tornova* bell towers as main reference points (Figure 2c, A and B points). Therefore, this second grid could have probably been used by the map-maker to draw on the map the azimuths measured in field and the sighting tracks toward all the noticeable points.

It is not possible to state whether the field survey strategy was the same for the other maps, as in them a similar large grid does not exist. If it was, the same baseline would have probably been used in P map (where it is clearly recognizable) but not in L map (where the *Rettinella* canal is roughly drafted in a marginal area of the map).

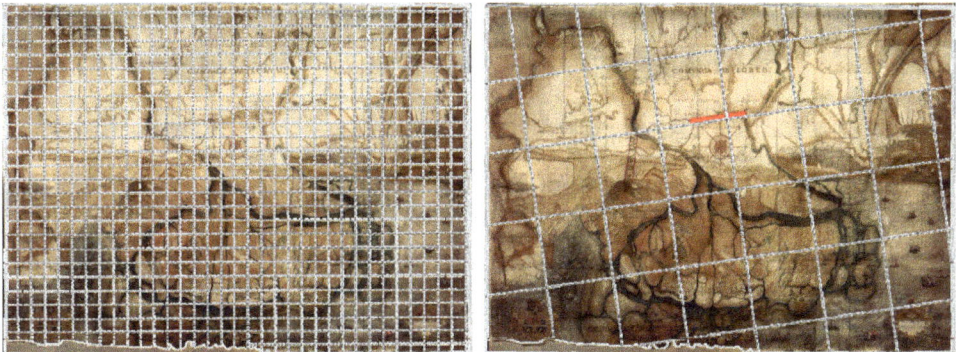

(a) The little-sized grid (b) The big-sized grid

Figure 6. Linedraw of the technical grids visible in F map, superimposed on the map. In (b) the *Rettinella* canal is highlighted in red.

4.3. Topographic measurements

Topographic measurements can be subdivided into three classes (Figure 7):

i.   explicit length measurements, which are clearly written as numbers of ancient Venice paces (1 Venice pace = 1.738674 m, as stated in Martini, 1883), e.g. in the area of the *Pertegado* ditch, whose name properly means "measured";
ii.   implicit length measurements, e.g. along the riverside where the points are graphically located in order to define the width and the geometry of the riverbanks;
iii. local bathymetric surveys and some selected riverbed transects, in particular at the river mouth of the *Tramontana* branch.

In the latter case, the mapmaker clearly highlighted which topographic reaches are the best surveyed, by means of red lines drawn onto the map. In those reaches pickets were certainly located and their related distances measured. According to that, each of these points had to be further sighted from a standpoint lying on the opposite riverside.

Therefore, this class of evidences states the fundamental technical character of the maps as a whole or for some selected subareas.

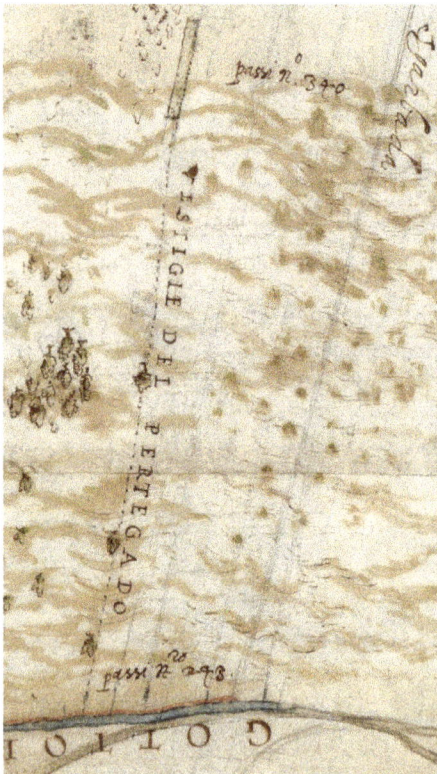

(b) Points along the riverside
and red reaches

(a) Explicit length measurements

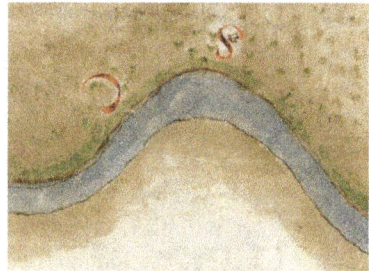

(c) Local bathymetric surveys

Figure 7. Topographic measurements in L map.

4.4. Sighting tracks

Some sighting tracks (sometimes identified by a number and a letter) are clearly visible, in particular in L map (Figure 8) and occasionally in P map. These tracks are preserved in sea areas lying between the Po delta and the continental coast reaches in the northern and southern part of the delta lobe. They try to define the correct distance existing between selected points lying on the fast-prograding delta coast and well-visible landmarks lying on the opposite coastal domain, e.g. *Mesola* defense towers and *Chioggia* bell towers. The length of the longest tracks is about 10 km, a plausible value for field surveys in marine environment free of the vegetative mantle (thus characterized by a good inter-visibility of the points).

This kind of information could suggest the author's need to control the existence of gross errors in the conclusive survey phase.

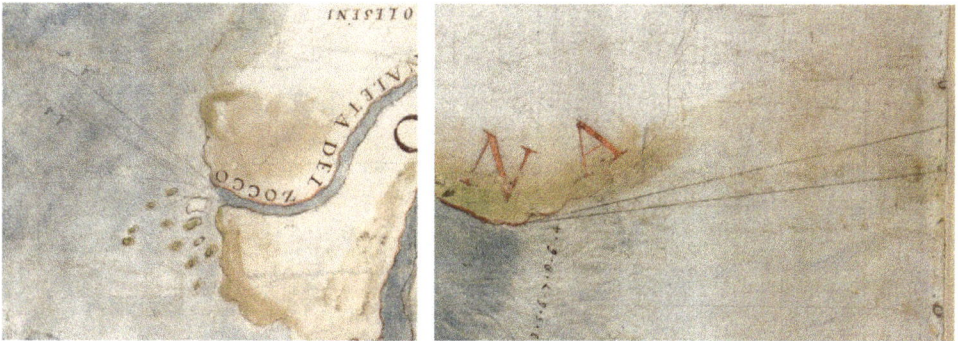

Figure 8. Examples of sighting tracks visible in L map.

4.5. "Lost landmarks"

Other signs of unclear value are visible in F map. They are not preserved in today landscape, thus they should have not been true *landmarks* (Figure 9): for this reason we prefer to use the term *lost landmarks* for their identification. They consist in few small crosses and a large one located at the diffluence point of two river branches of the *Po di Levante*. The large cross was probably a real reference-point as this object gave its name to a salt marsh (*Polesine de la Crose*) and to a still preserved building. The small crosses are located in desert areas of the E and SE realms of the delta lobe, along a red double curvilinear track. The meaning of the curvilinear tracks has been deduced from the comparison with another map of the same area dating back to the year 1608: they define the boundaries of the lands confiscated by the Venice Government in 1588.

This class of evidences highlights how difficult the ancient land surveying was, due to lacking of tall and fixed benchmarks in high-speed changing areas. On the other hand, it highlights how much important, from an economic point of view, those new generated areas were for the local government, thus suggests a possible reason to start the survey works. A question remains open: were those landmarks the only ones existing or are they simply a selected set?

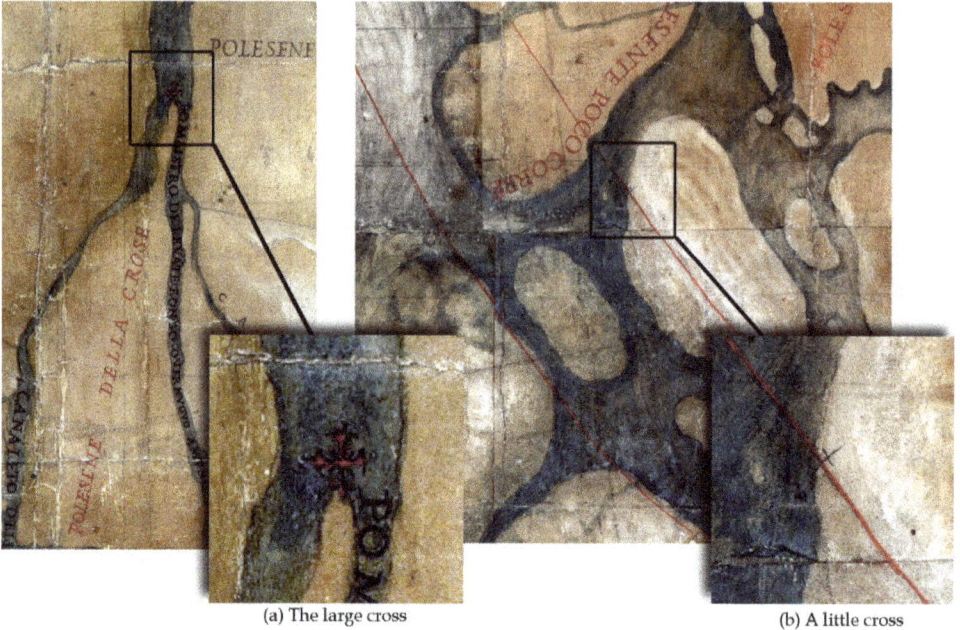

(a) The large cross

(b) A little cross

Figure 9. Signs of unclear value (*lost landmarks*) drawn in F map: a large cross at the diffluence point of two river branches, and a small cross lying on a red double curvilinear track.

4.6. Preparing/correcting/updating drawings

In each map, former drawings of topographical elements can be observed at various places. They represent details of riverbeds or local ways and were probably drawn as a first approximation of topographic elements. In the last definitive drawing, they appear corrected in their shape and/or location (Figure 10).

It may be believed that they represent the first attempt at defining the topographic design in those parts of the map derived from previous cartographic documents (as stated in the cartouche of F and L maps, see Paragraph 4.1). In some other cases, these kinds of correction probably referred to real environmental changes developed over time, as in the case of the minor branches of the *Scirocco Po* branch near *Mea* place (e.g. in P map, see Figure 10a).

It is difficult to conceal that these features represent an interpretative problem, as they suggest a high approximation degree in the drafting of the topographic details.

4.7. Additional iconography

In F map a very particular detail consists in a number of appliqués showing ancient vessels (Figure 11a). It was possible to notice such a detail only thanks to a direct consultation of the original analogical document in ASVe. We could ask ourselves why these symbols were stuck on the map and not directly drawn on it, whereas a very beautiful fish image (a sturgeon) was drawn in front of the *Tramontana* branch mouth (Figure 11b).

The question is not out of sense because it can suggest chronological problems, also related to tracking of the reference grid (Figure 11a).

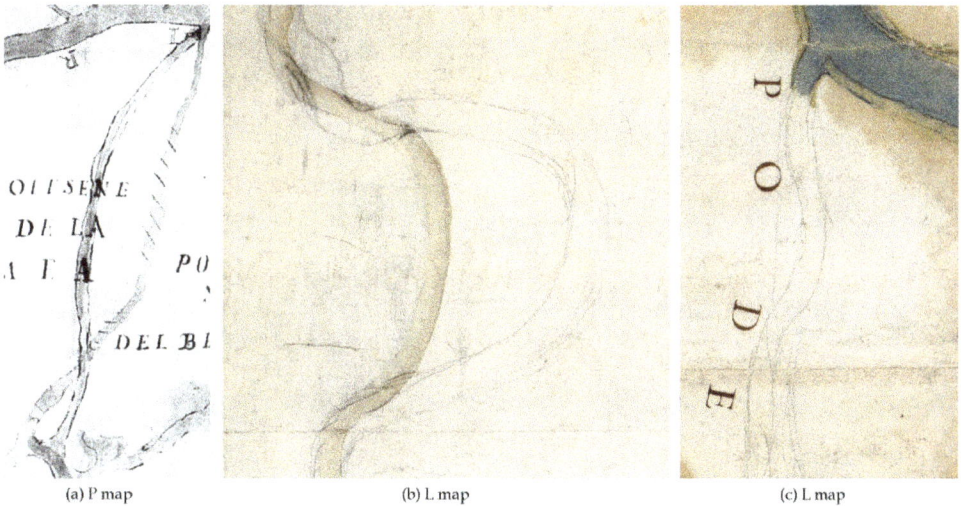

| (a) P map | (b) L map | (c) L map |

Figure 10. Examples of updating riverbed details in P and L maps.

| (a) Appliqués of ancient vessels | (b) Image of a sturgeon |

Figure 11. Examples of additional iconography in F map: appliqués showing ancient vessels and the image of a sturgeon.

4.8. Unresolved questions

Each map derives from the assemblage of single sheets on a canvas, clearly identifiable thanks to the visible seams (Figure 12). As a result, the homologue tracts of the drawing do not always fit well together. Hence, a problem arises concerning the age of the last sheet assemblage: does the assemblage date back to the original edition of the map or to a more recent one, being related to a series of restorations?

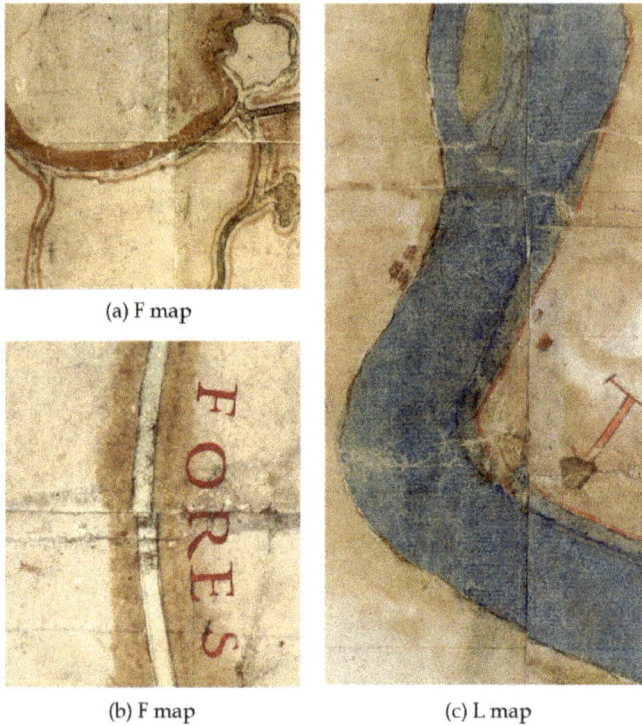

(a) F map

(b) F map   (c) L map

Figure 12. Visible seams in F and L maps, due to the assemblage of single sheets on the canvas: the homologue tracts of the drawing do not always fit exactly together.

As what we can see today is the product of the long lifespan of the maps, a fundamental question arises concerning type, number and entity of the restorations performed on them. Each restoration probably slightly changed the map almost imperceptibly. This may be true also for the smallest details, and it could partially deny the possibility of performing a correct diachronic study of the original document.

5. A discussion and a proposal

Notwithstanding some still open questions, the whole set of the discussed details highlights a very interesting class of information related to the hidden steps of pre-geodetic map construction. In particular, these details highlight the presence in the map of areas surveyed by means of different methodologies. This fact can explain the different entity of georeferencing residuals characterising different areas as well as the presence of high deformations in some particular areas rather than in others. The use of different survey methodologies can be due to various reasons, such as different levels of interest in the representation of different zones (e.g. areas confiscated by the Venice government, or areas planned for the forthcoming Po river diversion), or difficulties in surveying some others (e.g. areas too far from a selected reference baseline, or without well-recognizable landmarks), or, finally, the involvement of more than one person in the surveying works. As explicitly stated in some cartouches, the morphology of some depicted land areas can have been taken from other previously developed cartographies; those areas are not explicitly highlighted by the authors, but it is possible to infer their locations by excluding zones where explicit signs refer to specific surveying techniques (i.e. sighting tracks, length measurements, red reaches, bathymetric surveys). It can be hypothesized that some topographical elements characterized by former drawings (e.g. in L map, see Paragraph 4.6) were derived from previous cartographic documents. In fact, some visible former drawings (e.g. Figure 10b) are placed in peripheral areas where residuals from the georeferencing process are higher than those obtained in other areas of the map.

Other kinds of former drawings probably refer to corrections made by the authors over time: they can denote real environmental changes in shape and/or location of the topographic elements, or also intentionally-made corrections. The first ones can be highlighted by a roughly deletion of previously drawn elements (see Figure 10a), whereas the second ones seem to have been hidden by the authors by means of a brush stroke covering. This is the case of an astonishing map "correction", well recognizable in F map (Figure 1a), probably done at an early drawing stage. For that detail it was thought that the northern cluster of islands had been northward-shifted by the author for an unclear reason. This could be supported by both the comparison with the other maps and the existence of a wide area with no clear topographic detail lying immediately south of the delta coast. Hence, a possible verisimilar original map appearance could be that shown in Figure 13: it is a true "false case", generated by means of a simple "surgical" operation performed in a digital environment, in the opposite direction compared to the alleged author intervention. It must be noticed that the southward-shifted cluster of islands coincides perfectly in its morphology with the part of islands remained untouched.

The above described operation could now be attempted in all the other areas of F map with already evidenced anomalies (Cremonini & Samonati, 2009), allowing a complete regeneration of this ancient map. This could be an interesting way of approaching the study of pre-geodetic maps.

5.1. Georeferencing of the "false case"

The presence of intentionally-made corrections could explain why some areas, differently represented in F map in respect to P map (whereas in L map the represented land is smaller), maintain an unclear post-georeferencing deformation in F map. In order to demonstrate this assumption, a comparison of the new product (the "false case") with present cartography via modern georeferencing methods was made. In Figure 14 the overlay of the "false case" on present high resolution satellite images (*Bing Maps*™ environment) is reported; the map was georeferenced by means of a second order polynomial transformation, as done for the original map (see Figure 3b). An analysis of residuals of georeferencing could be the best way to check a real improvement in metric quality of the new cartographic product in respect to the original one, but in this particular case the test results impossible, due to the lack of still existing reference points in the deltaic area. Therefore, only a comparison between coeval maps is allowed. In Figure 15 the comparison between the "false case" and the georeferenced P map (by means of a second order polynomial transformation) is showed. In this case, the increase in correspondence for the northern cluster of islands details between the two maps is very high (see Figure 15b). For the purpose, a vectorization of P map was used in order to better visualize the overlay. Unfortunately, L map covers a smaller land area, and other coeval maps depicting the same deltaic area do not appear to exist: thus, the comparison with P map only is possible and significant.

Figure 13. A true "false case": the first attempt at restoring the possible original appearance of the former real map configuration before the alleged author intervention (in yellow the part translated and rotated from the white original location). North is right.

5.2. Towards possible future developments

The quite simple analysis above proposed could be considered as a stimulus toward a new way of approaching the study of pre-geodetic cartography: an approach capable to deeply enter the factual map genesis processes. In fact, a mere mathematical approach could not be satisfying for situations similar to the above discussed. As shown in Figures 16 and 17, the attempt of mutual comparison among the georeferenced coastline locations for the three studied maps could remain widely questionable. In these case, in fact, a gross discrepancy in the location of the same coastline (2 up to 3.5 km) in the same year 1592 remains between F and P maps, therefore neither reliable coastal erosion rates (Figure 16) nor a correct point-by-point response of the ancient beach line to the wind-induced long shore currents (Figure 17) can be believably proposed.

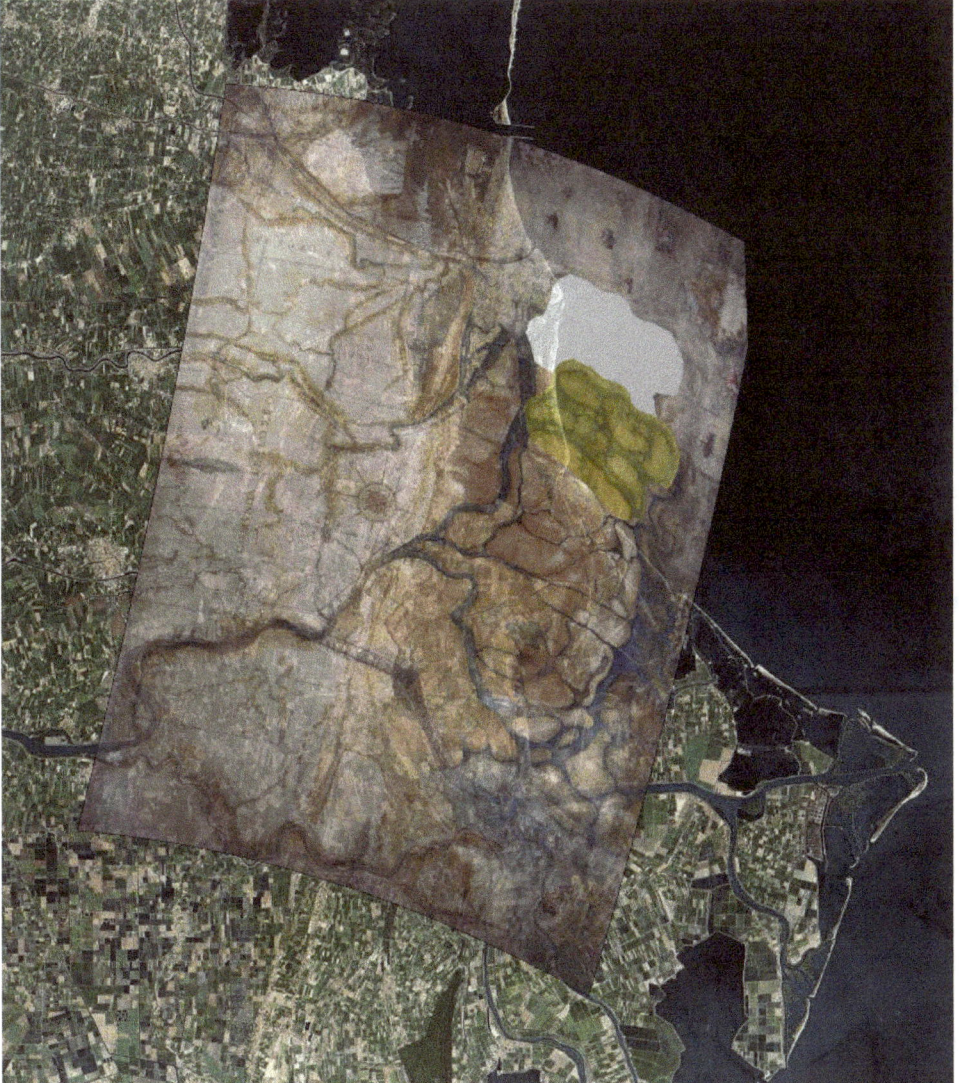

Figure 14. Overlay of the "false case", georeferenced by means of a second order polynomial transformation, on present high resolution satellite images (*Bing Maps*™ environment).

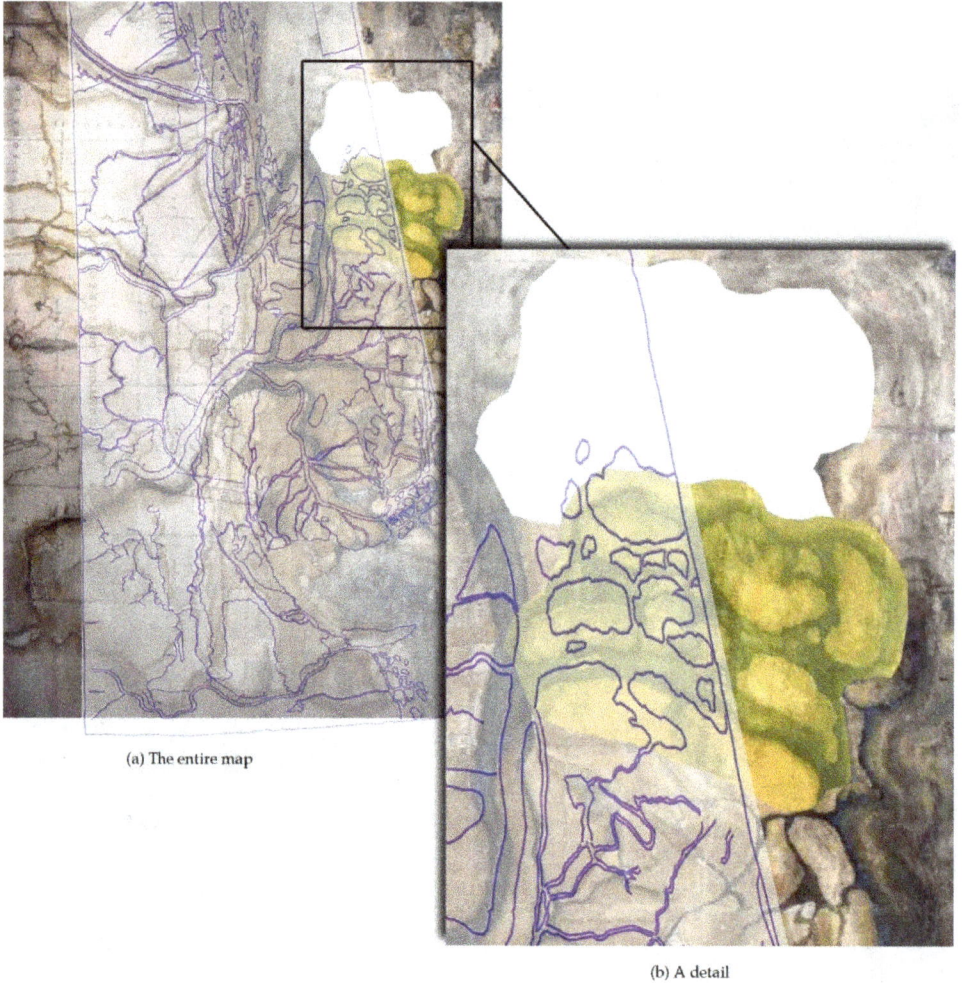

(a) The entire map

(b) A detail

Figure 15. Overlay of a georeferenced vectorization of the P map (in blue) on the "false case".

To overcome this impasse, a complex methodological approach would be adopted: it would be capable to check and take into account the metric information still preserved in the ancient maps, though partially hidden, and finally to regenerate the lacking topographic details or relocate them in a more reliable way. Only at that stage, a new phase of the study, based on a new and more reliable ancient map, will be triggered for various classes of researchers.

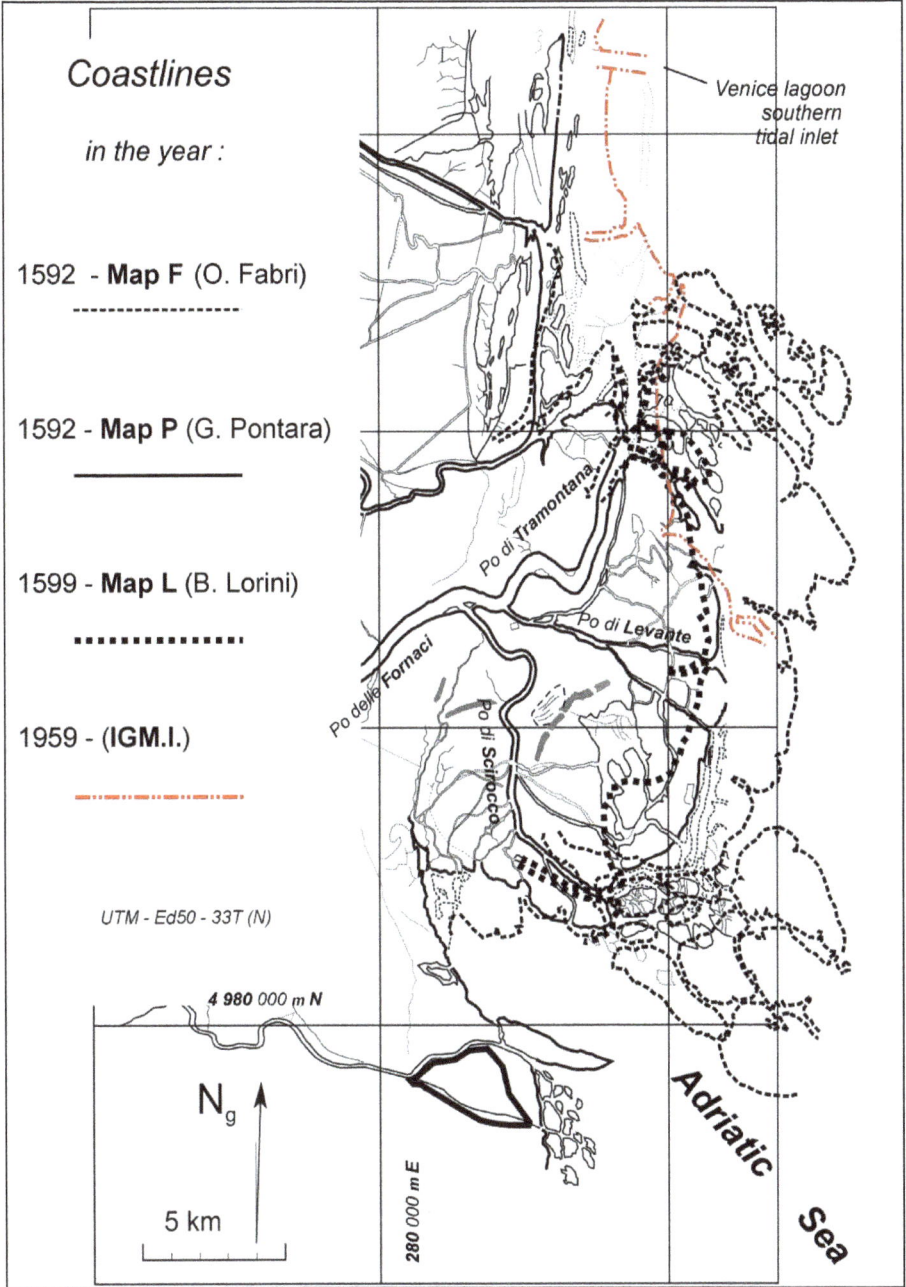

Figure 16. Coastline shape from the georeferencing of F, P and L maps on Italian modern cartography (Cremonini and Samonati, 2009; Cremonini 2010, Fig. 2).

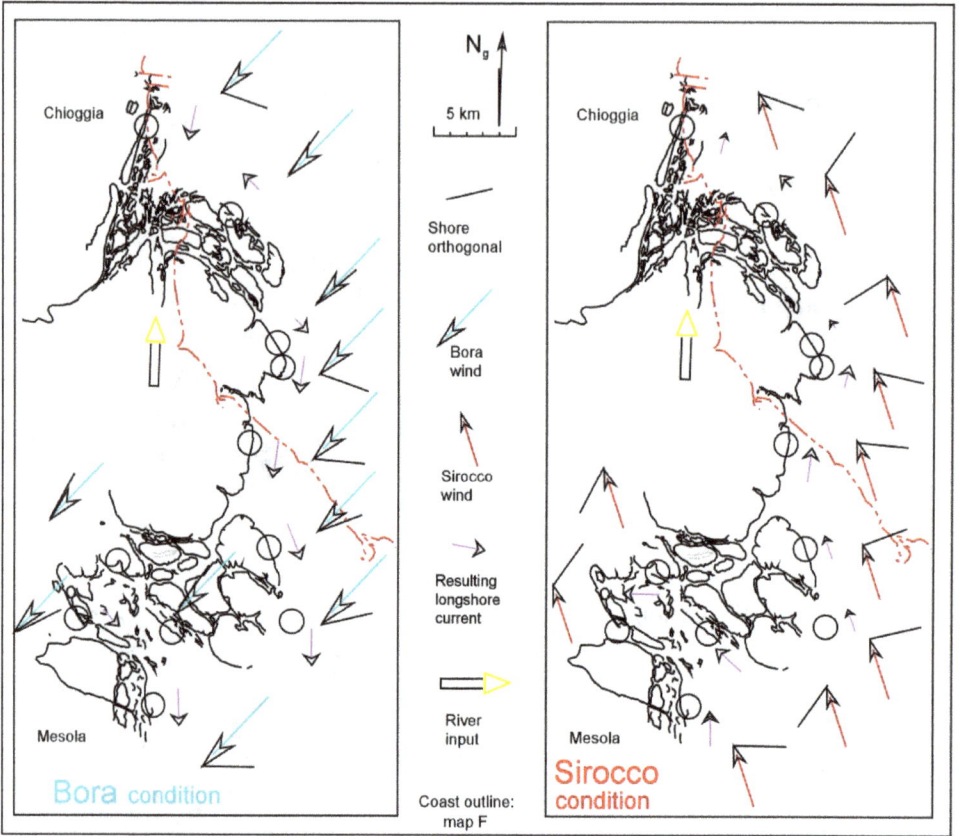

Figure 17. Long shore current directions induced by Bura and Sirocco winds on F map coastline profile (Cremonini and Samonati 2009; Cremonini 2010, Fig. 1).

6. Conclusion

In the study, careful recovery and analysis of some data concealed in three sixteenth century maps were done, and various technical signs preserved in them were identified and discussed. All these details show interesting information concerning the original survey performed to generate the maps, and help to comprehend the differences in the represented coastal areas among the analysed coeval maps. Notwithstanding the analysis cannot comprehensively explain all questions concerning the eroded coastal areas, due to the fact that existing landmarks cannot be found in disappeared areas, an attempt to explain the difference in the morphology of the coast was made. On the base of the recognition of a gross deformation in one map, a good example of a possible alleged author's intervention was given, attempting to restore the original appearance of the possible former map configuration. Finally, a comparison of the new product with present cartography and other coeval maps, via modern georeferencing methods, was able to support the hypothesis.

Finally, the study shows that pre-geodetic maps usually should not be regarded as completely original technical products but rather as an assemblage of various data-sets coming from previous sources. Thus, a study of ancient cartography should start from a philological analysis of the maps (i.e. search for origin, target, models, cultural atmosphere of that time) and it should subsequently become a "stratigraphic" reading of both relationships existing among the drawing details and accidents affecting the maps. A thorough knowledge of the ancient field surveying techniques (i.e. pathways, distances, contemporaneous use of various kinds of instruments, previous cartographic sources, etc.) is essential in order to understand genesis and size of the surveying errors recorded in the map, and therefore it is also essential to direct the georeferencing techniques towards the definition of the real topography and morphology of those areas that today no longer exist, e.g. due to natural environmental processes. In other words, the technical approach to ancient cartography cannot be reduced to a simple georeferencing attempt; the georeferencing should rather act as an essential support for a deep analysis of the maps, that in this way can become a more useful tool to various classes of researchers.

Author details

Gabriele Bitelli and Giorgia Gatta
Department of Civil, Environmental and Materials Engineering (DICAM),
Alma Mater Studiorum - University of Bologna, Bologna, Italy

Stefano Cremonini
Department of Earth and Geological-Environmental Sciences,
Alma Mater Studiorum - University of Bologna, Bologna, Italy

Acknowledgement

We are grateful to the Venice National Archive, for the consultation of the map by Fabri and the map by Lorini et al., and to the Engineering Faculty Library *Gian Paolo Dore* of the University of Bologna, for the consultation of the textbook by O. Fabri.

7. References

Adcock E.P, Varlamoff M.T, Kremp V (2004) *Principi dell'IFLA per la cura e il trattamento dei materiali di biblioteca.* Bari: IFLA (International Federation of Library Associations and Institutions).

Balletti C (2006) Georeference in the analysis of the geometric content of early maps. e-Perimetron, Vol. 1, No. 1. pp 32-42. Available: http://www.e-perimetron.org/Vol_1_1/Vol1_1.htm.

Benavides J., Koster E. (2006) Identifying surviving landmarks on historical maps, e-Perimetron, Vol. 1, No. 3. pp 287-296. Available: http://www.e-perimetron.org/Vol_1_3/Vol1_3.htm.

Bitelli G, Cremonini S, Gatta G (2010) Late Renaissance survey techniques revealed by three maps of the old Po river delta. E-perimetron, Vol. 5, No. 3. pp 172-175. Available: http://www.e-perimetron.org/Vol_5_3/Vol5_3.htm.

Bitelli G, Cremonini S, Gatta G (2009) Ancient maps comparisons and georeferencing techniques: a case study from the Po river delta (Italy). e-Perimetron, Vol. 4, No. 4. pp 221-228. Available:
http://www.e-perimetron.org/Vol_4_4/Vol4_4.htm.

Boutoura C, Livieratos E (2006) Some fundamentals for the study of the geometry of early maps by comparative methods. e-Perimetron, Vol. 1, No. 1. pp 60-70. Available: http://www.e-perimetron.org/Vol_1_1/Vol1_1.htm.

Cremonini S (2007a) *Questioni di geomorfologia costiera del delta del Po anteriormente al 1604. Evidenze dalla cartografia storica.* In: *Annali di Ricerche e Studi di Geografia*, 63, 3/4. pp. 53-67.

Cremonini S. (2007b) Some remarks on the evolution of the Po River plain (Italy) over the last four millennia. In: Marabini F., Galvani A., Ciabatti M. eds., China-Italy bilateral symposium on the coastal zone: evolution and safeguard, Bologna 4-8 November 2007. pp. 17-24.

Cremonini S, Samonati E (2009) Value of ancient cartography for geoenvironmental purposes. A case study from the Po river delta coast (Italy). In: *Geografia, Fisica e dinamica del Quaternario*, No. 32. pp 135-144.

Cremonini S. (2010) Climatic suggestions from the Po River delta coastline (Italy) at the beginning of the Little Ice Age. Proceedings of the China-Italy Bilateral Symposium on the coastal Zone and continental shelf evolutional trend, Bologna, October 5-8, 2010. pp. 114-119

Fabri O (1673) *L'uso della squadra mobile. Con la quale per teoria et per pratica si misura geometricamente ogni distanza, altezza, e profondità. S'impara à perticare, liuellare, et pigliare in dissegno le Città, Paesi, et Provincie. [...].* Padova: Gattella.

Jenny B, Hurny L (2011) Studying cartographic heritage: analysis and visualization of geometric distortions. Computers & Graphics 35-2. Elsevier. pp 402–411.

Livieratos E (2006) On the Study of the Geometric Properties of Historical Cartographic Representations. Cartographica, Vol. 41, issue 2. pp. 165-175.

Martini A. (1883) *Manuale di metrologia, ossia misure, pesi e monete in uso attualmente e anticamente presso tutti i popoli.* Torino: Loescher.

Panepinto E (2009) *Ottavio Fabri, Perito et Ingegnero Publico.* Degree thesis on European History and Geography. University of Verona, Faculty of Arts and Philosophy.

Chapter 3

Web Map Tile Services for Spatial Data Infrastructures: Management and Optimization

Ricardo García, Juan Pablo de Castro, Elena Verdú,
María Jesús Verdú and Luisa María Regueras

Additional information is available at the end of the chapter

1. Introduction

Web mapping has become a popular way of distributing online mapping through the Internet. Multiple services, like the popular Google Maps or Microsoft Bing Maps, allow users to visualize cartography by using a simple Web browser and an Internet connection. However, geographic information is an expensive resource, and for this reason standardization is needed to promote its availability and reuse. In order to standardize this kind of map services, the Open Geospatial Consortium (OGC) developed the Web Map Service (WMS) recommendation [1]. This standard provides a simple HTTP interface for requesting geo-referenced map images from one or more distributed geospatial databases. It was designed for custom maps rendering, enabling clients to request exactly the desired map image. This way, clients can request arbitrary sized map images to the server, superposing multiple layers, covering an arbitrary geographic bounding box, in any supported coordinate reference system or even applying specific styles and background colors.

However, this flexibility reduces the potential to cache map images, because the probability of receiving two exact map requests is very low. Therefore, it forces images to be dynamically generated on the fly each time a request is received. This involves a very time-consuming and computationally-expensive process that negatively affects service scalability and users' Quality of Service (QoS).

A common approach to improve the cachability of requests is to divide the map into a discrete set of images, called tiles, and restrict user requests to that set [2]. Several specifications have been developed to address how cacheable image tiles are advertised from server-side and how a client requests cached image tiles. The Open Source Geospatial Foundation (OSGeo) developed the WMS Tile Caching (usually known as WMS-C) proposal [3]. Later, the OGC released the Web Map Tile Service Standard (WMTS) [4] inspired by the former and other similar initiatives.

Most popular commercial services, like Google Maps, Yahoo Maps or Microsoft Virtual Earth, have already shown that significant performance improvements can be achieved by adopting this methodology, using their custom tiling schemes.

The potential of tiled map services is that map image tiles can be cached at any intermediate location between the client and the server, reducing the latency associated to the image generation process. Tile caches are usually deployed server-side, serving map image tiles concurrently to multiple users. Moreover, many mapping clients, like Google Earth or Nasa World Wind, have embedded caches, which can also reduce network congestion and network delay.

This chapter deals with the algorithms that allow the optimization and management of these tile caches: population strategies (*seeding*), tile pre-fetching and cache replacement policies.

2. Tiling schemes

Maps have been known for a long time only as printed on paper. Those printed cartographic maps were static representations limited to a fixed visualization scale with a certain Level Of Detail (LOD). However, with the development of digital maps, users can enlarge or reduce the visualized area by zooming operations, and the LOD is expected to be updated accordingly.

The adaptation of map content is strongly scale-dependent: A small-scale map contains less detailed information than a large scale map of the same area. The process of reducing the amount of data and adjusting the information to the given scale is called cartographic generalization, and it is usually carried out by the web map server [5].

In order to offer a tiled web map service, the web map server renders the map across a fixed set of scales through progressive generalization. Rendered map images are then divided into tiles, describing a tile pyramid as depicted in Figure 1.

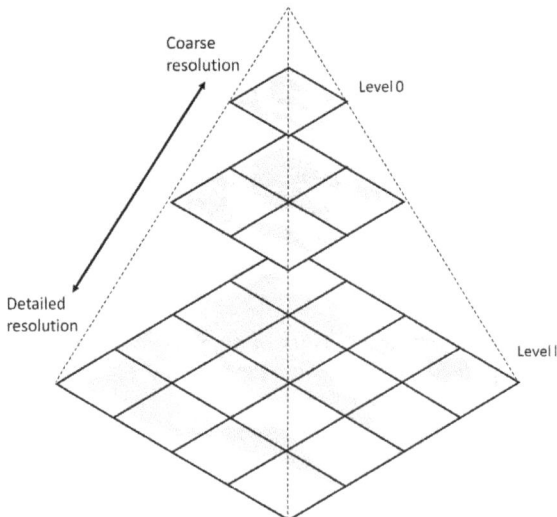

Figure 1. Tile pyramid representation.

For example, Microsoft Bing Maps uses a tiling scheme where the first level allows representing the whole world in four tiles (2x2) of 256x256 pixels. The next level represents the whole world in 16 tiles (4x4) of 256x256 pixels and so on in powers of 4. A comprehensive study on tiling schemes can be found in [2].

2.1. Simplified model

Given the exponential nature of the scale pyramid, the resource consumption to store map tiles results often prohibitive for many providers when the cartography covers a wide geographic area for multiple scales. Consider for example that Google's BigTable, which contains the high-resolution satellite imagery of the world's surface as shown in Google Maps and Google Earth, contained approximately 70 terabytes of data in 2006 [6].

Besides the storage of map tiles, many caching systems also maintain metadata associated to each individual tile, such as the time when it was introduced into the cache, the last access to that object, or the number of times it has been requested. This information can then be used to improve the cache management; for example, when the cache is out of space, the LRU (*Least Recently Used*) replacement policy uses the last access time to discard the least recently used items first.

However, the space required to store the metadata associated to a given tile may only differ by two or three orders of magnitude to the one necessary to store the actual map image object. Therefore, it is not usually feasible to work with the statistics of individual tiles. To alleviate this problem, a simplified model has been proposed by different researchers. This model groups the statistics of adjacent tiles into a single object [7]. A grid is defined so all objects inside the same grid section are combined into a single one. The pyramidal structure of scales is therefore transformed in some way in a prism-like structure with the same number of items in all the scales.

3. Web Map Server workload

In order to deal with this complexity some cache management algorithms have been created. However, the efficiency of the designed algorithms usually depends on the service's workload. Because of this, prior to diving into the details of the cache management policies, a workload characterization of the WMS services need to be shown. Lets take some real-life examples for such characterization: trace files from two different tiled web map services, Cartociudad[1] and IDEE-Base[2], provided by the National Geographic Institute (IGN)[3] of Spain, are presented in this chapter.

Cartociudad is the official cartographic database of the Spanish cities and villages with their streets and roads networks topologically structured, while IDEE-Base allows viewing the Numeric Cartographic Base 1:25,000 and 1:200,000 of the IGN.

Available trace files were filtered to contain only valid web map requests according to the WMS-C recommendation. Traces from Cartociudad comprise a total of 2.369.555 requests

---

[1] http://www.cartociudad.es
[2] http://www.idee.es
[3] http://www.ign.es/ign/main/index.do?locale=en

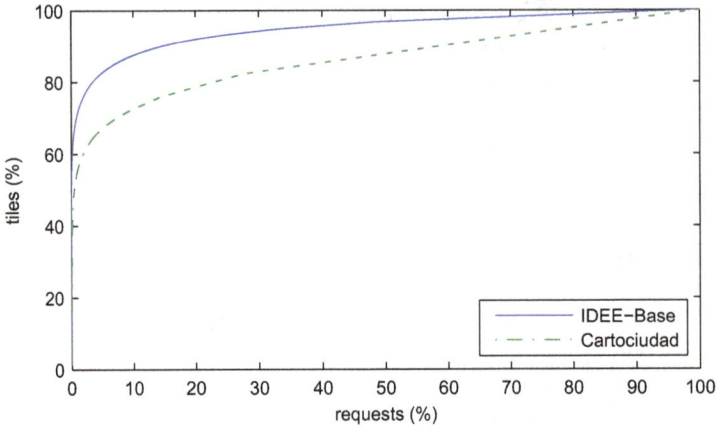

Figure 2. Percentile of requests for the analyzed services.

received from the 9th December of 2009 to 13th May in 2010. IDEE-Base logs reflect a total of 16.891.616 requests received between 15th March and 17th June in 2010.

It must be noted that the performance gain achieved by the use of a tile cache will vary depending on how the tile requests are distributed over the tiling space. If those were uniformly distributed, the cache gain would be proportional to the cache size. However, lucky for us, it has been found that tile requests usually follow a heavy-tailed Pareto distribution, as shown in Figure 2. In our example, tile requests to the Cartociudad map service follow the 20:80 rule, which means that the 20% of tiles receive the 80% of the total number of requests. In the case of IDEE-Base, this behaviour is even more prominent, where the 10% of tiles receive almost a 90% of total requests. Services that show Pareto distributions are well-suited for caching, because high cache hit ratios can be found by caching a reduced fraction of the total tiles.

Figure 3 and Figure 4 show the distribution of tile requests to each resolution level of the tile pyramid for the analyzed services. The maximum number of requests is received at resolution level 4 for both services. This peak is due to the fact that this is the default resolution on the initial rendering of the popular clients in use with this cartography, as it allows the visualization of the whole country on a single screen. As can be observed, the density of requests (requests/tile) is higher at low resolution levels than at higher ones. Because of this, a common practice consists in pregenerating the tiles belonging to the lowest resolution levels, and leave the rest of tiles to be cached on demand when they are first requested.

4. Tile cache implementations

With the standardization of tiled web map services, multiple tile cache implementations have appeared. Between them, the main existent implementations are: TileCache, GeoWebCache and MapProxy. A comparison between these implementations is summarized in Table 1.

As can be seen, TileCache and MapProxy are both implemented in Python (interpreted language), while GeoWebCache is implemented in Java (compiled language). These three

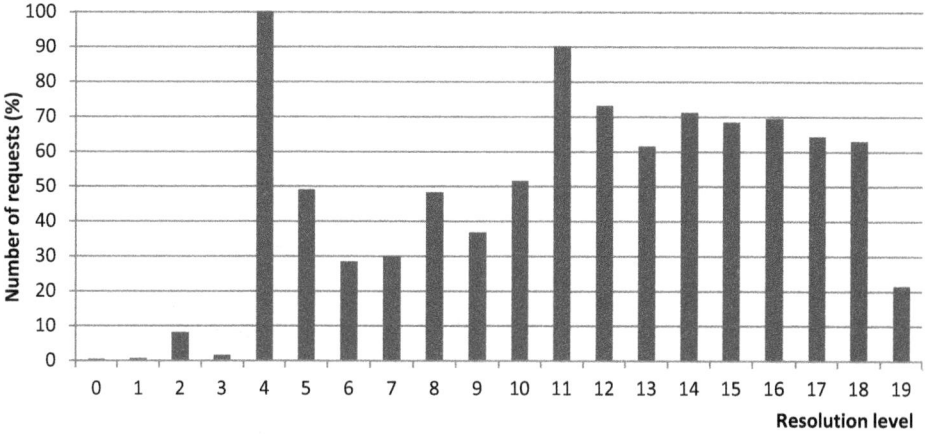

Figure 3. Distribution of requests along the different resolution levels for Cartociudad service.

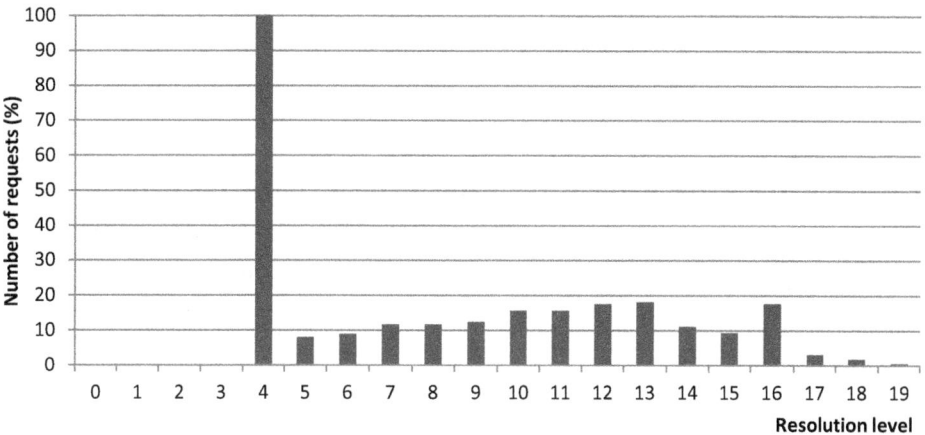

Figure 4. Distribution of requests along the different resolution levels for IDEE-Base service.

services implement the WMS-C, TMS and KML service interfaces. GeoWebCache and MapProxy also offer the WMTS service from OGC. In addition, GeoWebCache can recombine and resample tiles to answer arbitrary WMS requests, and can also be used to serve maps to Google Maps and Microsoft Bing Maps.

All these services offer the possibility of storing map image tiles directly in the file system. TileCache and GeoWebCache also support the MBTiles speficication[4] for storing tiled map data in a SQLite database for immediate use and for transfer. MapProxy supports the Apache CouchDB[5], a document-oriented database that can be queried and indexed in a MapReduce

---

[4] http://mapbox.com/mbtiles-spec/
[5] http://couchdb.apache.org/

	TileCache	GeoWebCache	MapProxy
Company	Metacarta Labs	OSGeo	Omniscale
Implementation	Python	Java	Python
Supported services	WMS-C, TMS, KML	WMS, WMS-C, TMS WMTS, KML, Google Maps, Bing Maps	WMS, WMS-C, TMS WMTS, KML
Tile storage	Disk, GoogleDisk, Memcached, Amazon S3, MBTiles	Disk	Disk, MBTiles CouchDB
Tile metadata storage	No	Yes	Yes
Replacement policies	LRU	LRU, LFU	None
Seeding regions	bounding box center and radius	bounding box	bounding box WKT polygons any OGR source
Supports Meta Tiles	Yes	Yes	Yes
Supports Meta Buffer	Yes	Yes	Yes
Reprojection on-the-fly	No	Yes (with Geoserver)	Yes (native)

Table 1. Comparison of features between different open-source tile cache implementations: TileCache, GeoWebCache and MapProxy.

fashion, as backend to store tiles. Moreover, TileCache can store map tiles in the cloud through Amazon S3[6] or to maintain them in memory using Memcached[7].

GeoWebCache maintain tile metadata, such as the last access time or the number of times that each tile has been requested. By using this metadata, it supports the LRU and LFU replacement policies. TileCache supports LRU by using the operating system's time of last access.

These services allow to specify a geographic region for automatically seeding tiles. For example, TileCache can be configured to seed a particular regions defined by a rectangular bounding box or a circle by specifying its center and radius. GeoWebCache supports only the

---

[6] http://aws.amazon.com/es/s3/
[7] http://memcached.org/

former. MapProxy offers three different ways to describe the extent of a seeding or cleanup task: a simple rectangular bounding box, a text file with one or more polygons in WKT format, or polygons from any data source readable with OGR (e.g. Shapefile, PostGIS).

These three services support both metatiling and meta-buffer methods. The meta-buffer adds extra space at the edges of the requested area.

When a request of a tile in an unsupported coordinate reference system (CRS) is received, both GeoWebCache and MapProxy supports the reprojection on the fly from one of the available CRSs to the specified one. The former achieves this using GeoServer, while the latter offers it natively.

5. Cache management algorithms

Significant improvements can be achieved by using a cache of map tiles, like the ones discussed above. However, adequate cache management policies are needed, especially in local SDIs with lack of resources. In this section, our contributions to the main cache strategies are presented: cache population (or *seeding*), cache replacement and tile prefetching.

5.1. Cache population

Anticipating the content that users will demand can guide server administrators to know which tiles to pregenerate and to include in their server-side caches of map tiles. With this objective in mind, a predictive model that uses variables known to be of interest to Web map users, such as populated places, major roads, coastlines, and tourist attractions, is presented in [8].

In contrast, we propose a descriptive model based on the mining of the service's past history [7]. Past history can be easily extracted, for example, from server logs. The advantage of this model is that it is able to determine in advance which areas are likely to be requested in the future based exclusively on past accesses, and it is therefore very simple.

In order to experiment with the proposed model, real-world logs from the IDEE-Base nation-wide public web map service have been used. Request logs were divided in two time ranges of the same duration. The first one was used as source to make predictions and the second one was used to prove the predictions created previously. Due to the difficulty of working with the statistics of individual tiles, the simplified model presented in Section 2 has been used. Concretely, the experiment was conducted with the simplified model to the grid cell defined by the level of resolution 12.

Figure5 shows the heatmaps of requests extracted from the web server logs of IDEE-Base service, propagated to level 12 through the proposed model. These figures demonstrate that some entities such as coast lines, cities and major roads are highly requested. These elements could be used as entities for a predictive model to identify priority objects, as explained in [8].

These figures show that near levels are more related than distant ones, but all of them share certain similarity. This relationships between resolution levels encourages the use of statistics collected in a level to predict the map usage patterns in another level with detailer resolution. For example, as shown in Figure5(c) and Figure5(e), resolution levels 14 and 16 are very

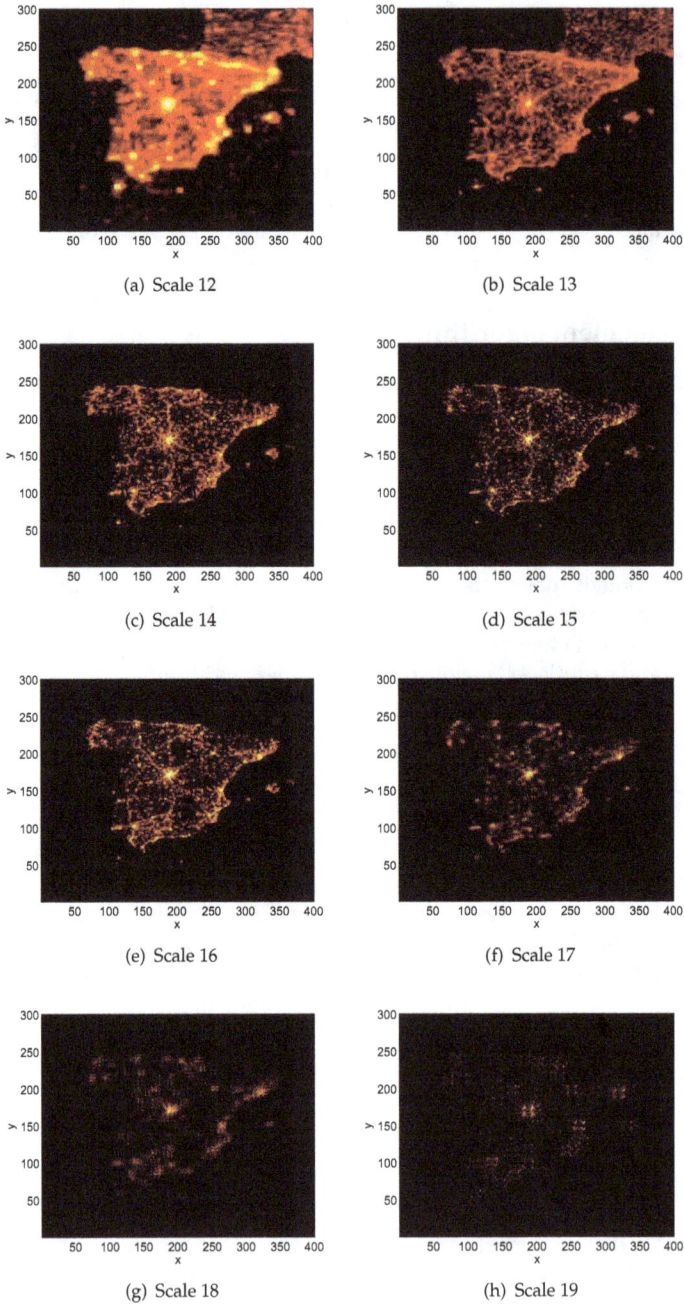

(a) Scale 12

(b) Scale 13

(c) Scale 14

(d) Scale 15

(e) Scale 16

(f) Scale 17

(g) Scale 18

(h) Scale 19

Figure 5. Heatmap of the requests to the *IDEE-BASE* service propagated from levels 12-19 to level 12.

correlated. It is easier to work with the statistics of level 14 than with those of level 16 which has much more elements.

Table 2 represents the hit percentage achieved by using this model for the IDEE-Base service. This table shows the percentage of hits obtained for the level identified by the column index from the statistics collected in the level identified by the row index. Last column shows the resources consumption, as a percentage of cached tiles. Last row collects the results of combining the statistics of all levels to make predictions over every level. Shadowed cell in Table 2 indicates that using retrieved statistics of level 13 as the prediction source, a hit rate of 92.1573% is obtained for predictions made in the level 18, being necessary the storage of a 25.8049% of the tiles in cache.

	12	13	14	15	16	17	18	19	resources
12	98.6417	98.9362	99.3573	99.4737	99.6901	99.2637	99.2993	94.7561	40.2172
13	87.8163	93.5760	95.8939	96.4146	97.4372	95.2686	92.1573	75.5073	25.8049
14	53.0529	61.5825	86.7783	88.2709	91.1807	81.6460	63.4527	43.9129	9.3302
15	37.1553	47.9419	77.9136	84.0861	83.7095	69.9489	57.0746	33.3348	5.2354
16	46.9387	57.5640	84.3110	86.7747	91.8272	78.7670	64.3781	41.8433	7.7686
17	30.2021	37.6348	57.0138	60.1330	62.2834	69.5106	55.5134	23.4647	3.2676
18	23.5791	25.5913	41.7535	46.1559	45.8693	41.4502	61.9763	33.3799	2.3291
19	8.8690	8.6848	12.4556	13.1338	14.3302	12.2756	13.6932	44.1113	1.2295
prop	98.9340	99.3080	99.6074	99.6321	99.7763	99.4244	99.4308	97.2315	41.3647

Table 2. Percentage (%) of cache hits through the simplified model obtained from *IDEE-BASE* logs, using the mean of the normalized frequencies as the probability threshold.

Nevertheless, it must be noted that the main benefit of using a partial cache is not the reduction in the number of cached tiles. The main benefits are the savings in storage space and generation time. As explained in [8], the amount of saved tiles is bigger than the storage saving. It reveals that the most interesting tiles come at a bigger cost. Mainly, popular areas are more complex, and it is necessary more disk space to store them.

Figure6 and Figure7 represent the cache hit ratios obtained by the simplified model for the *IDEE-BASE* service. This model bases its operation on the knowledge of past accesses, assuming a certain stationarity of requests; it assumes that map regions that have been popular in the past will maintain its popularity in the future. However, from a certain percentage of cached objects, identified by the continuous vertical line, the simplified model is not able to make predictions. Tiles situated at the right of this line correspond to objects that have never been requested so are not collected in server logs. To complete the model, these never-requested tiles have been randomly selected for caching, yielding a linear curve for this interval.

Results demonstrate that the simplified model obtains better results for predicting user behavior from near resolution levels. For low-resolution levels high cache hit ratios are achieved by using a reduced subset of the total tiles. However, descending in the scale pyramid, the requested objects percentage decreases, so the model prediction range and its ability to make predictions decrease too. For future work, instead of randomly selecting objects for caching in this interval, interesting features could be identified and used to define priority objects.

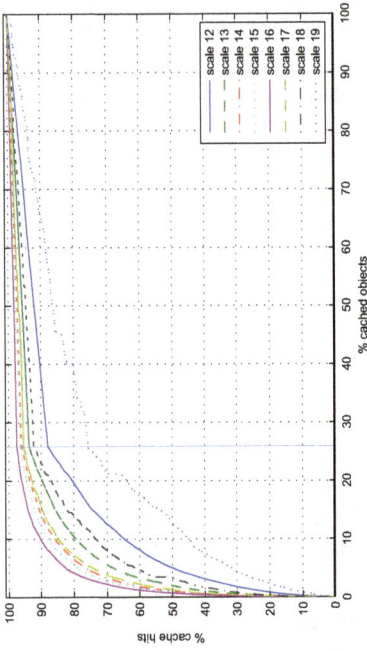

(a) Scale 12

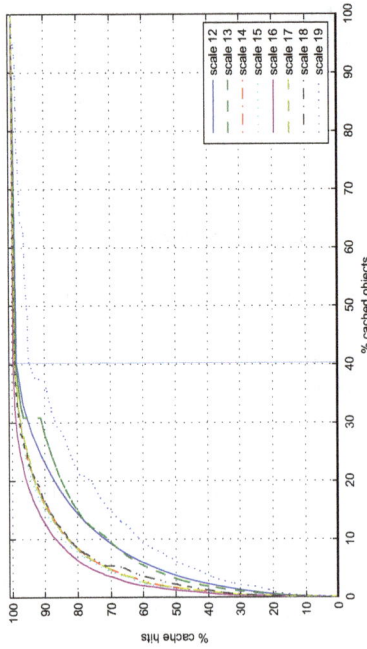

(c) Scale 14

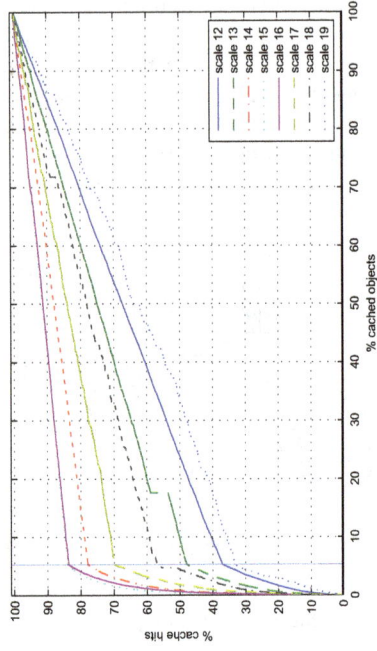

(b) Scale 13

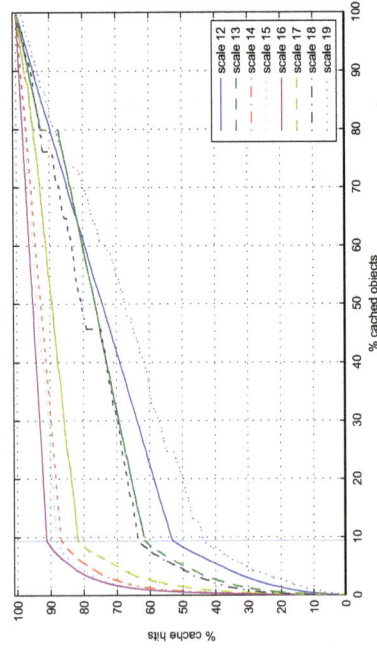

(d) Scale 15

Figure 6. Percentage of hits vs cached objects for *IDEE-BASE* service through the simplified model. Scales 12 to 15.

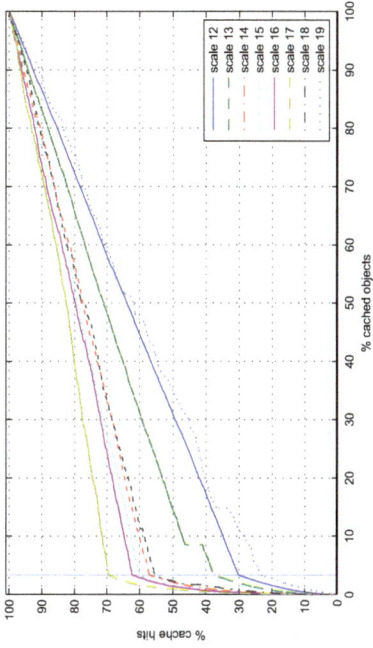

(a) Scale 16

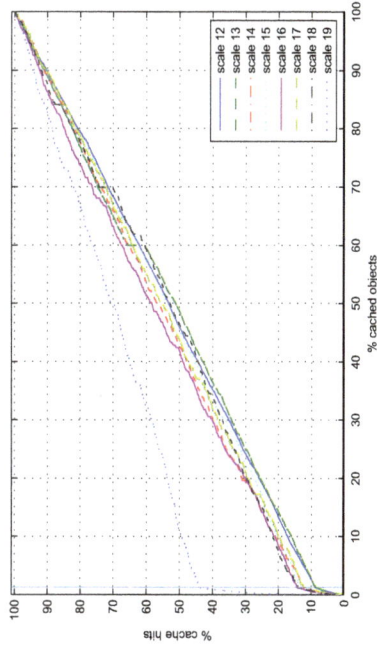

(b) Scale 17

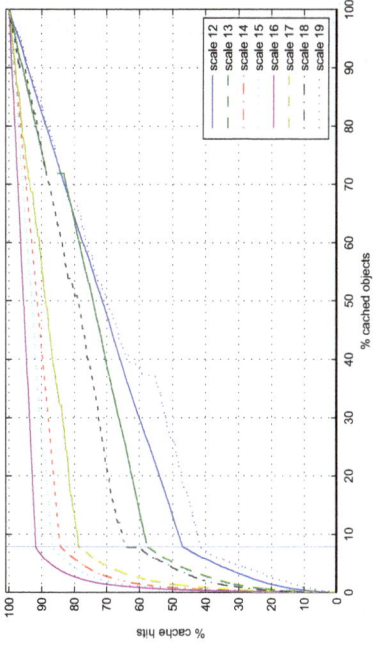

(c) Scale 18

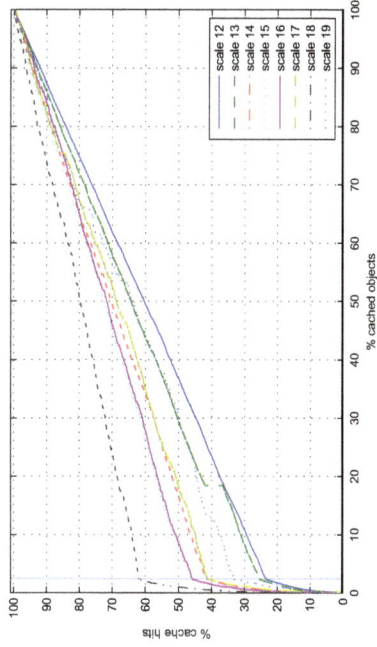

(d) Scale 19

Figure 7. Percentage of hits vs cached objects for *IDEE-BASE* service through the simplified model. Scales 16 to 19.

5.2. Tile pre-fetching

For a given tile request, tile pre-fetching methods try to anticipate which tiles will be requested immediately afterwards. There are several works in the literature that address object prefetching in Web GIS: [9, 10] approximate which tiles will be used in advance based on the global tile access pattern of all users and the semantics of query; [11, 12] use an heuristic method that considers the former actions of a given user.

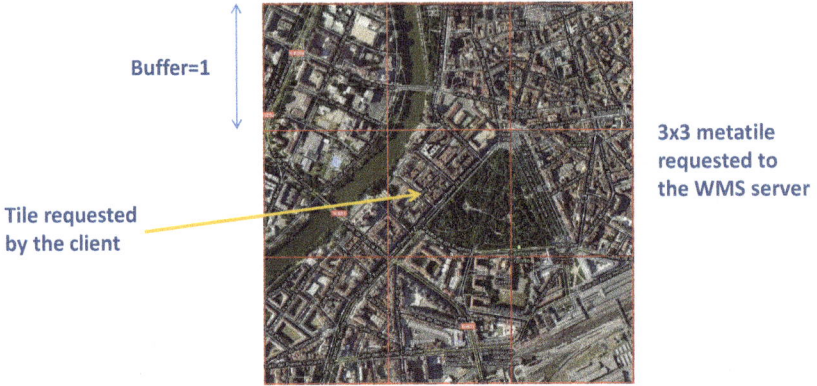

Figure 8. Metatile 3x3 centered in the requested tile.

We propose another pre-fetching strategy, known as metatiling, that works as follows [13]: when the proxy receives a tile request from a client and a cache miss is produced, it requests a larger image tile (called metatile) to the remote backend. This metatile includes the requested tile and also the surrounding ones contained in a specified buffer, as shown in Figure 8. Then, the proxy cuts the metatile into individual tiles, returns the requested tile to the client, and stores all these fragments into the cache, as shown in Figure 9. The main advantage of metatiling is that it can reduce the bottleneck between the proxy cache and the remote Web Map Server.

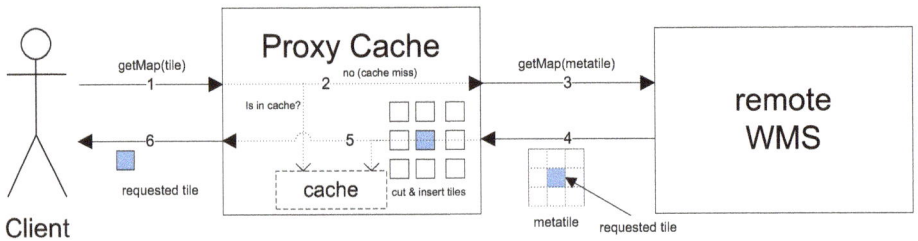

Figure 9. Tile request flow with metatiling.

Moreover metatiling reduces the problem of duplicating the labeling of features that span more than one tile. This problem is illustrated in Figure 10. Depending on the WMS server's configuration, this feature can be labeled once on each tile (Figure 10(a)). By increasing the geographic bounding box of tile requests, the WMS server avoids label duplicates (Figure 10(b)).

(a) Buffer=0

(b) Buffer=2

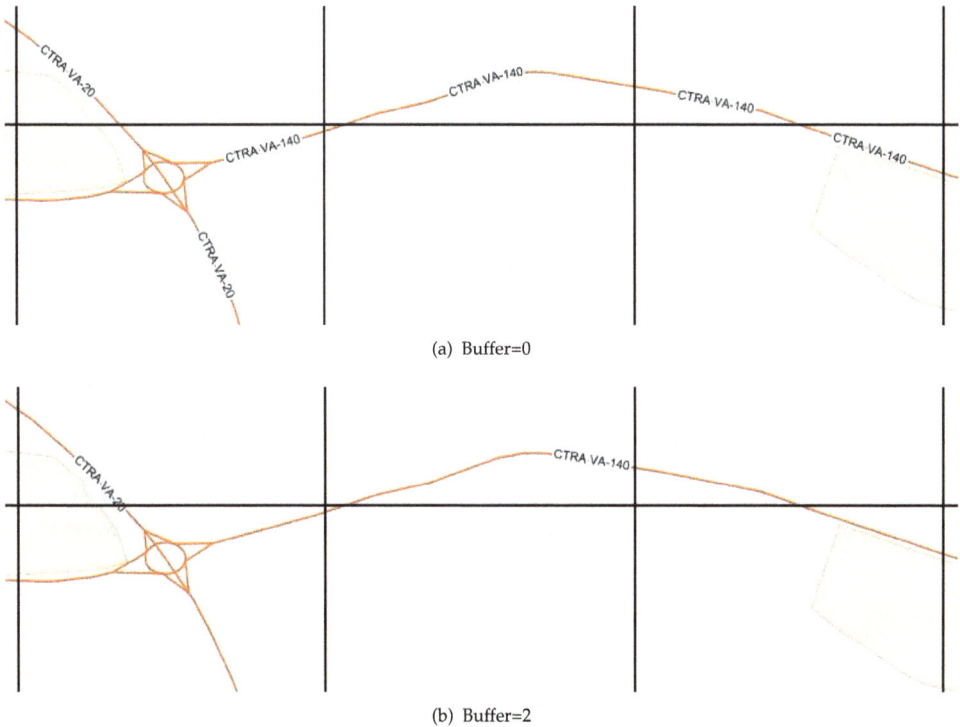

Figure 10. WMS labelling issues. (a) Requesting individual tiles yields duplicate labels between adjacent tiles. (b) With metatiling labels are not duplicated.

The analyzed tile cache implementations (see Section 4) allow users to configure the size of metatiles. For a given request, the cache orders a metatile of pre-configured size to the WMS server, centered on the requested tile. Considering a scenario where the cache is neither complete nor empty, this selection of the area to generate may not be very efficient, because it is probable that some of the tiles contained in the metatile would already be cached.

Under the assumption that the surrounding area of the requested tile is not uniformly cached, a novel algorithm for the optimal selection of the metatiles to generate has been developed. This procedure, illustrated in Figure 11, seeks to obtain, based on the current state of the cache, the metatile that contains the requested tile (but not necessarily centered in it) and that provides the system with the maximum new information.

In order to validate the hypothesis that a performance improvement can be achieved by using metatiles, the following experiment has been realized. A total of 2000 different tiles have been requested to the CORINE (*CoORdination of INformation of the Environment*) service[8] proxied by the *WMSCWrapper* tile cache. The experiment has been repeated for different metatile sizes, always starting from an empty-cache state. The mean latencies measured for each configuration are collected in Table 3.

---

[8] http://www.ign.es/ign/layoutIn/corineLandCover.do

buffer (B)	$\tau_{m,metatile}$	$\tau_{m,metatile\_n}$	$Gain_{metatiling}$
0 (no metatiling)	1454,10 ms	1454,10 ms	1
1 (metatile 3x3)	2933,94 ms	325,99 ms	4,46
2 (metatile 5x5)	5660,63 ms	226,42 ms	6,42
3 (metatile 7x7)	9561,54 ms	195,13 ms	7,45

Table 3. Mean latencies to obtain an object from the WMS original service for different metatile sizes. Proxy cache: *WMSCWrapper*; remote WMS: CORINE Land Cover.

The first column of the table shows the mean latency of a cache miss $\tau_{m,metatile}$ for different metatile sizes. This delay includes the transmisission and propagation delays in the network, the map image generation time in the remote web map service and the processing time in the proxy cache. The values of the second column $\tau_{m,metatile\_n}$ are computed by normalizing those of the first column by the number of tiles encompassed by each metatile ($[2B + 1]^2$). The last column shows the cache gain achieved by the use of metatiling, computed as the average acceleration in the delivery of a tile versus not using metatiling, as depicted in Equation 1.

$$Gain_{metatiling}(B) = \frac{\tau_{m,metatile\_n}(0)}{\tau_{m,metatile\_n}(B)} \qquad (1)$$

Results reflect that the latency involved in the request of a metatile increases with the buffer size. However, it increases in less proportion than the number of tiles it is compossed by. Therefore, the mean latency to obtain each individual tile decreases when increasing the size of the metatile requested to the remote web map service. In other words, it is faster to retrieve a metatile composed by $n \times n$ tiles than the n^2 tiles individually.

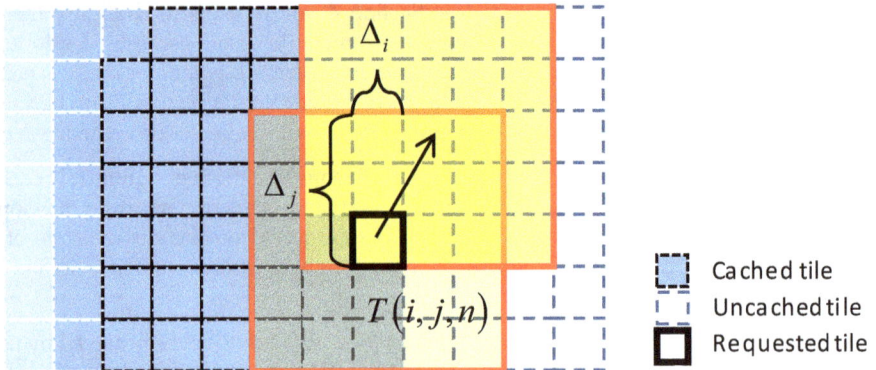

Figure 11. Metatile selection algorithm.

A limiting factor when choosing the metatile size is the overhead in memory consumption required to generate the map image. For example, by default the maximum amount of memory that a single request is allowed to use in *Geoserver* is 16MB, which are sufficient to render a 2048x2048 image at 4 bytes per pixel, or a 8x8 meta-tile of standard 256x256 pixel tiles.

Table 3 shows the maximum gain that can be achieved by the use of metatiling techniques. This maximum gain occurs when the whole metatile is used to cache new tiles that were not yet cached. While this is the case when automatically seeding tiles in sequential order with non-overlapping metatiles from an empty cache or in the early startup of the service, it would be useful to evaluate metatiling in the most general scenario where the cache is partially filled. In that case, each metatile is likely to add redundant information, since it is probable that some of the tiles encompassed by the metatile were already cached, thus reducing the effective gain of this method.

The performance of metatiling during dynamic cache population with users' requests has been evaluated using the *WMSCWrapper* tile cache, described in Section 4. Simulations were driven by trace files from the public WMS-C tiled web map service of Cartociudad. Being traces recorded in a real, working system, these logs represent a more realistic pattern of user behavior than a synthetic pattern. The CORINE WMS service was used as remote backend.

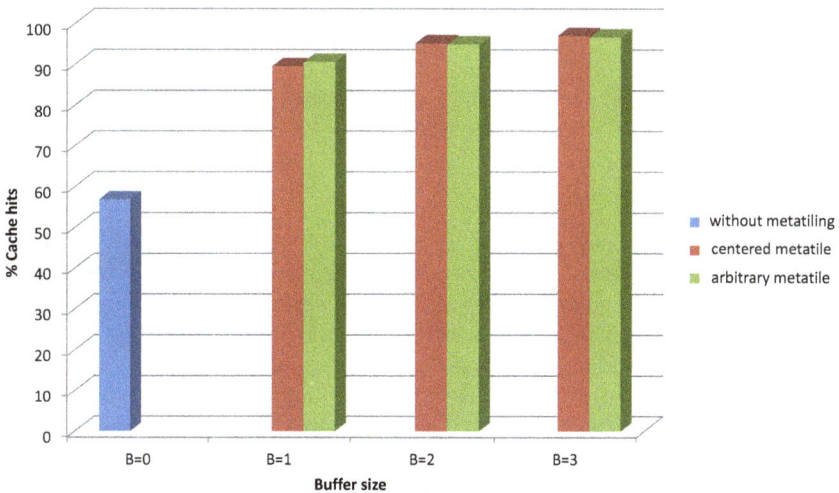

Figure 12. Cache-hit ratios obtained for different buffer sizes and metatile configurations.

A total of 1.000.000 requests were made to the cache. The experiment was repeated for different metatile configurations. For each configuration, the cache-hit ratio and the number of cached tiles after task completion have been collected, starting from an empty cache. Results are shown in Figure 12 and Figure 13.

As can be shown, both the cache-hit ratio and the number of cached tiles grow with the buffer size. For a fixed buffer size, both metatiling strategies (centered and minimum-correlation) obtain similar results. However, the number of cached objects is significantly improved with the minimum-correlation configuration. The improvement increases with the metatile size.

Thus, the advantage achieved with the minimum-correlation metatile configuration is that, maintaining the cache misses, and therefore maintaining the number of requests to the remote

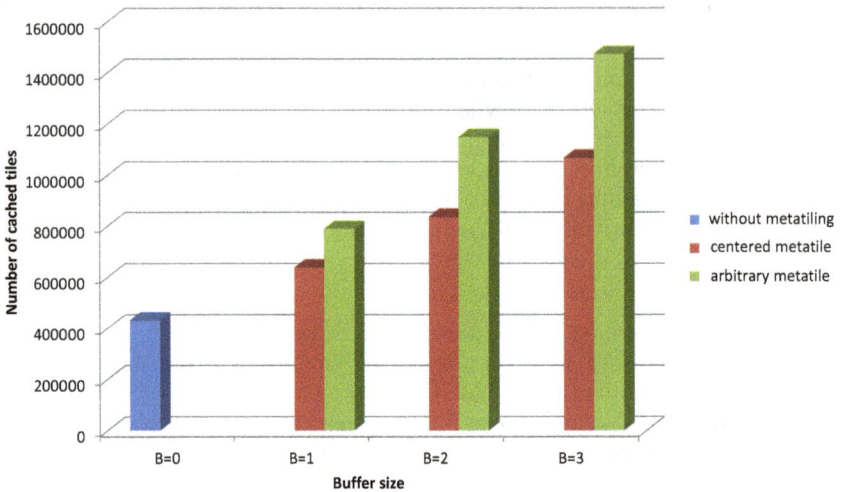

Figure 13. Number of cached tiles for different buffer sizes and metatile configurations.

WMS server, a broaden population of the cache is achieved. These extra pre-generated map image tiles stored in the cache will allow a faster delivery of future requests.

5.3. Cache replacement policies

When the tile cache runs out of space, it is necessary to determine which tiles should be replaced by the new ones. Most important characteristics of Web objects, used to determine candidate objects to evict in Web cache replacement strategies, are: *recency* (time since the last reference to the object), *frequency* (number of times the object has been requested), *size* of the Web object and *cost* to fetch the object from its origin server. These properties classifies replacement strategies as recency-based, frequency-based, recency/frequency-based, function-based and randomized strategies [14]. Recency-based strategies exploit the temporal locality of reference observed in Web requests, being usually extensions of the well-known LRU strategy, which removes the least recently referenced object. Another popular recency-based method is the Pyramidal Selection Scheme (PSS) [15]. Frequency-based strategies rely on the fact that popularity of Web objects is related to their frequency values, and are built around the LFU strategy, which removes the least frequently requested object. Recency/frequency-based strategies combine both, recency and frequency information, to take replacement decisions. Function-based strategies employ a general function of several parameters to make decisions of which object to evict from the cache. This is the case of GD-Size [16], GDSF [17] and Least-Unified Value (LUV) [18]. Randomized strategies use a non-deterministic approach to randomly select a candidate object for replacement.

For a further background, a comprehensive survey of web cache replacement strategies is presented in ([14]). According to that work, algorithms like GD-Size, GDSF, LUV and PSS were considered "good enough" for caching needs at the time it was published in 2003. However, the explosion of web map traffic did not happen until a few years later.

In this section, we propose a cache replacement algorithm that uses a neural network to estimate the probability of a tile request occurring before a certain period of time, based on the previously discussed properties of tile requests: recency of reference, frequency of reference, and size of the referenced tile [19, 20]. Those tiles that are not likely to be requested shortly are considered as good candidates for replacement.

5.3.1. Related work

The use of neural networks for cache replacement was first introduced by Khalid [21], with the KORA algorithm. KORA uses backpropagation neural network for the purpose of guiding the line/block replacement decisions in cache. The algorithm identifies and subsequently discards the dead lines in cache memories. It is based on previous work by [22], who suggested the use of a shadow directory in order to look at a longer history when making decisions with LRU. Later, an improved version of the former, KORA-2, was proposed [23, 24]. Other algorithms based on KORA were also proposed [25, 26]. A survey on applications of neural networks and evolutionary techniques in web caching can be found in [27]. [28–32] proposes the use of a backpropagation neural network in a Web proxy cache for taking replacement decisions. A predictor that learns the patterns of Web pages and predicts the future accesses is presented in [33]. [34] discusses the use of neural networks to support the adaptivity of the Class-based Least Recently Used (C-LRU) caching algorithm.

5.3.2. Neural network cache replacement

Artificial neural networks (ANNs) are inspired by the observation that biological learning systems are composed of very complex webs of interconnected neurons. In the same way, ANNs are built out of a densely interconnected group of units. Each artificial neuron takes a number of real-valued inputs (representing the one or more dendrites) and calculates a linear combination of these inputs. The sum is then passed through a non-linear function, known as *activation function* or *transfer function*, which outputs a single real-value, as shown in Figure 14.

In this work, a special class of layered feed-forward network known as multilayer perceptron (MLP) has been used, where units at each layer are connected to all units from the preceding layer. It has an input layer with three inputs, two-hidden layers each one comprised of 3 hidden nodes, and a single output (Figure 15). According to the standard convention, it can be labeled as a 3/3/3/1 network. It is known that any function can be approximated to arbitrary accuracy by a network with three layers of units [35].

Learning an artificial neuron involves choosing values for the weights so the desired output is obtained for the given inputs. Network weights are adjusted through supervised learning using subsets of the trace data sets, where the classification output of each request is known. Backpropagation with momentum is the used algorithm for training. The parameters used for the proposed neural network are summarized in Table 4.

The neural network inputs are three properties of tile requests: recency of reference, frequency of reference, and the size of the referenced tile. These properties are known to be important

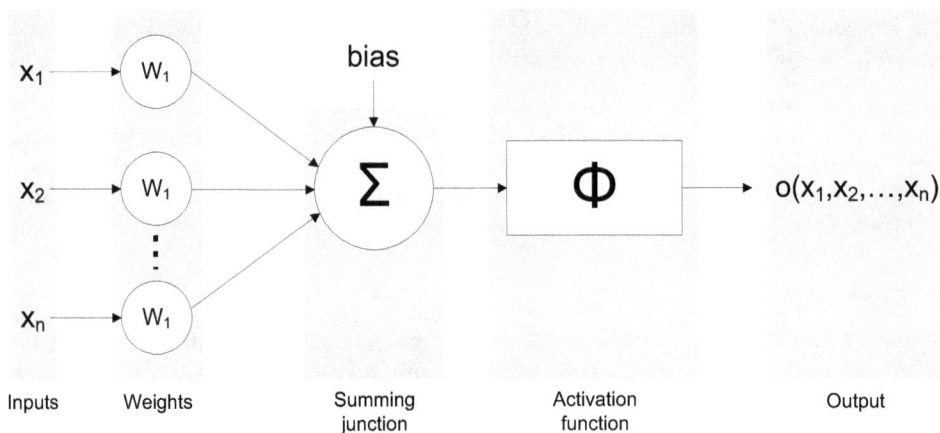

Figure 14. Artificial neuron.

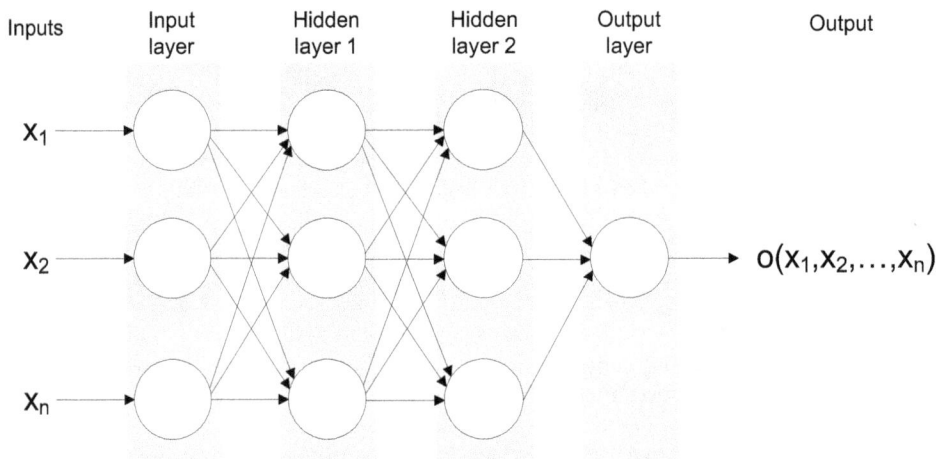

Figure 15. Proposed two-hidden layer feed-forward artificial neural network.

in web proxy caching to determine object cachability. Inputs are normalized so that all values fall into the interval $[-1, 1]$, by using a simple linear scaling of data as shown in Equation 2, where x and y are respectively the data values before and after normalization, x_{min} and x_{max} are the minimum and maximum values found in data, and y_{max} and y_{min} define normalized interval so $y_{min} \leq y \leq y_{max}$. This can speed up learning for many networks.

$$y = y_{min} + (y_{max} - y_{min}) \times \frac{x - x_{min}}{x_{max} - x_{min}} \tag{2}$$

Recency values for each processed tile request are computed as the amount of time since the previous request of that tile was made. Recency values calculated this way do not address the case when a tile is requested for the first time. Moreover, measured recency values could be too disparate to be reflected in a linear scale.

To address this problem, a sliding window is considered around the time when each request is made, as done in [28]. With the use of this sliding window, recency values are computed as shown in Equation 3.

$$recency = \begin{cases} max(SWL, \Delta T_i) & \text{if object } i \text{ was requested before} \\ SWL & otherwise \end{cases} \quad (3)$$

where ΔT_i is the time since that tile was last requested.

Recency values calculated that way can already be normalized as stated before in Equation 2.

Frequency values are computed as follows. For a given request, if a previous request of the same tile was received inside the window, its frequency value is incremented by 1. Otherwise, frequency value is divided by the number of windows it is away from. This is reflected in Equation 4.

$$frequency = \begin{cases} frequency + 1 & \text{if } \Delta T_i \leq SWL \\ max\left[\frac{frequency}{\frac{\Delta T_i}{SWL}}, 1\right] & otherwise \end{cases} \quad (4)$$

Size input is directly extracted from server logs. As opposite to conventional Web proxies where requested object sizes can be very heterogeneous, in a web map all objects are image tiles with the same dimensions (typically 256x256 pixels). Those images are usually rendered in efficient formats such as PNG, GIF or JPEG that rarely reach 100 kilobytes in size. As discussed in [8], due to greater variation in colors and patterns, the popular areas, stored as compressed image files, use a larger proportion of disk space than the relatively empty non-cached tiles. Because of the dependency between the file size and the "popularity" of tiles,

Parameter	Value
Architecture	Feed-forward Multilayer Perceptron
Hidden layers	2
Neurons per hidden layer	3
Inputs	3 (recency, frequency, size)
Output	1 (probability of a future request)
Activation functions	Log-sigmoid in hidden layers, Hyperbolic tangent sigmoid in output layer
Error function	Minimum Square Error (mse)
Training algorithm	Backpropagation with momentum
Learning method	Supervised learning
Weights update mode	Batch mode
Learning rate	0.05
Momentum constant	0.2

Table 4. Neural network parameters

tile size can be a very valuable input of the neural network to correctly classify the cachability of requests.

During the training process, a training record corresponding to the request of a particular tile is associated with a boolean target (0 or 1) which indicates whether the same tile is requested again or not in window, as shown in Equation 5.

$$target = \begin{cases} 1 \ \textit{if the tile is requested again in window} \\ 0 \ \textit{otherwise} \end{cases} \tag{5}$$

Once trained, the neural network output will be a real value in the range [0,1] that must be interpreted as the probability of receiving a successive request of the same tile within the time window. A request is classified as *cacheable* if the output of the neural network is above 0.5. Otherwise, it is classified as *non cacheable*.

The neural network is trained through supervised learning using the data sets from the extracted trace files. The trace data is subdivided into training, validation, and test sets, with the 70%, 15% and 15% of the total requests, respectivelly. The first one is used for training the neural network. The second one is used to validate that the network is generalizing correctly and to identify overfitting. The final one is used as a completely independent test of network generalization.

Each training record consists of an input vector of recency, frequency and size values, and the known target. The weights are adjusted using the backpropagation algorithm, which employs the gradient descent to attempt to minimize the squared error between the network output values and the target values for these outputs [36]. The network is trained in batch mode, in which weights and biases are only updated after all the inputs and targets are presented. The pocket algorithm, which saves the best weights found in the validation set, is used.

Neural network performance is measured by the correct classification ratio (CCR), which computes the percentage of correctly classified requests versus the total number of processed requests.

	CartoCiudad	IDEE-Base
training	76.5952	75.6529
validation	70.2000	77.5333
test	72.7422	82.7867

Table 5. Correct classification ratios (%) during training, validation and testing for Cartociudad and IDEE-Base.

Figure 16 shows the CCRs obtained during training, validation and test phases for Cartociudad and IDEE-Base services. As can be seen, the neural network is able to correctly classify the cachability of requests, with CCR values over the testing data set ranging between 72% and 97%, as shown in Table 5. The network is stabilized to an acceptable CCR within 100 to 500 epochs.

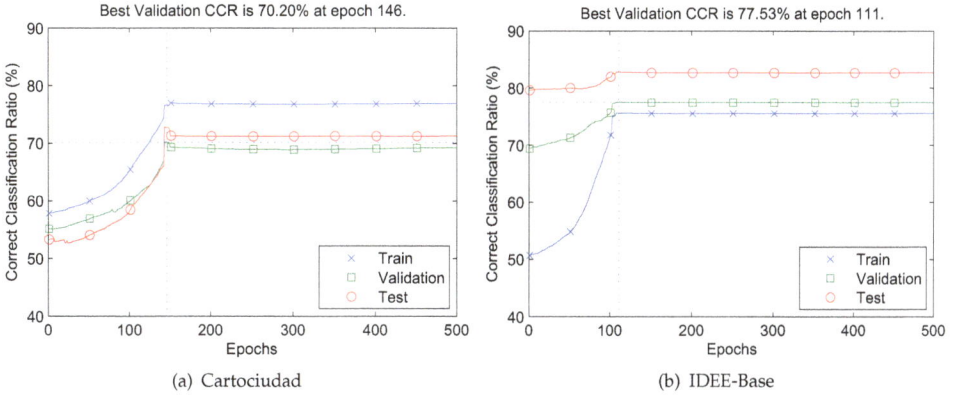

Figure 16. Correct classification ratios achieved with the neural network for CartoCiudad and IDEE-Base.

6. Conclusion

Serving pre-generated map image tiles from a server-side cache has become a popular way of distributing map imagery on the Web. However, in order to achieve an optimal delivery of online mapping, adequate cache management strategies are needed. These strategies can benefit of the intrinsic spatial nature of map tiles to improve its performance. During the startup of the service, or when the cartography is updated, the cache is temporarily empty and users experiment a poor Quality of Service. In this chapter, a seeding algorithm that populates the cache based on the history of previous accesses has been proposed. The seeder should automatically cache tiles until an acceptable QoS is achieved. Then, tiles could be cached on-demand when they are first requested. This can be improved with short-term prefetching; anticipating the following tiles that will be requested after a particular request can improve users' experience. The metatiling approach presented here requests, for a given tile request, a bigger map image containing adjacent tiles, to the remote WMS backend. Since the user is expected to pan continuously over the map, those tiles are likely to be requested. Finally, when the tile cache runs out of space, it is necessary to determine which tiles should be replaced by the new ones. A cache replacement algorithm based on neural networks has been presented. It tries to estimate the probability of a tile request occurring before a certain period of time, based on the following properties of tile requests: recency of reference, frequency of reference, and size of the referenced tile. Those tiles that are not likely to be requested shortly are considered as good candidates for replacement.

Acknowledgements

This work has been partially supported by the Spanish Ministry of Science and Innovation through the project "España Virtual" (ref. CENIT 2008-1030), a FPI research fellowship from the University of Valladolid (Spain), the National Centre for Geographic Information (CNIG) and the National Geographic Institute of Spain (IGN).

Author details

Ricardo García, Juan Pablo de Castro, Elena Verdú, María Jesús Verdú and Luisa María Regueras
Department of Signal Theory, Communications and Telematics Engineering, Higher Technical School of Telecommunications Engineering, University of Valladolid, Campus Miguel Delibes, Paseo Belén 15, 47011 Valladolid, Spain

7. References

[1] Jeff de la Beaujardiere, editor. *OpenGIS Web Map Server Implementation Specification*. Open Geospatial Consortium Inc, OGC 06-042, 2006.

[2] Elias Ioup John T. Sample. *Tile-Based Geospatial Information Systems, Principles and Practices*. Springer Science+Business Media, LLC, Boston, MA, online-ausg. edition, 2010.

[3] Open Geospatial Foundation. WMS-C wms tile caching - OSGeo wiki, 2008.

[4] Núria Juliá Juan Masó, Keith Pomakis, editor. *OpenGIS Web Map Tile Service Implementation Standard*. Open Geospatial Consortium Inc, OGC 07-057r7, 2010.

[5] J. C. Muller. Generalization of spatial databases. *Geographical Information Systems*, 1:457–475, 1991.

[6] Fay Chang, Jeffrey Dean, Sanjay Ghemawat, Wilson C. Hsieh, Deborah A. Wallach, Mike Burrows, Tushar Chandra, Andrew Fikes, and Robert E. Gruber. Bigtable: a distributed storage system for structured data. In *Proceedings of the 7th USENIX Symposium on Operating Systems Design and Implementation - Volume 7*, OSDI '06, pages 15–15, Berkeley, CA, USA, 2006. USENIX Association.

[7] R. García, J.P. de Castro, M.J. Verdú, E. Verdú, L.M. Regueras, and P. López. A descriptive model based on the mining of web map server logs for tile prefetching in a web map cache. *International Journal of Systems Applications, Engineering & Development*, 5(4):469–476, 2011.

[8] Sterling Quinn and Mark Gahegan. A predictive model for frequently viewed tiles in a web map. *T. GIS*, 14(2):193–216, 2010.

[9] Yong-Kyoon Kang, Ki-Chang Kim, and Yoo-Sung Kim. Probability-based tile pre-fetching and cache replacement algorithms for web geographical information systems. In *Proceedings of the 5th East European Conference on Advances in Databases and Information Systems*, ADBIS '01, pages 127–140, London, UK, 2001. Springer-Verlag.

[10] V. K Kang, Y. W Park, B. Hwang, and Y. S Kim. Performance profits of tile pre-fetching in multi-level abstracted web geographic information systems. In *Proceedings of the Int. Symposium on Internet and Multimedia*, Indonesia, December 2002.

[11] Dong Lee, Jung Kim, Soo Kim, Ki Kim, Kim Yoo-Sung, and Jaehyun Park. Adaptation of a neighbor selection markov chain for prefetching tiled web gis data. In Tatyana Yakhno, editor, *Advances in Information Systems*, volume 2457 of *Lecture Notes in Computer Science*, pages 213–222. Springer Berlin / Heidelberg, 2002.

[12] Serdar Yesilmurat and Veysi Isler. Retrospective adaptive prefetching for interactive web gis applications. *GeoInformatica*, pages 1–32, 2011. 10.1007/s10707-011-0141-8.

[13] R. Garcia, J.P. de Castro, M.J. Verdu, E. Verdu, L.M. Regueras, P. Lopez, and D. Garcia. Estrategias de metatiling para la aceleración de servicios de mapas teselados en las

infraestructuras de datos espaciales. In *The X Symposium on Telematic Engineering, JITEL 2011*, 2011.

[14] S. Podlipnig and L. Böszörmenyi. A survey of web cache replacement strategies. *ACM Computing Surveys (CSUR)*, 35(4):374–398, 2003.

[15] C. Aggarwal, J.L. Wolf, and P.S. Yu. Caching on the world wide web. *Knowledge and Data Engineering, IEEE Transactions on*, 11(1):94 –107, jan/feb 1999.

[16] Pei Cao and Sandy Irani. Cost-aware www proxy caching algorithms. In *Proceedings of the USENIX Symposium on Internet Technologies and Systems on USENIX Symposium on Internet Technologies and Systems*, pages 18–18, Berkeley, CA, USA, 1997. USENIX Association.

[17] Martin Arlitt, Ludmila Cherkasova, John Dilley, Rich Friedrich, and Tai Jin. Evaluating content management techniques for web proxy caches. *SIGMETRICS Perform. Eval. Rev.*, 27:3–11, March 2000.

[18] Hyokyung Bahn, Kern Koh, S.H. Noh, and S.M. Lyul. Efficient replacement of nonuniform objects in web caches. *Computer*, 35(6):65 –73, June 2002.

[19] R. García, J.P. de Castro, M.J. Verdú, E. Verdú, L.M. Regueras, and P. López. An adaptive neural network-based method for tile replacement in a web map cache. In Beniamino Murgante, Osvaldo Gervasi, Andrés Iglesias, David Taniar, and Bernady Apduhan, editors, *Computational Science and Its Applications - (ICCSA 2011)*, volume 6782 of *Lecture Notes in Computer Science*, pages 76–91. Springer Berlin / Heidelberg, 2011. 10.1007/978-3-642-21928-3_6.

[20] R. Garcia, J.P. de Castro, M.J. Verdu, E. Verdu, L.M. Regueras, and P. Lopez. A cache replacement policy based on neural networks applied to web map tile caching. In *The 2011 International Conference on Internet Computing - (ICOMP)*, 2011.

[21] Humayun Khalid. A new cache replacement scheme based on backpropagation neural networks. *SIGARCH Comput. Archit. News*, 25:27–33, March 1997.

[22] J. Pomerene, T. R. Puzak, R. Rechtschaffen, and F. Sparacio. Prefetching mechanism for a high-speed buffer store. *US patent*, 1984.

[23] H. Khalid and M.S. Obaidat. Kora-2: a new cache replacement policy and its performance. In *Electronics, Circuits and Systems, 1999. Proceedings of ICECS '99. The 6th IEEE International Conference on*, volume 1, pages 265 –269 vol.1, 1999.

[24] H. Khalid. Performance of the KORA-2 cache replacement scheme. *ACM SIGARCH Computer Architecture News*, 25(4):17–21, 1997.

[25] M.S. Obaidat and H. Khalid. Estimating neural networks-based algorithm for adaptive cache replacement. *Systems, Man, and Cybernetics, Part B: Cybernetics, IEEE Transactions on*, 28(4):602 –611, August 1998.

[26] H. Khalid and MS Obaidat. Application of neural networks to cache replacement. *Neural Computing & Applications*, 8(3):246–256, 1999.

[27] P. Venketesh and R. Venkatesan. A Survey on Applications of Neural Networks and Evolutionary Techniques in Web Caching. *IETE Technical Review*, 26(3):171–180, 2009.

[28] H. ElAarag and S. Romano. Training of nnpcr-2: An improved neural network proxy cache replacement strategy. In *Performance Evaluation of Computer Telecommunication Systems, 2009. SPECTS 2009. International Symposium on*, volume 41, pages 260 –267, july 2009.

[29] S. Romano and H. ElAarag. A neural network proxy cache replacement strategy and its implementation in the Squid proxy server. *Neural Computing & Applications*, pages 1–20.

[30] H. ElAarag and J. Cobb. A Framework for using neural networks for web proxy cache replacement. *SIMULATION SERIES*, 38(2):389, 2006.

[31] H. ElAarag and S. Romano. Improvement of the neural network proxy cache replacement strategy. In *Proceedings of the 2009 Spring Simulation Multiconference*, pages 1–8. Society for Computer Simulation International, 2009.

[32] J. Cobb and H. ElAarag. Web proxy cache replacement scheme based on back-propagation neural network. *Journal of Systems and Software*, 81(9):1539–1558, 2008.

[33] W. Tian, B. Choi, and V. Phoha. An Adaptive Web Cache Access Predictor Using Neural Network. *Developments in Applied Artificial Intelligence*, pages 113–117, 2002.

[34] R.A. El Khayari, M.S. Obaidat, and S. Celik. An Adaptive Neural Network-Based Method for WWW Proxy Caches. *IAENG International Journal of Computer Science*, 36(1):8–16, 2009.

[35] G. Cybenko. Approximation by superpositions of a sigmoidal function. *Mathematics of Control, Signals, and Systems (MCSS)*, 2:303–314, 1989. 10.1007/BF02551274.

[36] Tom M. Mitchell. *Machine Learning*. McGraw-Hill, New York, 1997.

Chapter 4

Use of Terrestrial 3D Laser Scanner in Cartographing and Monitoring Relief Dynamics and Habitation Space from Various Historical Periods

Gheorghe Romanescu, Vasile Cotiugă and Andrei Asăndulesei

Additional information is available at the end of the chapter

1. Introduction

Nowadays, modern cartography employs various techniques and methods, which were deemed inconceivable just two decades ago. The GIS technology is particularly important among these, a technology that advanced cartography to the highest standards of graphical (visual) representation. 3D laser scanning is one of the most notable new cartography techniques that is part of this "new wave". This instrument, specifically a Leica LIDAR scanner, capable of monitoring land dynamics with an accuracy of 1 mm/1 mm from a distance of 300 m, as well as its use, are the subject of the present paper. In the field of physical geography and geo-archaeology, the 3D laser scanner can be used to map and monitor soil degradation processes (ravine-creation, surface erosion, landslides, etc.), the erosion and aggradation of a river's channel and valley, shore and slope processes, beaches and escarpments, dunes, caves and man-made excavations (including salt mines), iceberg dynamics, archaeological sites and the spatial position of the archaeological finds, heritage buildings and structures, works of art (sculptures, reliefs, etc.) etc.

Even though the 3D laser scanning technique is relatively old, i.e. from the '90s, it has enjoyed increased use in physical geography and in geo-archaeology only during the last few years. Most of the cartography projects employing the 3D laser scanner focus on the morphology and dynamics of the slope geomorphological processes, as well as on river bed and shoreline dynamics. Up until the advances in technology allowed the 3D scanners to become widely available and practical enough for cartographic use, countless attempts to use "modern" cartographic methods were made, but, in most cases, they proved to be inexact and cumbersome. Therefore, most of the studies employed the classical, rudimentary, methods which rely on wooden or metal markers and analog data collecting.

This study will present several examples from eastern Romania: the Moldavian Plateau, the Bend Subcarpathians, the Danube floodplain, etc. (Figure 1). The geomorphological processes that are active throughout the Moldavian Plateau are carefully watched by specialised institutions but, unfortunately, the necessary equipment is lacking or is unsuitable, and the results are therefore unsatisfactory. By using a 3D scanner, many of the results expected from dynamic geomorphology and geo-archaeology will be accurate and the database much enriched.

Figure 1. The location of the main sites on Romanian territory scanned with a 3D scanner

The eastern part of Romania is extremely rich in Neolithic, Chalcolithic, Ancient and Medieval remains, and it is strongly affected by surface geomorphological processes and, wich is critical, by powerful floods. Elaborating a study on the dynamics of the relief from Eastern Romania implies the development of a specific methodology for the 3D scanner, as well as the acquisition of correct data, on a millimetre or centimetre scale, on the changes that can occur in very short time intervals.

The correlation between the data acquired via 3D scanning and that obtained through classical methods will certainly lead to improvements in the manner in which geomorphologic risk phenomena are prevented and combated, and how the historical conditions of the evolution of human habitation can be better interpreted (Figure 2). Concurrently, the aim is to make the technique versatile enough as to be employed in various domains: geomorphology (land degradation), agronomy (landscaping, irrigations), environmental protection, land management, architecture, civil engineering, etc.

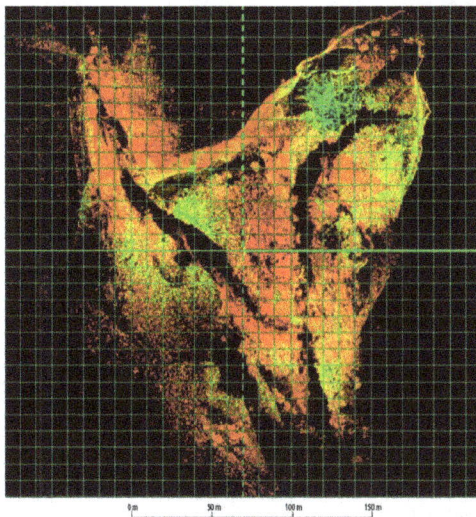

Figure 2. The point-cloud obtained from the measurements conducted in the gully impacting a Chalcolithic archaeological site from Cucuteni culture

Since 2000, the international literature in the field of laser-based 3D technologies grew continuously. At the present, certain fields of research, such are geo-archaeology, geomorphology, photogrammetry etc. created their own methodologies and the papers published in these fields are greatly sought after. From this point of view, it is important to mention the use of the 3D scanner in geomorphology [1-25], hydrology [26-29], architecture [30-32], archaeology [33-35], cartography and topography [36-37], methodology [38-54], etc.

The relief dynamic from Eastern Romania is extremely active and most of the major archaeological sites are affected by gullying or landslide processes [55-67] (Figure 3). The ultimate objective of the present undertaking is to elaborate cartographic material (approximately 25 colour and B&W illustrations) representative and suggestive enough to demonstrate the dynamic nature of the environment and its impact on relevant and significative locations. To this purpose, a series of cartographical representations will be produced, depicting the dynamics of a gully and the effects of several landslides which affected some Chalcolithic sites belonging to the Cucuteni culture (specifically, Cucuteni-Baiceni and Habasesti-Holm, Iasi County, Romania), and a river (the Danube) which affected the morphology of a Chalcolithic tell (an archaeological mound created by successive human occupation and abandonment of a location over several centuries), belonging to the Gumelnita culture, from Harsova (Constanta County).

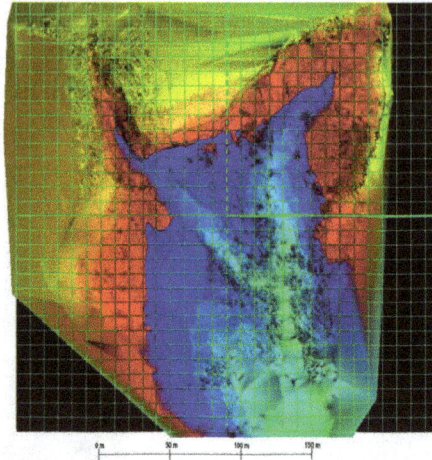

Figure 3. The final cartographic product for the investigated gully

For a better monitoring of relief dynamics, the use of the GPS is imperiously necessary. This tool can delimit with accuracy the spatial limits of the phenomenon. These coordinates constitute the backbone of the morphological analysis of a dynamic land form. Unfortunately, the 3D scanner cannot precisely identify by itself the border between one area and another. By using the 3D scanner we are able to move to the next step of the analysis of the dynamics of the relief and man-made forms. The accuracy of the obtained data and the exceptional cartographical representation means that the 3D has a long use-life and a greatly-diverse usability (Figure 4, 5).

For the study of the geomorphological processes we selected an area in the Moldavain Plateau that is affected by intense gullying. The complex analysis of the Cucuteni Ravine aims also at the impact it has on the Chalcolithic site. For all the other examples we presented only the fields of research they pertain to, while the results will be included in several future papers.

2. Materials and methods

The phenomena of gullying are affecting, on a global scale, large areas in the tropical and temperate regions [68-76]. In Romania, these are specific to the Moldavian Plateau, the Transylvanian Depression and the Getic Plateau [60-61, 65-66, 16, 77]. Apart from the archaeology and architecture applications, which are more visible to the public, the 3D ground laser scanning is successfully used in monitoring the processes of ravine-creation in eastern Romania. This technology was also used with promising results in monitoring the riverine valleys and the landslides [78, 5, 79, 38, 80-83]. As a result of its performance, it can be used successfully to assess the state of the environment, especially in land mapping and with limited extension in measuring the rate of erosion in certain land surfaces [33, 84-85].

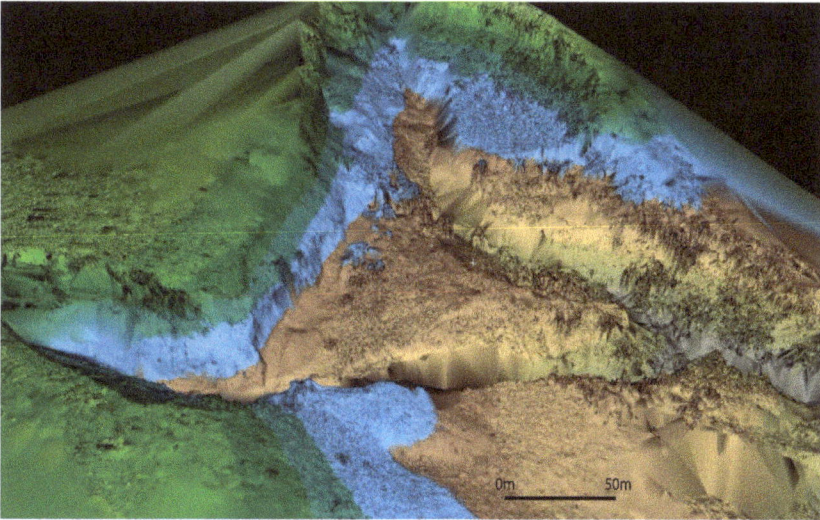

Figure 4. The detailed cartographic representation of the upper sector or the gully

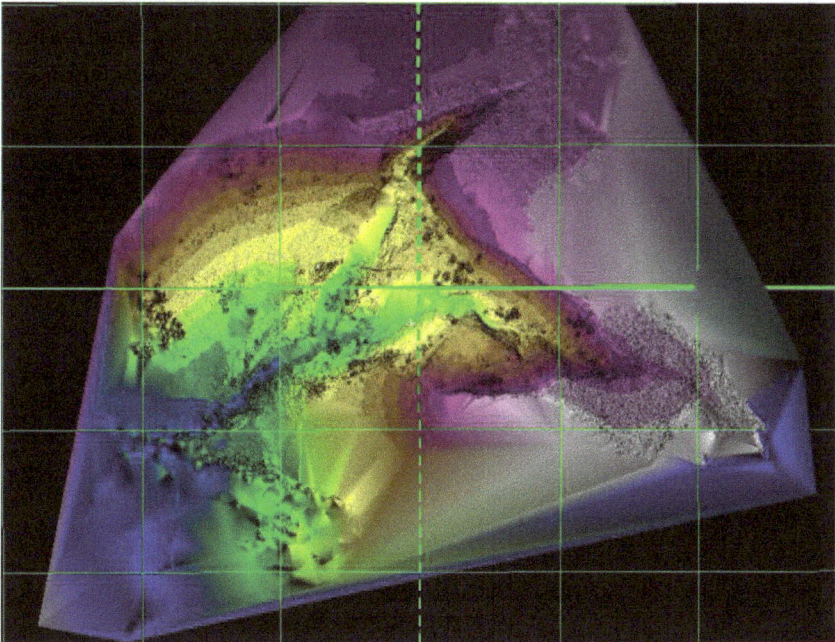

Figure 5. The final cartographic representation of the gully (100 m grid)

In terms of definition, the 3D laser scanning is one of the new technologies, by which the geometry of an object or surface can be automatically measured, without the aid of a reflector, with speed and precision well above the ones of the classic solutions. The groundbased laser scanner records the tridimensional points by measuring the vertical and horizontal angles, as well as the distance to each point. Even though there are largely different technologies, the field-use of the 3D scanner uses elements of the work methodology of the total station. In case the morphology of the scanned object is complex, more than one station points will be used. Thus, all its surfaces will be scanned. In case the shadowed surfaces are not reached by the laser beam, the software captures such areas automatically then integrates them to the scanned object. The process relies on registering distances and angles, and the data thus produced is used to compute the points' coordinates. The ability to register a massive amount of 3-D information in a relatively short time is the main advantage of this instrument, in contrast with classical equipment such as the total station [24, 10].

The methodology used for the analysis of Cucuteni Ravine is strongly influenced by its impressive size and the local topography, which required, in the end, the use of 17 station points. To merge the 17 positions, 6 reflective tie-points (Figure 6) were used for each of them, except for the last one. Nevertheless, for the entire model we used 24 tie-points, since some of the scanning positions were referenced to tie-points also used by other station points, where the physical distance allowed. Terrestrial laser scanning (TLS) generates several point clouds, with local coordinates and additional info (the light intensity in the reflected beam, and the RGB values obtained from an external or internal photographic camera). The point clouds, after having been registered from different positions, must be merged as to obtain a complete model of the scanned target. This procedure is called "registration" and involves merging the point clouds through the use of reflective tie-points, specially built and delivered by the manufacturer, which are automatically recognized by the scanner when a very fine scan is performed.

Figure 6. The round target used as tie-point

For the current project, in all of the three scanning sessions, we employed a Leica ScanStation HDS 3600 3D scanner. It is a time-of-flight active scanner, which works by timing the round-trip time of a pulse of light. The operation range is 270º horizontally and 360º vertically (Figure 7), and the active distance is 300 m. With a resolution of 6 mm at a distance of 50 m, and due to its ability to register approximately 2000 points per second, the ScanStation HDS scanner is among the most effective equipment of its type. The average resolution for all of the scans was of approximately 6 mm, and the registered points numbered millions, despite the fact that the majority of positions overlapped. Although an external photographic camera can be attached to the scanner, we thought it to be no better suited for the task than the internal camera [86].

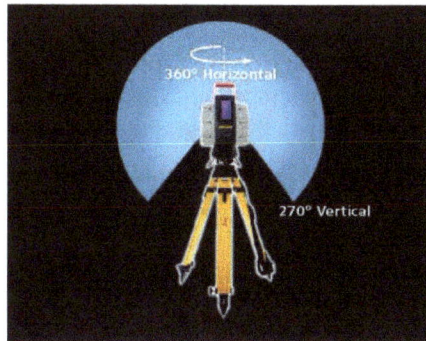

Figure 7. The visual field of the scanner

A particularly important phase of the fieldwork facet of the project was the geo-referencing of the point clouds. Using a reference station positioned on fixed known spot and a Leica 1200 GPS receiver, we referenced the point cloud established as the basis for the 3D model to the national coordinates system (Stereographic 1970). In fact, our methodology was based, for all of the three sessions, on computing the differences between the obtained geo-referenced models using CAD and GIS. In respect to the raw data processing, it was carried out by filtering the data using the Cyclone dedicated software program, registering the data (see above), reducing the point cloud, creating a mesh by triangulation, and texturing the model. The final results of the analysis were produced by exporting sections, transverse and longitudinal, of the three tridimensional models obtained in each session.

As a case study, the selection of the Cucuteni-Baiceni Ravine for the present paper was based on its high level of activity and its lack of arboreal vegetation, which could have partially impeded the measuring of the volumetric parameters. The undergrowth vegetation had then to be erased using the technical methods allowed by the software. Furthermore, for increased accuracy, the edge of the gully was outlined during each session by using the two "traditional" instruments mentioned above (the reference station and the 1200 GPS receiver, both produced by Leica). The operation was somewhat cumbersome, because in such cases the data must be collected from extremely numerous positions, as to take into account all of the inflexions [44, 10]. All of the positions were geo-referenced and corroborated with older measurements. In this way, we were able to estimate the rate of soil erosion in the gully, for each of the measurements taken. The conjoint use of these two types of measurements (GPS and 3D scanner) means that the risk of error was much diminished.

Another stage worth mentioning is, in the lab phase of the project, the filtering and modeling of data, to the aim of producing results compatible with the complementary software we used (CAD, GIS). Data filtering is a compulsory stage of the analysis, as the vegetation present at the scanning site, together with the very large amount of data, could result in errors in the Digital Terrain Model export-for-GIS process, as well as during the volume calculation of the whole resulting model. In the first step, the tridimensional model was trimmed by removing the points lying outside the analyzed area, in order to reduce its originally large size. Concurrently, the model was scaled down in order to be more easily handled by the software. All these steps were carried out within the Cyclon platform, the proprietary software of the scanner, which was specially designed to handle the point-clouds in a 3D environment. The large number of axial and cross-sections of the model were produced at the same time, together with the various graphic, image and video exports.

The stage of interpolating the scanner-produced points, to the aim of achieving a Digital Elevation Model, was carried out within the GIS platform (using the ArcGIS suite). Based on the resulting DEM, several derivates were produced (contour lines, volume calculations, various graphs and diagrams etc.) The proprietary software of Leica is also providing export functionality, but the options are more oriented toward GIS and CAD formats. In order to reduce to minimum the error margin of the volume calculations for the three models (of three successive years), which is the definitive element for outlining the timeline of change between the three scans, we used the Cyclon software. Following this procedure, there were no major, visible differences between the employed calculation methods.

The Cucuteni gully was selected for the present research because it is extremely active. The area occupied by the gully is very sparsely covered by vegetation, and the trees are virtually absent; therefore, nothing prevented volumetric measuring. The very sparse shrub vegetation was removed using the techniques made available by the dedicated software. The reason behind the selection of the gully for our investigation was due to the fact that the gullying process is affecting a very important archaeological site dating back to circa 5000 B.P. This made it easier to assess the rate of erosion over a period of great lengths.

Three consecutive measurements were performed at relatively equal intervals in 2008, 2009 and 2010 (Figure 8, 9, 10, 11). The last measurement was made in spring 2010 after a solid

Figure 8. The 3D scanner in action on the Cucuteni-Baiceni Ravine

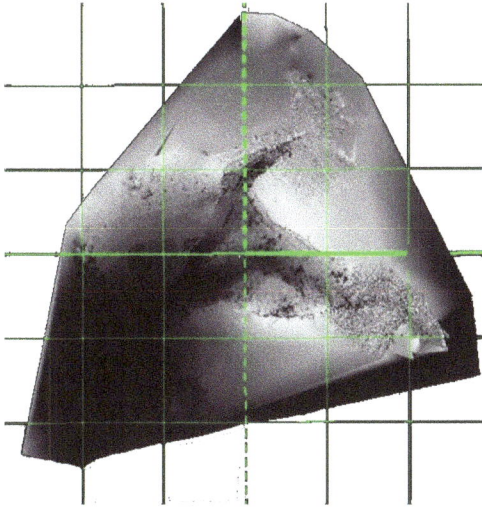

Figure 9. The 3D model and the map of the Cucuteni Ravine in 2008 (100 m grid)

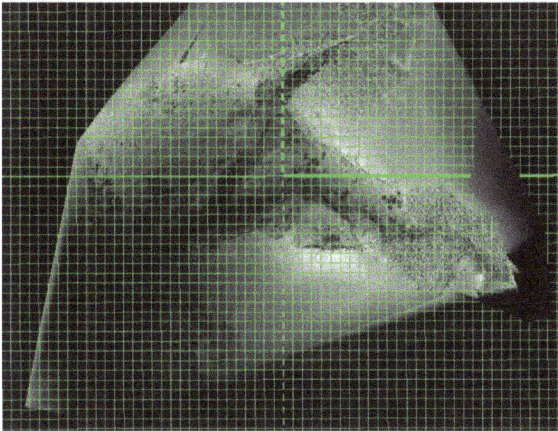

Figure 10. The 3D model and the map of the Cucuteni Ravine in 2008 (10 m grid)

winter precipitation and high rainfall in spring. For the historical evolution of gullies, topographic maps and military plans of the Romanian Army were consulted. During the Second World War the Army had placed a battery of guns in the area of the unit studied. Unfortunately, it has only been possible to make use of zoning maps from after 1950. The maps drawn earlier are not accurate and they are often for orientation only, with a high degree of generalization.

Meteorological data on precipitation, daily and monthly, were provided by the Meteorological Centre, Iasi, Moldavia. They were focused on Cotnari Meteorological Station, located near the Cucuteni-Baiceni ravines. The most important stations were rather uniformly distributed on Moldavian territory (Eastern Romania) [87]. Certain old cartographic material was provided by the Military Topography Directorate in Bucharest. The orthophoto maps and the military maps dating from WW II were supplied by the ANCPI (The National Agency for Surveying and Real Estate Publicity) in.

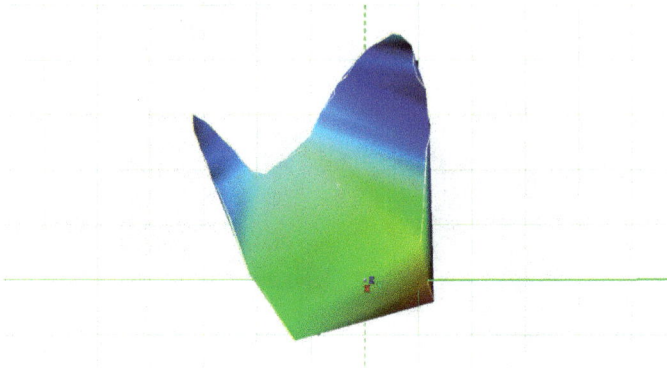

Figure 11. The model used for volume calculation for the eroded and transported material (50 m grid)

3. Results

There were several thematic maps elaborated for the Cucuteni-Baiceni Ravine, emphasizing its dynamics. The morpho-graphic and morpho-metric maps were built on the basis of the 3D scans (Figure 12), which were further used as basis for the spatial yearly analysis of the ravine. In this case the erosion and accumulation sectors were outlined on the walls of the ravine or on the bottom of the cut (Figure 13, 14). The measurements included three consecutive years: 2008, 2009 and 2010.

The Cucuteni Ravine was extremely active during the years 2008-2010, due to the high frequency of torrential rainfall. The selected site can be regarded as representative for the whole Moldavian Plateau, as its relief energy is high and the loess deposits of its subbasement are very friable [16, 58].

Along the gullying processes, the monitoring could include also the gravitational processes, such as landslides (Figure 15). Due to the loess subbasement and the general deforestation in the Moldavian Plateau, large tracts of farmlands are subject to large-area landslides [62, 64]. As in the case of gullying, the landslides can be scanned successively, monitoring the reference points in order to assess the dynamics of the phenomenon. In the present paper, the case of Holm Hill, near the village of Habasesti (Iasi County), presents a complex situation due to the presence of an archaeological site (Cucuteni culture). The application of the 3D scan-based analysis proved to be salutary, as the combined derivates produced using the point-cloud enabled a successful monitoring of the interest area. The use of 3D graphic modeling software (3D Studio Max) resulted in the creation of models useful for studying the topography and planimetry of the prehistoric settlement. All the modeling processes were based on the available archaeological data and the tridimensional model produced by scanning (Figure 16).

Use of Terrestrial 3D Laser Scanner in Cartographing and Monitoring Relief Dynamics and Habitation
Space from Various Historical Periods

83

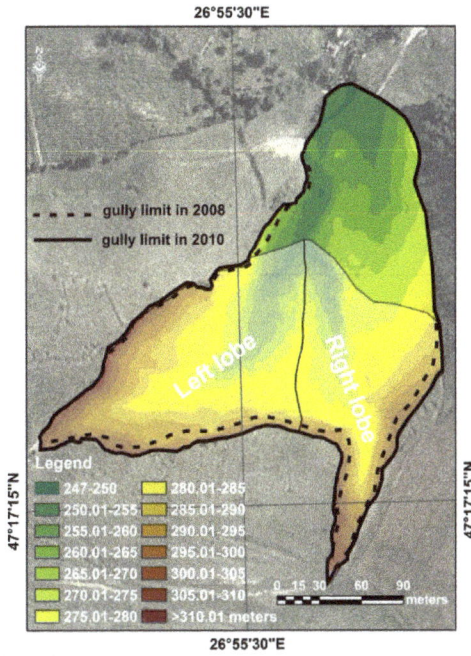

Figure 12. Making the morphologic map of the Cucuteni-Baiceni Ravine

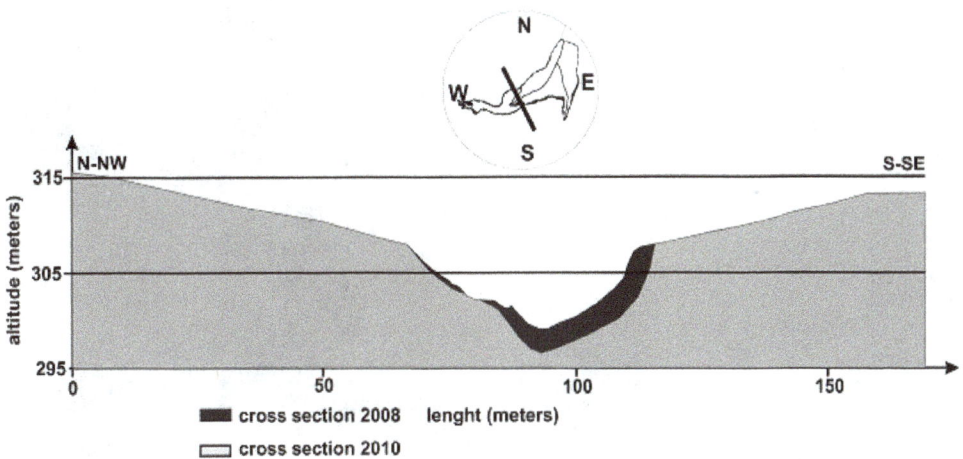

Figure 13. Cross-section through the Cucuteni-Baiceni Ravine, with the indication of the 2008 and 2010
measuring sessions (erosion only)

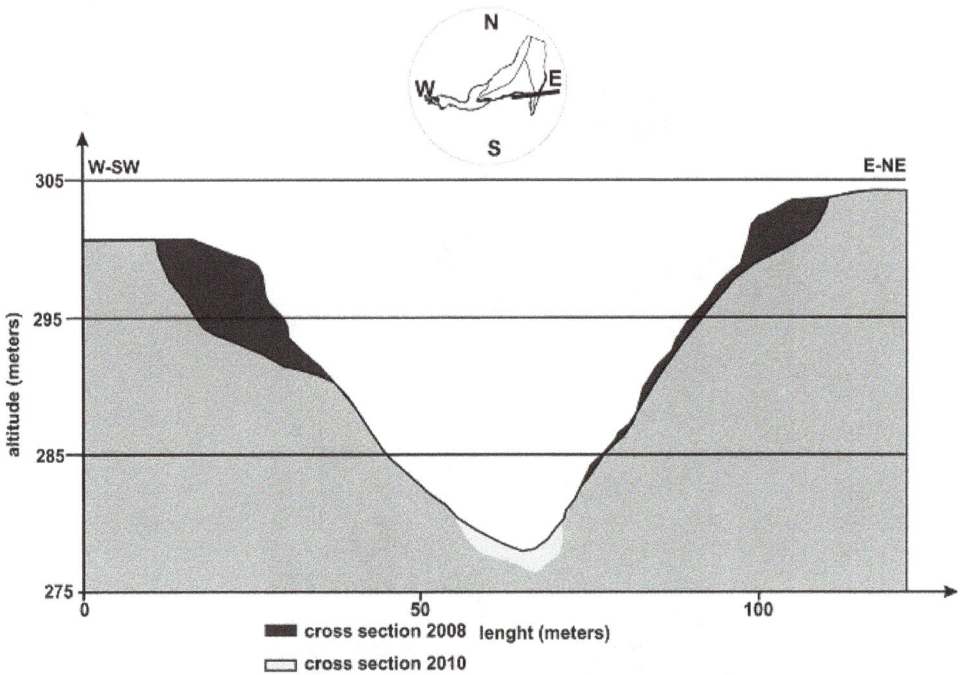

Figure 14. Cross-section through the Cucuteni-Baiceni Ravine, with the indication of the 2008 and 2010 measuring sessions (erosion and accumulation)

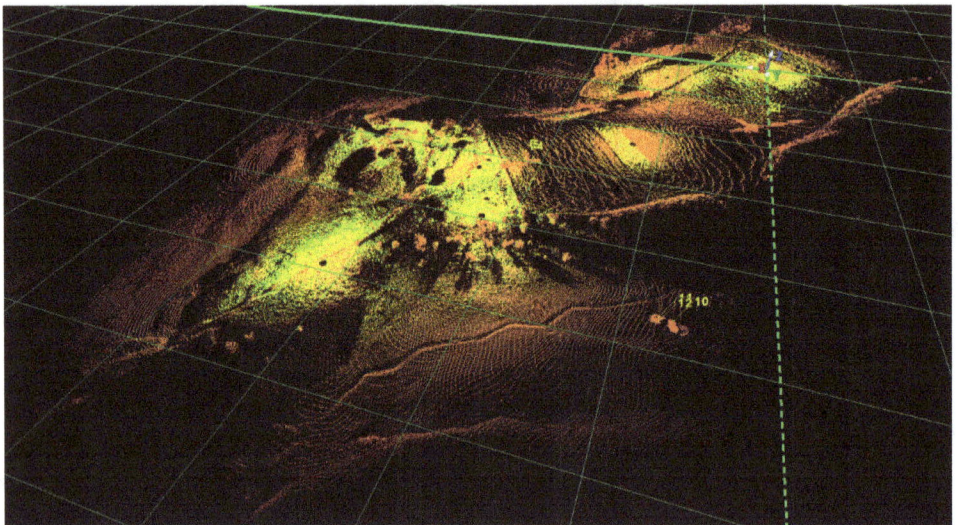

Figure 15. Landslide on the Habasesti-Holm (50 m grid)

Use of Terrestrial 3D Laser Scanner in Cartographing and Monitoring Relief Dynamics and Habitation
Space from Various Historical Periods

85

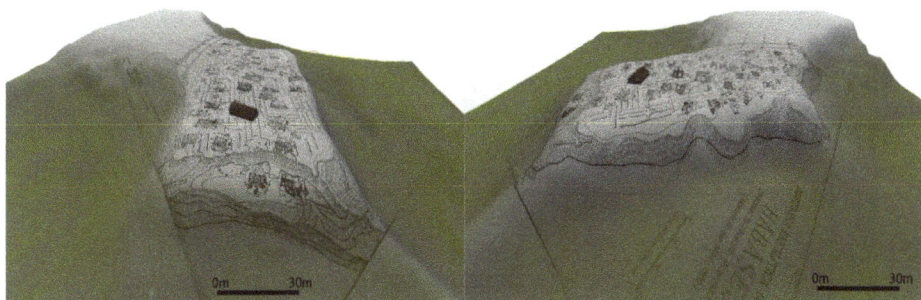

Figure 16. The prehistoric site of Habasesti following point-cloud processing

The 3D scans are also instrumental in monitoring the riverbanks and in analyzing the
dynamics of the riverine valleys and channels (Figure 17). The most useful measurements
include the lateral erosion, especially at flood times, as well as the accumulation processes
along the riverbed and at river-mouths. In the latter case, the measurements should be taken at
much shorter intervals, as the accumulation and erosion processes can affect the navigation. It
is recommended that the measurements be carried out after each flash-flood. In this case, the
riverbank and riverbed processes can be successfully monitored, especially the thalveg ones.

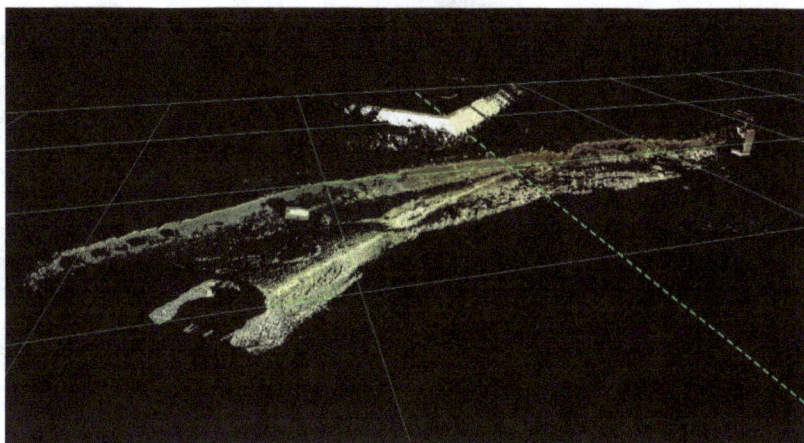

Figure 17. The channel of Trotus river at Onesti (30 m grid)

The steep riverbaks are easier to monitor, particularly in the case of highly friable
subbasements (Figure 18, 19). The spasmodic character of rivers in the Moldavian Plateau
bestows a high level of dynamicity to their channels. The solid alluvia of the eastern
Romanian rivers are volumetrically high, while the geomorphologic processes result in
bank-collapse and large accumulations. The use of a large palette of colours in process-

modeling may increase the accuracy of interpretation of the hydro-geomorphologic phenomena, as it outlines the erosion and accumulation areas, the wetter or drier layers, the finer or coarser grained strata, etc.

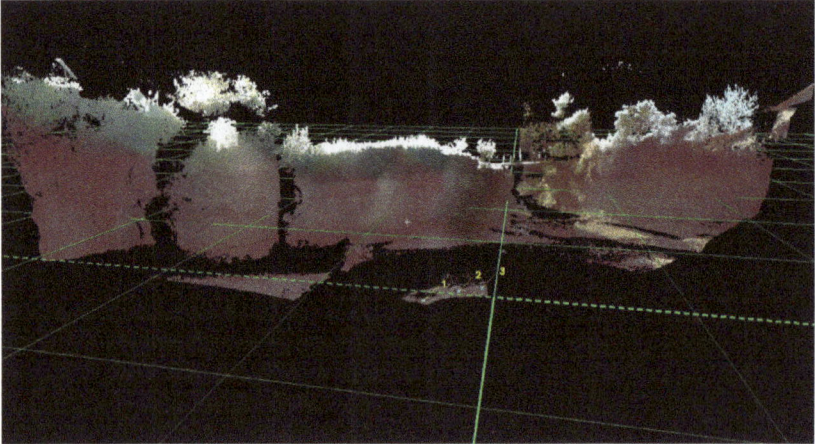

Figure 18. Detail of Dofteana river banks – 1 (10 m grid)

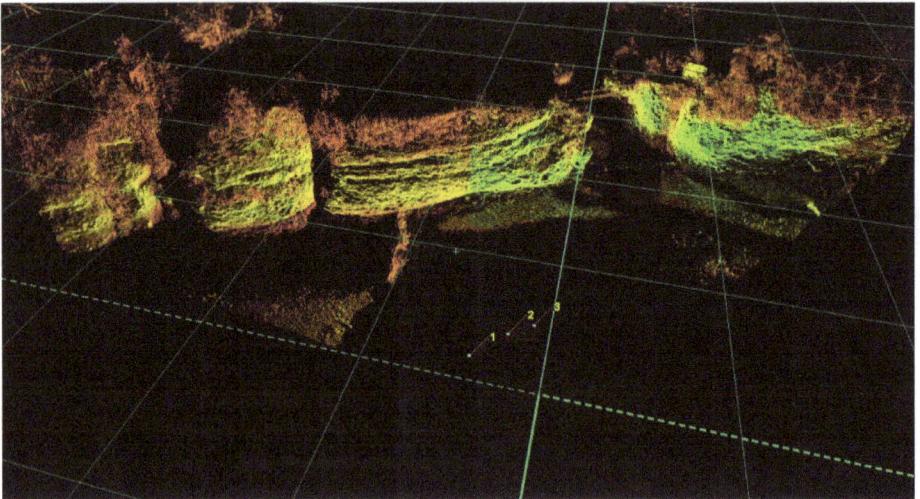

Figure 19. Detail of Dofteana river banks – 2 (10 m grid)

One of the first uses given to the 3D scanner was the study of archaeological sites. As such, the excavations and the tracking of the finds were recorded accurately in time and space (Figure 20, 21). The archaeological use of the 3D scanner included the sites of Bucsani, Ibida, the tells of Harsova, and Tangaru, the walled town of Ulpia Traiana and the defensive ditch of Silistea, Neamt County. For the future, we plan to monitor in this way all the archaeological excavations in Moldova and Dobruja.

Use of Terrestrial 3D Laser Scanner in Cartographing and Monitoring Relief Dynamics and Habitation
Space from Various Historical Periods

87

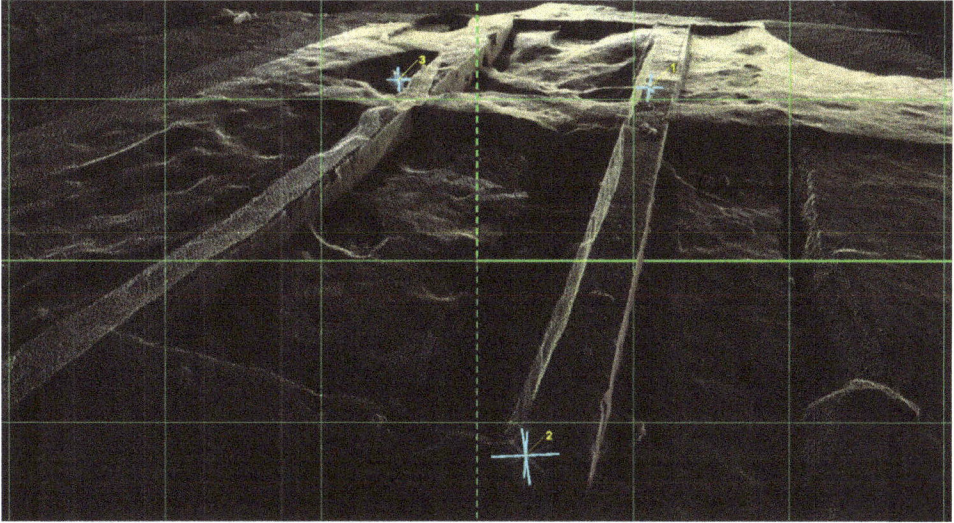

Figure 20. Bucsani archaeological site – during the archaeological excavations (1 m grid)

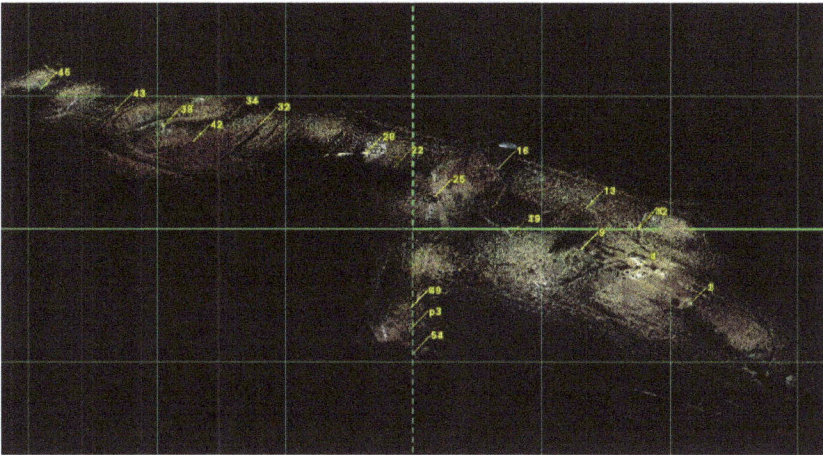

Figure 21. The walled town of Ibida – the archaeological excavations (200 m grid)

The 3D scanning on archaeological sites applies not only to the excavation area but to the location and tracking of finds as well. The most remarkable results were achieved in the field of architectural reconstruction of the built space. The Romanian territory conceals a veritable archaeological treasure, with several hundreds of major sites. In spite of this richness, most of the excavation projects are still at the level of intention or slow startup.

In a first phase, we carried out the measurements required to outline the features of a tell (ex: Tangaru) (Figure 22). To delineate the excavation stages and to accurately locate the finds, several successive scans are carried out on the site, as is the case of the sample research stage at the tell of Harsova, on the right bank of the lower course of Danube (Figure 23). The outstanding importance of the Harsova tell resulted in its close and adequate monitoring, as well in its promotion on the international heritage scene.

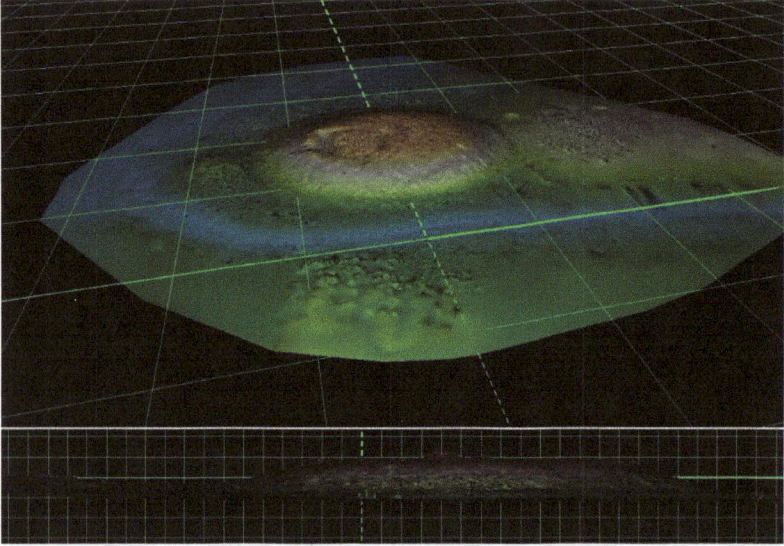

Figure 22. The morphologic features of the tell of Tangaru (during the data-modeling) (50 m grid up, 5 m grid bottom)

Figure 23. The tell of Harsova, on the right bank of the Danube (5 m grid)

The 3D scans carried out at the site of the ancient walled town of Ulpia Traiana in the Orastie Mountains unveiled a large and well-provisioned urban settlement, au par with the better known contemporary Roman towns. The tridimensional image can help in filling the unknown areas and in the clarification of certain issues of urban zoning (Figure 24). The complete image of a settlement is useful in carrying out a succesfull synchronic comparative study.

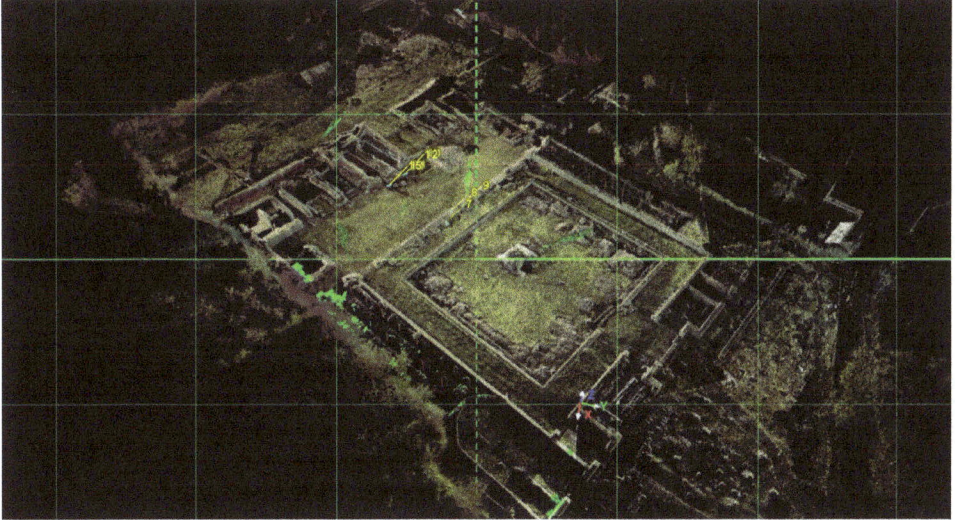

Figure 24. The walled town of Ulpia Traiana in Orastie Mountains (30 m grid)

On the archaeological site of Silistea (Neamt County), the 3D scanning was directed on the defensive ditch of the settlement (Figure 25), making important contributions to the monitoring of the cultural landscape.

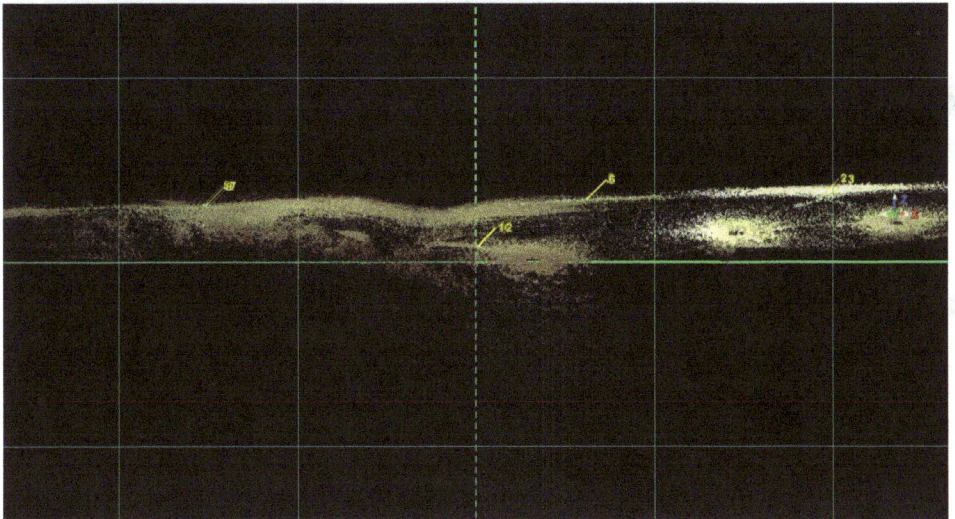

Figure 25. Silistea, Neamt County – the defensive ditch (30 m grid)

The architecture and art history benefit also from the 3D scanning, which opens new directions of research for them. In this regard, the scans are useful for studying the relation between a statue and its environment (Figure 26) or the makings of a building (Figure 27, 28) (on the small scale), or for analyzing the center of a city (on the large scale) (Figure 29).

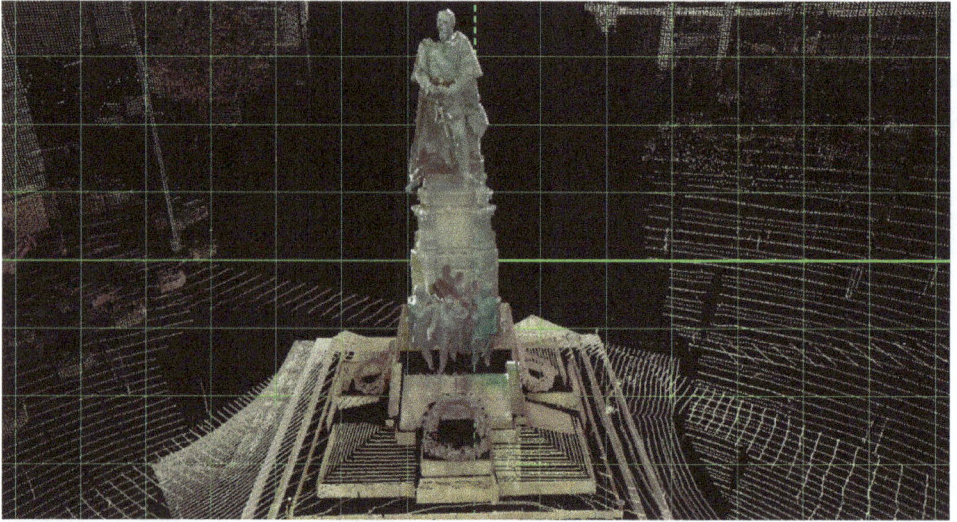

Figure 26. Prince Alexandru Ioan Cuza statue, in Union Square, Iasi (2 m grid)

Figure 27. The Cantacuzino Palace, in Bucharest (5 m grid)

Use of Terrestrial 3D Laser Scanner in Cartographing and Monitoring Relief Dynamics and Habitation
Space from Various Historical Periods

91

Figure 28. The Cantacuzino Palace, in Bucharest – the final product

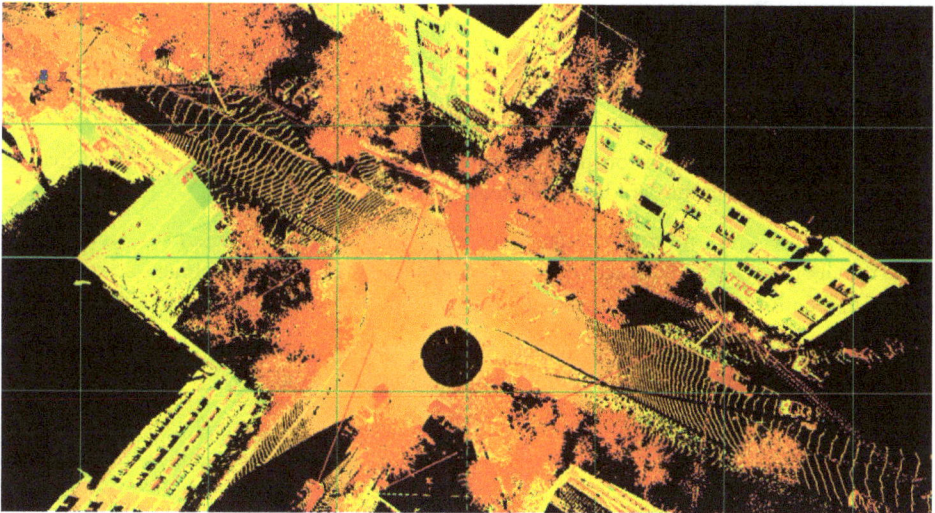

Figure 29. Iasi city-center (10 m grid)

The 3D scanner is also an instrumental aid in designing the land networks (Figure 30): forestry roads, county and national roads and motorways, or other land-based facilities: ski tracks, bobsleigh courses, golf fields, power, gas and oil transport networks, etc.

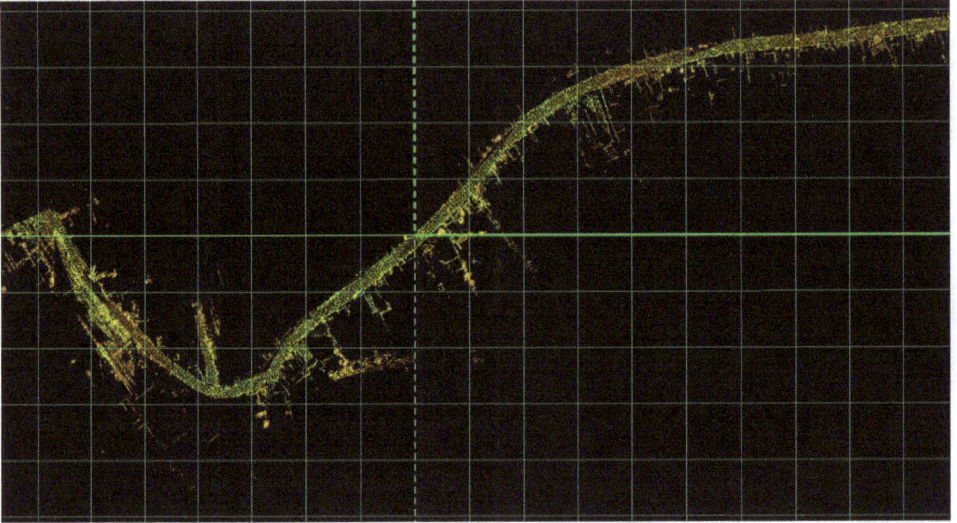

Figure 30. Road-building in Rarau Mountains (Suceava County) (50 m grid)

4. Conclusions

The 3D scanner is the most powerful and versatile instrument used in monitoring the natural and man-related processes on short term scale. The processing/production times are short and the mapping products are flawless.

On a large scale, it can be useful for a wide range of activities. Up to the present, it was used for dynamicity analyses in geomorphology (gullying, landsliding, bank- and slope- collapsing, riverine dynamics etc.), archaeology, architecture, civil and military engineering etc.

The first practical application on the Romanian territory was carried out on the Cucuteni Ravine (Iasi County), in the Moldavian Plateau.

For small items, with minute change over time, the most recommended resolution is 1mm/1mm. For lare areas, with active dynamics over time, the most recommended resolutions are 5-6mm/5-6 mm (reducing significantly the scanning time – two or three times faster).

The mapping products resulted following the interpretations of the 3D scanning results are much more accurate that the ones produced using classic methods.

Author details

Romanescu Gheorghe

Department of Geography, "Alexandru Ioan Cuza" University of Iasi, Romania

Cotiugă Vasile
Departament of History, "Alexandru Ioan Cuza" University of Iasi, Romania

Asăndulesei Andrei
ARHEOINVEST Platform, "Alexandru Ioan Cuza" University of Iasi, Romania

Acknowledgement

We extend our thanks to the Geo-archaeology Laboratory within the Faculty of Geography and Geology, "Alexandru Ioan Cuza" University of Iasi, which provided the instruments and carried out the data-processing.

5. References

[1]   Abellan A, Vilaplana JM, Martinez J (2006) Application of a long-range terrestrial laser scanner to a detailed rockfall study at Vall de Nuria (Eastern Pyrenees, Spain). Eng. Geol. 88(3-4): 136-148.

[2]   Abellan A, Jaboyedoff M, Oppikofer T, Vilaplann JM (2009) Detection of millimeric deformation using a terrestrial laser scanner: experiment and application to a rockfall event. Natural Hazards and Earth System Sciences 9: 365-372.

[3]   Barnett T, Chalmers A, Diaz-Andreu M, Ellis G, Longhurst P, Sharpe K, Trinks I (2005) 3D Laser Scanning for Recording and Monitoring Rock Art Erosion. International Newsletter an Rock Art (INORA) 41: 25-29.

[4]   Bauer A, Paar G, Kaltenbock A (2005) Mass Movement Monitoring Using Terrestrial Laser Scanner for rock Fall Management. In: van Oosterom P, Zlatanova S, Fendel E M, editors. Geo-information for Disaster Management. Berlin: Springer. pp. 393-406.

[5]   Bitelli G, Dubbini M, Zanutta A (2004) Terrestrial laser scanning and digital photogrammetry techniques to monitor landslide bodies. In: Altan M O, editor. Proceeding of the XXth ISPRS Congress Geo-Imagery Bridging Continents. Istanbul: ISPRS. pp. 246-251.

[6]   Bouike M, Viles H, Nicoli J, Lyew-Ayee P, Ghent R, Holmlund J (2008) Innovative applications of laser scanning and rapid prototype printing to rock breakdown experiments. Earth Surface Processes and Landforma 33: 1614-1621.

[7]   Bryer A (2003) Technologie pour le leve des mouvements historiques. La photogrammetrie digitale comparee au laser Scanner 3D. Levee de l'Arc d'Auguste, a Aoste, en Italie, Memoire pour le titre d'ingenieur diplome par l'etat en batiments et travaux publics, geometrie et topographie. Paris: Conservatoire National des Arts et Metiers. 81 p. Available:
http://www.esgt.cnam.fr/documents/dpe/ memoires/03_Bryer_mem.pdf. Accessed 2011 Nov 23.

[8]   Kottke J (2009) An Investigation of Quantifying and Monitoring Stone Surface Deterioration Using Three Dimensional Laser Scanning. University of Pennsylvania Scholarly Commons. Available:
http://repository.upenn.edu/cgi/viewcontent.cgi?article=1126&context=hp_theses. Accessed 2011 Nov 6.

[9]   Kwong A K L, Tham L G, King B A (2005) Application of 3D Laser Scanning Technique to Slope Movement Monitoring. Journal of Testing and Evaluation 33(4): 266-273.

[10] Milan D G, Heritage G L, Hetherington D (2007) Application of a 3Dlaser scanner in the assessment of erosion and deposition volumes and channel change in a proglacial river. Earth Surface Processes and Landforma 32: 1657-1674.

[11] Nagihara S, Mulligan K R, Xiong W (2004) Use of a three-dimensional laser scanner to digitally capture the topography of sand dunes in high spatial resolution. Earth Surface Processes and Landforms 29: 391-398.

[12] Nasermoaddeli M H, Pasche E (2008) Application of terrestrial 3D laser scanner in quantification of the riverbank erosion and deposition. In: Altinakar M S, Kokpinar M A, Aydin I, Cokgar S, Kirkgoz S, editors. River flow 2008, Proceedings of International Conference on fluvial Hydraulics, Cesme-Ismir, Turkey, Sep. 3-5, 2008. Ankara: Kubaba Congress Department and Travel Services. pp. 2407-2416.

[13] Oppikofer T, Jaboyedoff M, Blikra L H, Derren M H (2008) Characterization and monitoring of the Aknes rockslide using terrestrial laser scanning. In: Locat J, Perret D, Turmel D, Demers D, Leroueil S, editors. Proceedings of the 4th Canadian Conference on Geohazards: Forum Causes to Management. Quebec: Press de l'Universite Laval. pp. 211-218.

[14] Perroy R L, Bookhagen B, Asner G P, Chadwick O A (2010) Comparison of gully erosion estimates using airborne and ground-based LiDAR on Santa Cruz Island, California. Geomorphology 118(3-4): 288-300.

[15] Poulton C V L, Lee J R, Hobbs P R N, Jones L, Hall M (2006) Preliminary investigation into monitoring coastal erosion using terrestrial laser scanning: case study at Happisburgh, Norfolk. Bulletin of the Geological Society of Norfolk 56: 45-64.

[16] Romanescu G, Cotiuga V, Asandulesei A, Stoleriu C (2011) Use of the 3-D scanner in mapping and monitoring the dynamic degradation of soils. Case study of the Cucuteni-Baiceni Gully on the Moldavian Plateau (Romania). Hydrology and Earth System Sciences 16(3): 953-966.

[17] Rosser N J, Petley D N, Lim M, Dunning S A, Allison R J (2005) Terrestrial laser scanning for monitoring the processes of hard rock coastal cliff erosion. Quarterly Journal of Engineering Geology and Hydrogeology 38(4): 363-375.

[18] Schaefer M, Inkpen R (2010) Towards a protocol for laser scanning of rock surface. Earth Surface Processes and Landforms 35(4): 147-423.

[19] Sui L, Wang X, Zhoo D, Qu J (2008) Application of 3D laser scanner for monitoring of landslide hazards. International Archives of the Photogrammetry, Remote Sensing and Spatial Information Science 37: 277-281.

[20] Teza G, Galgaro A, Zaltron N, Genevois R (2007) Terrestrial laser scanner to detect landslide displacement fields: a new approach. International Journal of Remote Sensing, 28(16): 3425-3446.

[21] Thoma D P, Gupta S C, Bauer M A (2001) Quantifying River Bank Erosion with Scanning Laser Altimetry. International Archives of Photogrammetry and Remote Sensing, Annapolis, MD. XXXIV-3/W4: 169-174.

[22] Thoma D P, Gupta S C, Bauer M A, Kirchoff C E (2005) Airborne laser scanning for riverbank erosion assessment. Remote Sensing of Environment 95:493-501.

[23] Travelletti J, Oppikofer T, Delacourt C, Malet J P, Jaboyedoff M (2008) Monitoring landslide displacements during a controlled rain experiment using a long-range terrestrial laser scanning (TLS). International Archives of the Photogrammetry, Remote Sensing and Spatial Information Science 37: 485-490.

[24] Wasklewicz T, Stanley D, Volker H, Whitley D S (2005) Terrestrial 3D laser scanning: A new method for recording rock art. INORA 41: 16-25.

[25] Ylmaz H M, Yakar M, Yildiz F, Karabork H, Kavurmaci M M, Mutlouoglu O, Goktepe A (2010). Determining rates of erosion of an earth pillar by terrestrial laser scanning. The Arabian Journal for Science and Engineering 35(2A): 163-172.

[26] Alba M, Giussani A, Roncoroni F, Scaioni M, Valgoi P (2006) Geometric Modelling of a Large Dam by Terrestrial Laser Scanning, Shaping the Change. In: Shaping the change: XXIII International FIG Congress, 8-13 October 2006, Munich, Germany: proceedings. Copenhagen: FIG. pp. 1-15.

[27] Alba M, Fregonese L, Prandi F, Scaroni M, Valgoi P (2006) Structural Monitoring of a Large Dam by Terrestrial Lasser Scanning. International Archives of the Photogrammetry, Remote Sensing and Spatial Information Science 36(5): 1-6.

[28] Playan E, Zapata N, Burguete J, Salvador R, Serreta A (2010) Application of a topographic 3D scanner to irrigation research. Irrigation Science 28(3): 245-256.

[29] Schneider D (2006). Terrestrial Laser Scanning for Area Based Deformation Analysis of Towers and Water Dams. In: Proceedings 3rd IAG / 12th FIG Symposium, Baden, May 22-24, 2006. 10 p. Available:
http://www.fig.net/commission6/baden_2006/PDF/LS2/Schneider.pdf.   Accessed   2011
Nov 11.

[30] Barber D M, Ross W A D, Jon P M (2006) Laser Scanning for Architectural Conservation. Journal of Architectural Conservation 12(1): 35-52.

[31] Chang J R, Chang KT, Chen D H (2006) Application of 3D Laser Scanning on Measuring Pavement Roughness. Journal of Testing and Evaluation 34(2): 83-91.

[32] Ioanidis C, Valani A, Georgopoulos A, Tsiligiris E (2006) 3D Model Generation of Deformation Analysis using Laser Scanning Data of a Cooling Tower. 3rd IAG / 12th FIG Symposium, Baden, May 22-24, 2006. 10 p. Available:
http://www.fig.net/commission6/baden_2006/PDF/LS2/ Ioannidis.pdf. Accessed 2011 Nov 11.

[33] Böhler W, Marbs A (2006) 3D Scanning Instruments; CIPA, Heritage Documentation - International Workshop on Scanning for Cultural Heritage Recording, Corfu, Greece. Available: http://www.i3mainz.fh-mainz.de/publicat/korfu/p05_Boehler.pdf. Accessed 2011 Nov 11.

[34] Böehler W, Marbs A (2006) Investigating laser scanner accuracy. Available: http://www.uagp. net/images/stories/news/2007/10/PDF/1.pdf. Accessed 2011 Nov 11.

[35] Wang D Z, Edirisinghe M J, Jayasinghe S N (2007) A novel 3D pattering technique for forming advance ceramics. Key Engineering Materials 336-338(11): 977-979.

[36] Vaaja M, Hyyppa J, Kukko A, Kaartinen H, Hyyppa H, Alho P (2011) Mapping Topography Changes and Elevation Accuracies Using a Mobile Laser Scanner. Remote Sensing 3: 587-600.

[37] Zhou Y, Cui M, Yang L (2009) Application of 3D laser Scanner in topographic change monitor and analysis. In: Electronic Measurement & Instruments, 2009. ICEMI '09. 9th International Conference on. pp. 382-385.

[38] Bornaz L, Rinaudo F (2004) Terrestrial laser scanner data processing. Available: http://citeseerx.ist.psu.edu/viewdoc/download?doi=10.1.1.64.8938&rep=rep1&type=pdf. Accessed 2011 Nov 8.

[39] Bu L, Zhang Z (2008) Application of point clouds from terrestrial 3D Laser Scanner for deformation measurements. In: The International Archives of the Photogrammetry, Remote Sensing and Spatial Information Sciences. Vol. XXXVII. Part B5. Beijing. pp. 545-548. Available: http://www.isprs.org/proceedings/XXXVII/congress/5_pdf/95.pdf. Accessed 2011 Nov 8.

[40] Carlson W D (2006) Three-dimensional imaging of earth and planetary materials. Earth and Planetary Science Letters 249(3-4): 133-147.

[41] Gordon S J, Lichti D D (2004) Terrestrial laser scanners with a narrow field of view: the effect on 3-D reaction solutions. Survey Review 37(292): 22.

[42] Guarnieri A, Pirotti F, Pontin M, Vettore A (2006) 3D Surveying for Structural analysis Applications, In: Proceedings 3rd IAG / 12th FIG Symposium, Baden, May 22-24, 2006. Available:
http://www.fig.net/commission6/baden_2006/PDF/LS1/Guarnieri.pdf.   Accessed   2011 Nov 01.

[43] Kaasalainen S, Krooks A, Kukko A, Kaartinen H (2009) Radiometric Calibration of Terrestrial Laser Scanners with External Reference Targets. Remote Sensing 1(3): 144-158.

[44] Kern F (2002) Precise Determination of Volume with Terrestrial 3D-Laser scanner. In: Kahmen H. Niemeier W, Retscher G, editors. 2nd Symposium on Geodesy for Geotechnical and Structural Engineering: papers presnted at the symposium in Berlin, Germany, May 21-24, 2002. Vienna: Department of Applied and Engineering Geodesy, Institute of Geodesy and Geophysics, Vienna University of Technology. pp. 531-534.

[45] Lichti D D, Goerdon S J, Steward M P (2002) Groundbased laser scanners: operations, systems and applications. Geomatica 56: 21-33.

[46] MacMillan R A, Martin T C, Earle T J, McNabb D H (2003) Automated analysis and classification of landform using high-resolution digital elevation data: applications and issues. Canadian Journal of Remote Sensing 29: 592-606.

[47] Monserrat O, Crosetto M (2008) Deformation measurement using terrestrial laser scanning data and least squares 3 D surface matching, ISPRS Journal of Photogrammetry and Remote Sensing 63(1): 142-154.

[48] Pfeifer N, Lichti D D (2004) Terrestrial Laser Scanning. GIM International 12: 50-53.

[49] Pflipsen B (2006) Volume Computation - a comparison of total station versus laser scanner and different software. Master's Thesis in Geomatics. Available: http://hig.diva-portal.org/smash/ get/diva2:120447/FULLTEXT01. Accessed 2011 Nov 3.

[50] Schmid T, Schack-Kirchner H, Hildebrand E (2004) A case study of terrestrial laser-scanning in erosion research: calculation of roughness indices and volume balance at a logged forest site International Archives of Photogrammetry, Remote Sensing and Digital Information Sciences 36: 114-118.

[51] Slob S, Hack H R G K (2004) 3-D Terrestrial Laser Scanning a New Field Measurement and Monitoring Technique. In: Hack R, Azzam R, Charlier R, editors. Engineering Geology for Infrastructure Planning in Europe. A European Perspective. Berlin, New York: Springer. pp. 179-190.

[52] Soudarissanane S, Lindenbergh R, Gorte B (2008) Reducing the error in terrestrial laser scanning by optimizing the measurement set-up. In: Chen J, Jiang J, Maas H-G, editors. The International Archives of the Photogrammetry, Remote Sensing and Spatial Information Sciences. Vol. XXXVII. Part B5. Beijing 2008. Beijing. pp. 615-620. Available: http://www.isprs.org/proceedings/XXXVII/ congress/5_pdf/182.pdf. Accessed 2011 Nov 5.

[53] Tsakiri M, Lichti D, Pfeifer N (2006). Terrestrial Laser Scanning for Deformation Monitoring, In: Proceedings 3rd IAG / 12th FIG Symposium, Baden, May 22-24, 2006. Available: http://www.fig.net/commission6/baden_2006/PDF/LS2/Tsakiri.pdf. Accessed 2011 Nov 23.

[54] Voegtle T, Schwab I, Landes T (2008) Influences of different materials on the measurement of a Terrestrial Laser Scanner (TLS). In: Chen J, Jiang J, Maas H-G, editors. The International Archives of the Photogrammetry, Remote Sensing and Spatial Information Sciences. Vol. XXXVII. Part B5. Beijing 2008. Beijing. pp. 1061-1066. Available:
http://www.isprs.org/proceedings/XXXVII/ congress/5_pdf/107.pdf. Accessed 2011 Nov 5.

[55] Bacauanu V (1968) Campia Moldovei. Studiu geomorphologic. București: Editura Academiei. 220 p.

[56] Bacauanu V, Barbu N, Pantazica M, Ungureanu A, Chiriac D (1980) Podisul Moldovei. Natura, om, economie. Bucuresti: Editura Stiintifica si Enciclopedica. 350 p.

[57] Boghian D (2004) Comunitatile cucuteniene din bazinul Bahlui. Suceava: Editura Bucovina Istorica. 246 p.

[58] Branzila M (1999) Geologia partii sudice a Campiei Moldovei. Iasi: Editura Corson. 221 p.

[59] Cotiuga V, Cotoi O (2004) Parcul arheologic experimental de la Cucuteni. In: Petrescu-Dimbovita, Valeanu M C. Cucuteni-Cetatuie. Monografie arheologică. Piatra Neamt: Editura Constantin Matasa. pp. 337-351.

[60] Ionita I (2000) Geomorfologie aplicata. Iasi: Editura Universitatii "Al.I.Cuza". 230 p.

[61] Ionita I (2006) Gully development in the Moldavian Plateau of Romania. Catena 68(2-3): 133-140.

[62] Motoc M, Taloescu I, Negut N (1979) Estimarea ritmului de dezvoltare a ravenelor. Buletinul Informativ ASAS 8: 183-193.

[63] Pantazica M (1974) Hidrografia Cimpiei Moldovei. Iasi: Editura Junimea. 319 p.

[64] Poghirc P (1972) Satul din Colinele Tutovei. Bucuresti: Editura Stiintifica. 240 p.

[65] Radoane M, Ichim I, Radoane N (1995) Gully distribution and development in Moldavia, Romania. Catena 24: 127-146.

[66] Radoane M, Radoane N, Ichim I, Surdeanu V (1999) Ravenele. Forme, procese, evolutie. Cluj-Napoca: Editura Presa Universitara Clujeana. 266 p.

[67] Romanescu G, Romanescu G, Stoleriu C, Ursu A (2008) Inventarierea si tipologia zonelor umede si apelor adanci din Podisul Moldovei. Iasi: Editura Terra Nostra. 242 p.

[68] Barnolas M, Rigo T, Llasat M C (2010) Characteristics of 2-D convective structures in Catalonia (NE Spain): an analysis using radar data and GIS. Hydrology and Earth System Science 14: 129–139. Doi:10.5194/hess-14-129-2010.

[69] Xie H, Jia H (2010) The Development of 3D Laser Scanning Technique and Its Applications in Land Reclamation. Information Engineering and Electronic Commerce 2: 1-4. Doi:10.1109/IEEC.2010.5533250.

[70] Rinaudo F, Nex F (2011) LIDAR e Fotogrammetria Digitale verso una nuova integrazione. Geomedia 3: 14-22.

[71] Lee I, Schenk T (2001) 3D Perceptual Organization of Laser Altimetry Data. International Archives of Photogrammetry and Remote Sensing 34(3/W4): 119-127.

[72] Marzolff I, Poesen J (2009) The potential of 3D gully monitoring with GIS using high-resolution aerial photography and a digital photogrammetry system. Geomorphology 111(1-2): 48-60.

[73] Mikoš M, Vidmar A, Brilly M (2005) Using a laser measurement sydtem for monitoring morphological changes on the Strug rock fall, Slovenia. Natural Hazards and Earth System Sciences 5: 143-153.

[74] Bretar F, Chauve A, Bailly J-S, Mallet C, Jacome A (2009) Terrain surfaces and 3-D land10 cover classification from small footprint full-waveform lidar data: application to badlands. Hydrology and Earth System Science 13: 1531–1544. Ddoi:10.5194/hess-13-1531-2009.

[75] Leica HDS3000 data sheet. 2006. Available: http://www.leica-geosystems.com/hds/en/lgs_5574.htm. Accessed 2011 Apr 22.

[76] James T D, Carbonneau P E, Lane S N (2007) Investigating the effects of DEM error in scaling analysis. Photogrammetric engineering and remote sensing 73(1): 67–78.

[77] Erhan E (2001) Consideratii privind resursele climatice ale Moldovei. Lucrarile Seminarului Geografic "Dimitrie Cantemir" 19-20: 211-226.

[78] Blong R J, Graham O P, Veness J A (1982) The role of side wall processes in gully development; some N.S.W. examples. Earth Surface Processes and Landforms 7: 381-385.

[79] Bradford J M, Piest R F, Spomer R G (1978) Failure sequence of gully headwalls in Western Iowa. Soil Science Society of America Journal 42(2): 323-328.

[80] Bull W B (1997) Gully processes and modeling. Progress in Physical Geography 21(3): 354-374.

[81] De Oliveira M A T (1990) Slope geometry and gully erosion development: Bananal, Sao Paulo, Brazil. Zeitschrift. für Geomorphology N.F. 34(4): 423-434.

[82] Harvey A M (1992) Process interactions, temporal scales and the development of hillslope gully system: Howgill Fells, northwest England. Geomorphology 5: 323-344.

[83] Heede, B.H. 1976. Gully development and Control: The Status of our Knowledge. Fort Collins, Colo.: Rocky Mountain Forest and Range Experiment Station. 42 p.

[84] Heritage G, Hetherington D (2007) Towards a protocol for laser scanning in fluvial geomorphology. Earth Surface Processes and Landforms 32: 66-74.

[85] Piest R F, Bradford J M, Wyatt G M (1975) Soil Erosion and Sediment transport from gully. Journal of the Hydraulics Division 101(1): 65-80.

[86] Vandaele K, Poesen J, Govers G, Wesenael van B (1996) Geomorphic threshold conditions for ephemeral gully incision. Geomorphology 16(2): 161-173.

[87] Ursulescu N (2006) Donées récentes concernant l'histoire des communautés énéolithique de la civilisation Cucuteni. Acta Terrae Septemcastrensis 5: 79-113.

Chapter 5

Advanced Map Optimalization Based on Eye-Tracking

Stanislav Popelka, Alzbeta Brychtova, Jan Brus and Vít Voženílek

Additional information is available at the end of the chapter

1. Introduction

Mapmakers and cartographers perceive maps differently than the target group of users. In the case map is created by someone else than a professional cartographer the process of making maps suffers by considerable degree of subjectivity. Despite the several-hundred-years effort of objectification of map making process, cartographers sometimes have difficulties to imagine the way how the user will use the map, especially the way of map perception, reading, analysis and interpretation [1]. Cartographers often lack the reasoning for the decision-making in balancing the map design and layout, designing a map symbology, choosing the cartographic method of visualization or level of generalization [2]. For these reasons it is necessary to carry out a research on user perception of maps.

According to Golledge and Stimson [3] the perception is function of a cognition (thinking), which can be understood as a way of encoding, storing and integrating information into existing knowledge. In order to study reading and using maps the perception is very important, because it helps to structure the area depicted on the map.

Several approaches of the research on user perception and evaluation of the applicability and effectiveness of maps exist. The eye-tracking is one of the rarely used. Based on results of eye movements analyses, many questions, that were not yet been discussed in cartography adequately, can be answered. For example, how users obtain information from the map, what is the strategy of map reading, how often users look to the map legend, how easily can be map symbols interpreted, etc. Analyses of maps usability can help in optimizing the map symbology, composition or design, so the new maps can be created in order to respect the specific user's requirements.

The main sensory channel for cognitive processes is vision [4]; therefore the research on map visual perception is necessary

2. Usability studies

The term usability is defined as "the effectiveness, efficiency, and satisfaction with which specified users achieve specified goals in particular environments" [5]. Satisfaction quantifies subjective users' impressions dealing with such indicators as operability and learnability of a given task. Efficiency and effectiveness metrics are objective performance measures of speed and accuracy. [6]

Nielsen [7] defines usability as a quality attribute that assesses how are user interfaces easy to use. Usability can reveal qualities of the product as well as lack of its functionality, which usually arises during the design phase of a product. [8] The assessed product can be an image, web page, text or a map. To be able to derive qualitative or quantitative measures of the user experience (usability), a number of evaluation methods is possible to use. Li et al. [9] mentions these methods of usability evaluation:

- focus group studies,
- interview,
- direct observation,
- think-aloud protocol,
- retrospective think-aloud protocol,
- screen capture,
- eye-tracking.

Each method has its advantages and disadvantages. Frequently a combination of methods is used in the research. For example, see [10, 11, 12].

The method of focus group studies and interview are based on direct contact with the user. They are based on a targeted questioning and recording of discussions and responses of individuals or groups.

The method of direct observation leads to the detection of subject's behaviour in its natural environment without any interference by the observer. For observation various technical utilities, especially recorders, cameras and camcorders, are used. Direct observation sometimes leads to problems in professional ethics, especially when observed people are not acquainted with the fact that they are the subject of observation.

A frequently used method is "Think-Aloud". Its principle lies in the verbalization of the process performed by the user during solving a specific problem. Participating test subjects verbally describe the process of solving specific tasks and also their feelings [13]. This method is very quick and inexpensive, nevertheless, participant is not aware of all cognitive processes, and not all processes can be simply expressed in words. This method is very subjective in the term of observed subjects, who describe their experiences, and also in the term of the evaluation of their response. Detail usage of this method is discussed by Somersen [14]. Similar to the "Think-Aloud" method is a retrospective variant, when the subject describes a workflow after the task is completed.

The screen capture method of usability study was in the field of cartography and GIS used for example in Haklay et al. [15]. They assessed the usability of GIS software using screenshots that users had posted. On these images the windows of application were captured in the middle of the workday and composition of opened toolbars was analysed.

The last mentioned method, which is in the focus of this chapter, uses a device to monitor eye movements. Eye-tracking method can be considered to be objective, because it is not influenced by the opinion of the monitored person [16].

3. Eye-tracking technology

Eye-tracking technology is based on the principles of tracking movements of human eye while perceiving the visual scene. The measurement device used for measuring eye movements is commonly known as eye-tracker. [17]

When users are searching for the information in unknown environment (text articles, web pages, maps etc.), typically two types of processes occur: a perceptual one (the user should locate/notice the target) and a cognitive one (the user cognitively computes the visual input and understands the function of the target). Eye movement analysis provides valuable quantitative and qualitative information on both stages of visual search. [16]

Qualitative information describes the way in which user explores the stimuli. They can reveal areas of maximum interest, disruptive elements and strategy of searching for a specific element. Quantitative information describe the time spent by observing a particular phenomenon, speed of identifying information and several derived gaze data metrics.

3.1. Methods of eye-tracking

In general, there are two types of eye movement monitoring techniques: those that measure the position of the eye relative to the head, and those that measure the orientation of the eye in space, or the "point of regard". [18]

According to Duchowski [6], methods of eye movement tracking can be categorized into three main groups:

- electrooculography (EOG),
- scleral contact lens/search coil
- remote eye-tracking.

EOG is a method which was popular about 40 years ago. Its principle lies in measuring of skin's electric potential differences of electrodes placed around the eye. By recording quite small differences in the skin potential around the eye, the position of the eye can be tracked. [19]

The method of scleral contact lens or search coil uses an attachment to the eye, such as a special contact lens with an embedded mirror or magnetic field sensor, and the movement of the attachment is measured with the assumption that it does not slip significantly as the eye rotates. Both mentioned methods measure eye position relative to the head and they are not generally suitable for point of regard measurements.

Currently the most exploited eye movement measurement method is remote eye-tracking, also called the Pupil and Corneal Reflexion method. It relies on the measurement of visible features of eye, e. g. pupil, iris-sclera boundary and corneal reflection of a closely situated direct light source (often infra-red). The reflected light is recorded by a video camera or specially designed optical sensor. The information is then analyzed to extract eye rotation from changes in corneal reflections. The resulting corneal reflexion is also known as "glint" or the 1st Purkinje reflexion (P1) [20].

There are at least four Purkinje images (figure 1). The first Purkinje image (P1) is the reflection from the outer surface of the cornea. The second one (P2) is the reflection from the inner surface of the cornea. The third one (P3) is the reflection from the anterior surface of the lens and the last one (P4) is the reflection from the posterior surface of the lens. [21]

Eye position and gaze direction are estimated using information from an image sensors picking up reflection patterns on the cornea and other information points. By image analysis and mathematics a gaze point on a reference plane can be calculated.

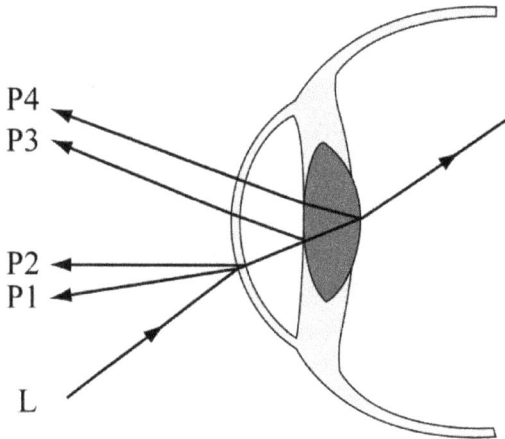

Figure 1. Four Purkinje images - the reflection (L) on different parts of the eye: P1 from the outer surface of the cornea, P2 from the inner surface of the cornea, P3 from the anterior surface of the lens, P4 from the posterior surface of the lens. [22]

3.2. Eye movements and algorithms of their detection

Human eyes can only perceive a limited fraction of the visual world at one point in time. Both eyes together provide a roughly elliptical view of the world which is approximately 200° of visual angle wide and 130° high. [23]

However not all parts of this view are perceived with equal acuity because the retina of the eye has a varying structure and composition. The fovea, part of the retina, is responsible for sharp central vision (also called foveal vision), which is necessary during reading, watching television or movies, driving, and any activity where visual detail is of primary importance. The really high-resolution area covers only about 2° of the visual field. The fovea is surrounded by the parafovea belt and the perifovea outer region [24]. The vision supported by this part of the eye is so called peripheral vision, which in comparison with foveal vision seem to be blurred [25].

Eye movement is not smooth. The eye moves in spurts and rests between each movement. During a fixation, eyes are relatively steadily looking at one spot in the visual scene. In order to achieve the most accurate visual impression of a visual scene, eyes move rapidly in mostly ballistic jumps (i.e., saccades) from one spot to another. Among those rather large saccadic eye movements that an attentive person can easily observe from his or her own experience, there are three other, much shorter eye movements, i.e. tremor, drift, and microsaccades. Their purpose is to avoid saturation effects of the visual receptors on the retina which would lead to fading perception. However, people are unaware of those tiny movements and they can be hardly detected by state-of-the-art unobtrusive eye-trackers. [23]

The analysis of fixations and saccades requires some form of identification that results from the processing of raw eye-movement data. Fixation and saccades identification is an inherently statistical description of observed eye movement behaviours.

It is important to define the exact detection algorithm for eye movement analysis, because different parameterizations of an algorithm might lead to different results [26]. Plenty of algorithms exist, but mostly used are I-VT and I-DT. In the case of the I-VT (Velocity-threshold fixation identification) algorithm, the eye-velocity value is compared to the threshold. If the sampled velocity is smaller than the threshold, the corresponding eye-position is marked as a part of a saccade, otherwise the eye-position sample is assigned to be a part of a fixation. The I-DT (Dispersion-Threshold Identification) algorithm takes into account the close spatial proximity of the eye position points in the eye movement trace. [27]

Based on statistical analysis of fixations, saccades, their mutual relationship and other characteristics, it is possible to identify certain attributes of respondent behaviour.

For example, long average fixation durations can be interpreted in two different ways as either: (1) the user has difficulties extracting information; or (2) the user is more engaged with interpreting a representation [28]. Hence, distinguishing between the two is case specific.

Saccade/Fixation ratio describes the ratio between search activity (represented by the number of the saccades) and processing activity (represented by the number of the fixations). A small saccade/fixation ratio indicates that the user is spending more "cognitive resources" on the task and less cognitive resources on gathering important background information [16].

A large number of saccades indicate a low degree of search efficiency or poor interface layout. User roams from place to place finding no satisfactory answer. Saccadic amplitude together with ScanPath duration and ScanPath length can refer to a strategy of user cognition style, or quality of examined layout [23].

3.3. Eye-tracking device

According to fundamentals of examined problem it is necessary to use a proper eye-tracker. Individual devices differ from one another by precision given by the spatial resolution and accuracy of the point of view. An important parameter is the time resolution, which is expressed in hertz (Hz). Department of Geoinformatics, Palacký University in Olomouc owns static SMI RED 250 Eye-tracker (figure 2) with sample frequency of 120 Hz, so data are recorded approximately every 8 ms.

Device parameters (resolution, mobility) must respect the purpose of the application for which the device is used. Research on sportsmen or driver concentration, or arrangement of goods on the shelves requires a mobile device mounted on the head of the subject (the headset). Evaluation of stimuli on the computer screen or television will rather use the static device.

Figure 2. Laboratory setup with static remote SMI RED 250 eye-tracker at Department of Geoinformatics, Palacký University in Olomouc.

3.4. Current eye-tracking based research on cartographical issues

Anatomy of the human eye is known for hundreds of years, but scientific interest in the processes of visual perception began only during the 19th century [29]. After the Second World War, one of the first measures of gaze direction was done in 1947. The research was focused on the behaviour of military pilots during aircraft landing. It was carried out by analyzing the video with more than 500 000 film frames [30].

With the improvements of the eye-tracking technologies, eye-tracking tools gain their impacts on the usability field and nowadays they are accepted as a tool to improve computer interfaces [11]. Currently eye-tracking is utilizable in many areas of human activity - psychology, medicine, marketing, commercial, etc.

One of the research goals of contemporary cartography is the investigation of perception processing of maps, not only from the commercial point of view, but also in planning, crisis management and rescue operations.

On the one hand, map is very important carrier of information that readers need to assimilate as quickly as possible and undistorted, so cartographers labour for its highest possible accuracy. On the other hand map design and its visual appearance are determinants of user popularity.

In both cases, the key to success is to answer a number of highly debated issues - for example how readers follow the information in the map, in which order and how fast they read the information, which compositional elements they read earliest, how many times they look back to the map legend, which map elements are easy and which difficult to handle, what affects the legibility of the map, etc. These findings can facilitate to evaluate the quality of the map composition, symbology and map content, and thus define the methodology for creating maps that will correspond with requirements of users.

With respect to the investigatory device, maps evaluation using eye-tracking technology is available both for analog maps and the digital cartographic outputs.

One of the first publications focused on the application of eye-tracking methods in cartography is Eye Movement Studies in Cartography and Related Fields [29], in which the author summarizes the results of various studies in the late 80's of the 20th century. It deals with the general knowledge of tracking of the human eye, studies on evaluation of specific graphic outputs, emphasizing the impossibility to generalize the findings in the behalf of dissimilar studies. He described several universal conclusions and highlighted the importance of distinguishing between user groups according to their age and education.

It is possible to separate the evaluation of information content of maps from the map design. However complex evaluation is more logical, because the information value of maps (e. g. content) can be increased or degraded by technical or artistic design.

An example of a complex evaluation of maps was presented by Alacam and Dalci [11], who compared four map portals (Google Maps, Yahoo Maps, Live Search Maps and MapQuest). Results of eye-tracking experiment revealed considerable variance in the strategy of solving particular quest in different map portals environment. Basic assumption of this study was that the lower the average duration of fixation, the more intuitive the environment. It was found, that users average fixation duration at the Google Map stimuli is statistically significantly lower, than in the case of other evaluated portals.

Coltekin et al. [31] in their research deal with the evaluation of user interface of cartographic software. Test subjects were ask to create a complex map in two different map applications. The study was designed as a between-subject experiment and eye movement analysis was coupled with traditional usability metrics to identify possible design issues. Initial analyses included statistical tests for satisfaction, effectiveness (accuracy of response), and efficiency (response speed).

In a different study the same authors [32] deal with a generic evaluation approach combining theory and data-driven methods based on sequence similarity analysis. The approach systematically studies users' visual interaction strategies when using highly interactive map interfaces. The result was that the participants generally follow a sequence that agrees with the hypothetical sequence representing user's strategies.

Another application of eye-tracking in cartography appears in the study of Opach and Nossum [33] where authors have explored the suitability of eye-tracking on two different semistatic and traditional cartographic animations of temperature and weather. Contrary to the author's previous web based experiment, analysis of the eye-tracking data revealed that the viewing behaviour were surprisingly similar. Three of the metrics used (fixation counts, observation length and time to first fixation) indicated very similar viewing strategies and behaviour during viewing different kind of cartography animations.

Fuhrman, Tamir and Komogortsev [34] have dealt with an assumption that three-dimensional topographic maps provide more effective route planning, navigation, orientation, and way-finding results than traditional two-dimensional representations. The eye-tracking metrics analysis indicates with a high statistical level of confidence that three-dimensional holographic maps enable more efficient route planning.

Popelka and Brychtova [35] used eye-tracking together with questionnaire investigation for evaluation user's attitudes toward interactive methods of virtual geovisualisation of changes in the city built-up area. Five approaches of visualization were assessed - textual description of changes, comparison of historical and recent pictures or photos, overlaying historical maps over the orthophoto, enhanced visualization of historical map in large scale using the third dimension and photorealistic 3D models of the same area in different ages.

Technologies and methods of eye-tracking have not yet been fully utilized in cartography, even though the possibilities are wide. Cartographic research in the field of eye-tracking currently focuses explicitly on improving the user quality of maps. Future potential expansion of eye-tracking technology can be seen in the activity of the International Cartographic Association, especially the Commission on Use and User Issues [36] and the newly established Commission on Cognitive Issues in Geographic Information Visualization [37].

4. Methods of eye-tracking data visualization

Results of eye-tracking measurements are presented as a text file containing a timestamp and a number of specifications describing coordinates of the point of regard, the pupil size, the angle of the eye position etc. First of all, it is necessary to classify the data, with a specified algorithm, as fixations and saccades (see chapter Eye movements and algorithms of their detection). Then, the data are even visualized in a suitable way, or can be statistically analysed.

There are several basic methods of the eye-tracking data visualization. Holmqvist et al. [20] present the main techniques of gaze data visualization as follows: ScanPath (GazePlot), Attention (Heat) maps and the AOI (Area of Interest) Analysis.

Following chapters will present different possibilities of visualization and eye-tracking measurements on concrete examples from cartography. In this way, methods of eye-tracking will be shown. All examples are based on source data, which are results of several authentic experiments on cartography rules evaluation.

All presented case studies were performed using the SMI RED 250 remote device with the sampling frequency of 120 Hz. Eye position was measured every 8ms.

Respondents were chosen from university students of Geoinformatics and Cartography and also from other studying fields which are not related to cartography.

4.1. ScanPath

ScanPath is defined as a route of oculomotor events through space within a certain time-span [20]. Thanks to ScanPath, it is possible to display raw data as well as calculated fixations and saccades. Circles of different sizes represent fixations (their radius corresponds with their length) and lines which connect the circles represent saccades [38].

When a larger amount of data is displayed, this method becomes restraining. Overlapping parts of individual fixations cause that it is not possible to identify their number visually. Figure 3 shows an example of a ScanPath. Respondents were asked to find the highest peak on one of the maps. The aim of this experiment was to compare two types of visualization - perspective 3D display and a classical orthogonal map supplemented with the shading. Raw data are displayed on the left of the picture, fixations and saccades on the right. From both pictures, it is evident that this particular respondent preferred the three-dimensional visualization. His answer is displayed by a red dot which represents the mouse-click.

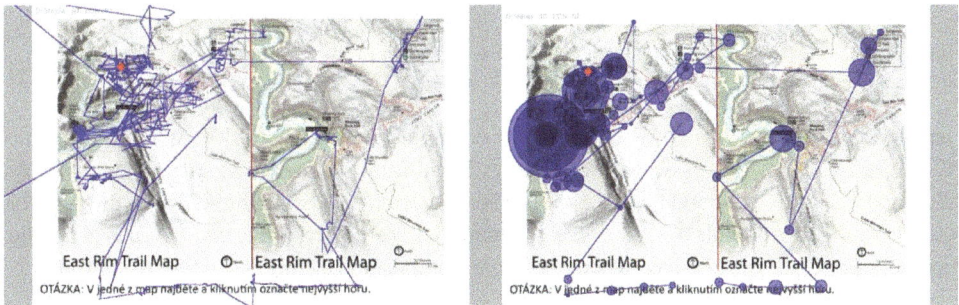

Figure 3. ScanPath showing raw data (left) and fixation and saccades (right)

4.2. ScanPath comparison

There is a great need for robust and general method for ScanPath comparison existing in many fields of eye-tracking research [39]. Privitera and Stark [40] introduced ScanPath comparison based on string editing. Fixations are replaced with characters standing for the AOI's they hit and the ScanPath is represented as character string. It is one of the first methods comparing not only the loci of fixations, but also their order. The principle of this method is the transformation of two-dimensional data (X, Y coordinates of fixations) to one-dimensional data (character string). Each ScanPath is recorded as a string of letters where each letter corresponds to the area of a current fixation location.

Two or more character strings are then compared and their similarity is measured. Defined by an optimization algorithm, string editing assigns a unit value to three different character operations: deletion, insertion, and substitution. Characters are then manipulated in order to transform one string into another and character manipulation values are tabulated [41]. String edit algorithm determines the number of operations needed to transform one sequence to another - the operations being insertions, deletions and substitutions. The calculated metric will be a measure of how different two sequences are. This method uses the Levenshtein algorithm to produce a string-edit distance between each sequence [42]. Example of the Levenshtein distance measure is in the figure 4. Character string comparison methods are widely used in bioinformatics to align DNA and protein sequences.

☐ ⊞ **Similarity discovery**
Search within sequences S1, S2, S3, S4, S5, S6, S7, S8, S9, S10, S11

	S1	S2	S3	S4	S5	S6	S7	S8	S9	S10	S11
S1	-	11.0	13.0	13.0	13.0	15.0	12.0	13.0	13.0	13.0	14.0
S2	11.0	-	8.0	8.0	8.0	11.0	10.0	10.0	11.0	6.0	10.0
S3	13.0	8.0	-	8.0	5.0	8.0	9.0	7.0	9.0	7.0	7.0
S4	13.0	8.0	8.0	-	8.0	9.0	6.0	8.0	9.0	8.0	9.0
S5	13.0	8.0	5.0	8.0	-	7.0	9.0	6.0	9.0	9.0	6.0
S6	15.0	11.0	8.0	9.0	7.0	-	9.0	9.0	12.0	10.0	9.0
S7	12.0	10.0	9.0	6.0	9.0	9.0	-	10.0	10.0	8.0	10.0
S8	13.0	10.0	7.0	8.0	6.0	9.0	10.0	-	9.0	7.0	8.0
S9	13.0	11.0	9.0	9.0	9.0	12.0	10.0	9.0	-	12.0	10.0
S10	13.0	6.0	7.0	8.0	9.0	10.0	8.0	7.0	12.0	-	8.0
S11	14.0	10.0	7.0	9.0	6.0	9.0	10.0	8.0	10.0	8.0	-

Scores shown are the Levenshtein distance between collapsed sequences.
This is a **distance** measure

Figure 4. Result of the Levenshtein distance measure between group of eleven geoinformatics while observing a map. Gridded AOI with grid of 5*5 cells was used

Depending on specific tasks, it is important to distinguish between gridded AOI and semantic AOI approaches. When using the Gridded AOI, stimulus is split into areas of equal size (rectangles) with no relation to semantics. The second approach, Semantic AOI uses the Area of Interest which corresponds to specific areas in stimuli. In cartography, Semantic AOI is generally more advantageous because it corresponds to map composition elements like title, legend, etc.

4.3. Attention maps

Attention maps, also called HeatMaps, are used for visualization of quantitative characteristics of the user's gaze. Thanks to attention maps, it is possible to identify to which area user pay attention and which are rather neglected. In eye-tracking, HeatMaps enable the creation of a brief summary of areas which are in the spotlight and so they are needed to be analyzed more thoroughly. Except of the function of visualization, they might be used as a background for AOI creation. When plotting AOI around a small object, fixations of some of the respondents could be noted outside the created AOI because of inaccurate measurement or calibration deflection. By means of HeatMaps, it is possible to get an overview of possible measurement inaccuracies and so to adapt the Area of Interest.

HeatMaps produced by SMI software BeGaze are created in two steps. The software first scales each pixel in proportion to the durations of all fixations landing on it. Typically, this results in a very sparse "Fixation Hit Map", since only a small proportion of the pixels have been "hit". In next step, the hit map is convolved with a Gaussian kernel with certain width. A wider kernel gives a smoother, less pointy appearance to the attention map [20].

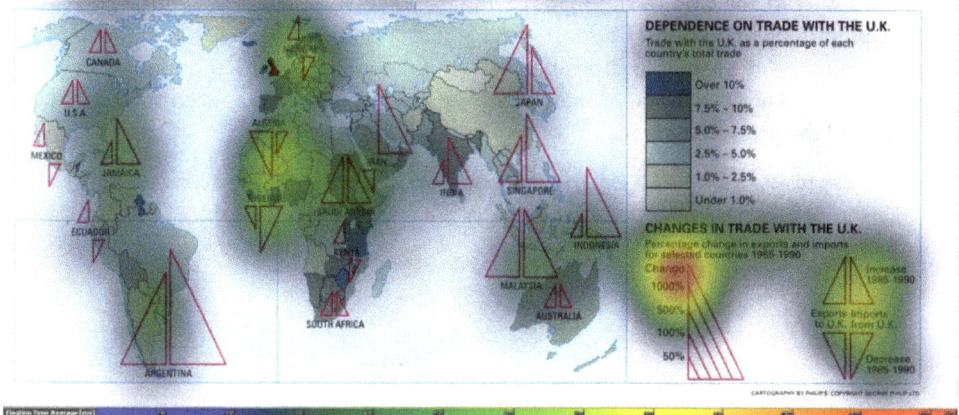

Figure 5. HeatMap created from fixations of seven respondents

4.4. Area of Interest

Areas of Interest (AOI) are regions in the stimulus which the researcher is interested in. The most important AOI metric is the dwell time defined as one visit in AOI, from entry to the exit. The dwell has its own duration, the starting point, the ending point, dispersion etc. In several ways, it is similar to a fixation, but it is of much larger entity, both in space and time [20]. It is also possible to follow the order in which the respondent looked at particular areas or the transition, the movement from one AOI to another etc.

In cartography the Areas of Interest analysis can be used advantageously. AOI analyses are based on evaluation of concrete parts of a map (legend, scale, title, specific phenomena in the map, etc.). When evaluating influence of a composition on the map reading, the application of AOI is very useful. By indicating and evaluating particular compositional elements as AOI, several characteristics can be find out - e.g. for how long the respondent was observing the given area, in which order he visited them etc.

The results can be visualized with use of the Sequence Chart, which displays observed areas in different colours on a timeline. Figure 6 shows results of an experiment, whose objective was to reveal differences in reading of a simple map by a group of students of cartography and cartography amateurs. Three different map compositions are displayed in figure 6. Maps (presented in the first row of figure 6) were projected in 5s intervals during which the respondents have to observe maps without answering any question. The second raw in the Figure 6 shows a sequence chart for a group of students of Geoinformatics and cartography, who have attended several cartography courses. The last row represents data given by cartographic amateurs, students of psychology, zoology etc.

Each stimulus was preceded by a short cross used to locate beginning of all trajectories at the same place (in the middle of the picture). That is why the AOI representing the map field is always pictured in first 500 ms. After this time; most geoinformatics students automatically read the title of the map, or rather noted fixations representing it in AOI. Cartographic amateurs did not do so. It is evident especially in the first column, where the stimulus was the "ideal" map composition [43]. In following columns, the composition was not in accordance with cartographic rules. Despite this fact, students of geoinformatics were trying to find the title of the map.

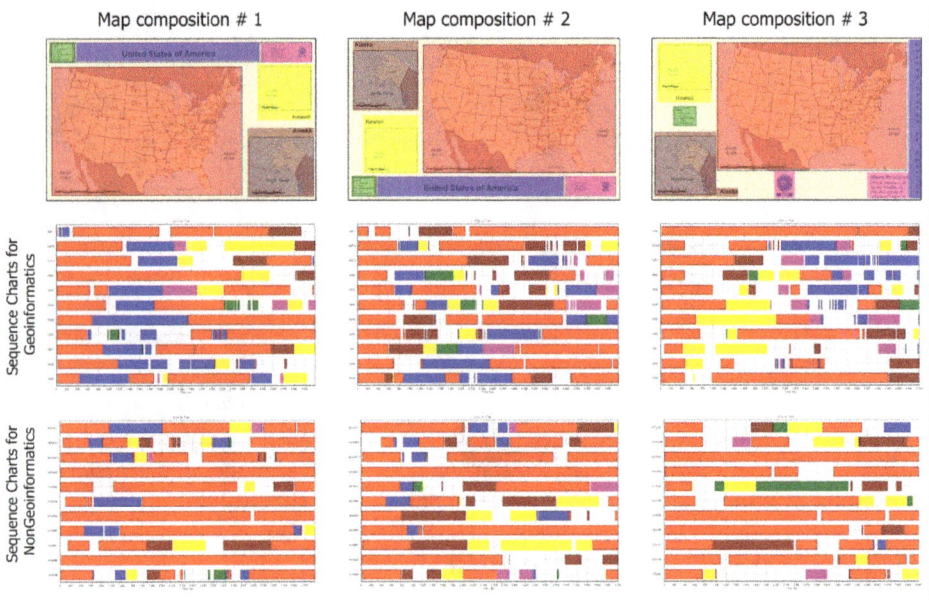

Figure 6. Sequence chart visualization. Sample data of the Esri were used for the creation of stimulus maps

The Sequence Chart is illustrative and easy to interpret, but it is necessary to evaluate the data by statistical approach.

The objective of the statistical hypotheses testing is to evaluate the data gained from experiments and the suitability of the purpose given before the testing. Statistical hypothesis is a certain purpose about the distribution of accidental quantities of a basic file.

In statistics, the result is called statistically significant when it is unlikely to have occurred by chance alone, according to a pre-determined threshold probability, the significance level. Critical tests of this kind may be called tests of significance. When such tests are available, we can say whether the second sample is/ is not significantly different from the first one [44].

From a more detailed evaluation of different types of basic eye movements (fixations and saccades), or gaze data metrics such as dwell time, it is possible to deduce a series of numeral metrics suitable for other statistical testing.

Different composition perception with two different groups of map users (experienced cartographers and non-cartographers) was verified by the testing of measured results by a two sample t-test, which is a method of mathematical statistics making possible to verify the null hypothesis that the means of two normally distributed populations are equal.

Differences of the mean dwell time of two groups of users were tested on particular Areas of Interest of the map list - title of the map, the map, the legend, the imprint and secondary maps. A zero hypothesis H0 was tested: mean values of particular choices are the same. Concrete results are illustrated in table 1.

	t	df	p-value	alpha	mean of carto [ms]	mean of non-carto [ms]	statement
Main map	-2,2189	33,374	0,0334	0,05	2375,58	3080,11	Rejecting H0
Map heading	3,7546	51,501	0,0004	0,05	631,27	230,78	Rejecting H0
Additive map 2	0,0963	36,715	0,9238	0,05	387,84	376,47	Fail to reject H0
Additive map 1	-0,1874	40,321	0,8523	0,05	260,59	284,08	Fail to reject H0
Masthead	1,1999	36,275	0,2380	0,05	127,27	57,57	Fail to reject H0
Map legend	0,2211	45,702	0,8260	0,05	99,50	84,33	Fail to reject H0

Table 1. Results of two sample t - test of dwell time on particular compositional map elements of two different user groups.

By comparing the mean dwell time in particular AOI it is evident that both groups spent most time on the main map field. Extremely long time was noticed at students of cartography in AOI covering the map title. The same AOI was on the 4th position with non-cartographers.

The t-test result disproved the zero-hypothesis saying that the values of the mean dwell time were the same with AOI map heading and the main map field. The visit rate of the main map field was significantly higher with non-cartographers. On the contrary, the map title visit rate was higher with cartographers. There was no difference of the dwell time statistically proved with other observed AOI.

4.5. Space-Time-Cube

Data sets created as a result of the eye-tracking are often very large, which limits their visualization possibilities by means of methods mentioned above. When displaying larger data sets with the ScanPath visualization, overlapping parts are created and the follow-up interpretation of results is not possible. The cause of this problem is displaying of three-dimensional data (X, Y, time) into two-dimensional space (X, Y).

With relatively smaller data sets, it is possible to use colours to differentiate fixations according to their order. During the visualization, transparency can be used to identify overlapping fixations.

Another possible solution is to neglect the time. Two dimensional data (X, Y) can be then displayed by means of HeatMap method which displays only the number of fixations, without their order. However, the loss of the information can make the analysis of results impossible in many cases.

Thanks to Space-Time-Cube (STC) which is the most important element of the Hägerstrand time-space model, the data can be effectively visualized without neglecting any of the data files [45].

In its basic appearance, the cube has on its base a representation of the geography (X, Y), while the cube's height represents time (Z) [46]. As it is evident from figure 7, if the location of the observed object or phenomenon does not change in time, the line is always perpendicular to the base of the cube. The steeper the line between two vertices, the slower the change in the position of observed object/phenomenon. Today there are softwares that automatically creates a Space-Time-Cube from the data in the database. It is also important that it is possible to interactively rotate the cube and select the best perspective for data analysis.

By means of Space-Time-Cube, it is possible to portray any space-time data. These might be for example data recorded by a GPS device, statistic data containing information about location and time, or data detected with eye-tracking.

In this case, coordinates X and Y describe the distribution of fixations in space, and time is described by the axis Z. Thanks to Space-Time-Cube it is possible to reveal different behaviour of particular users. On the other hand, ScanPath cannot identify in which direction the user moved when reading the picture. Up to now, application of Space-Time-Cube with analysis of eye-tracking data was examined only by [9, 47, 48].

Space-Time-Cube visualization is presented in this chapter on testing the user's perception of the map legend. Respondents were given the task to mark flax growing areas on the map. The aim of the test was to find out the proportion of respondents (%), which use the map legend to fulfil the task. Trajectories of eye movements of two respondents are displayed in figure 8. It is evident that during the first two second of solving the task, the respondents followed almost the same gaze trajectories. One of the first fixations of both respondents are localised in the legend. Then, the respondents tried to answer the given question and started to explore the map.

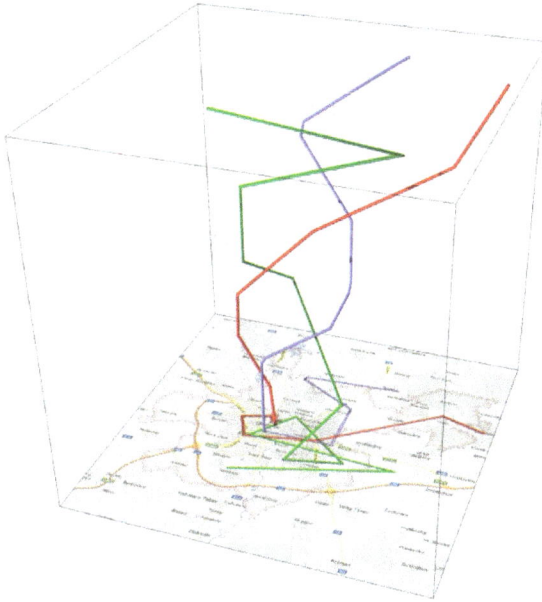

Figure 7. Time-space data displayed by means of Space-Time-Cube.

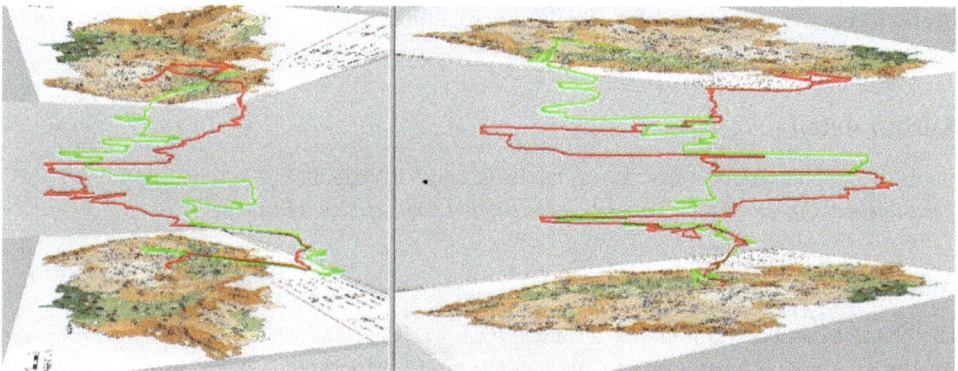

Figure 8. Use of Space-Time-Cube visualization for investigation of ScanPaths from two respondents

5. Conclusion and prospects

Up to now, technologies and methods of eye-tracking in cartography were not fully utilized despite their great possibilities in cartography. A cartographical research in the field of eye-tracking recently focuses on the improvement of the user quality of a map, particularly on the map composition improvement. However, there is a question how to define the user quality of a map or a "good" map composition. In the main, the user has to be able to interpret the content of the map correctly and accurately. A correct but a too long interpretation of a map cannot be considered as a sign of high user quality of the map. A method of the map content interpretation or the way of internal recording and later recalling of the information are related to its structure of cognitive and mental maps [49, 50, 51]. That is why the improvement of the user quality of a map is considered necessary if we want to perceive into cognitive processes going on during work with maps. In this field, eye-tracking can enable the user to do a research of cognitive maps. Nevertheless, it is necessary to respect the fact that maps have its own special dimension which cannot be neglected during the research because it is essentially connected with the user's map-content interpretation.

Thanks to easier (but not easy) access to high-performance eye-trackers, we can expect, in a short time period, more numerous and deeper researches on different aspects of map reading. In the field of map creation, there exist certain short and long-term rules. Many of them are respected without any international convention, for example a blue colour used for waters [52].

High initial investments on high quality equipment and a non-existence of a single methodology for preparing and evaluation of tests limit the implementation of the described technology in cartography research. It is also necessary to cooperate with a professional psychologist.

Cartographic research with eye-tracking methods will considerably contribute to argumentation of a high number of empirically based rules and instructions for map creation and the map language will be internationalized. By implication, it will enable geographers to present better results of their researches and studies.

Author details

Stanislav Popelka, Alzbeta Brychtova, Jan Brus* and Vít Voženílek
Department of Geoinformatics, Palacký University in Olomouc, Czech Republic

Acknowledgement

The chapter has been completed within the project CZ.1.07/2.4.00/31.0010 Supporting the creation of a national network of new generation of Cartography – NeoCartoLink which is co-financed from European Social Fund and State financial resources of the Czech Republic and the project The small format photography in the study of the effect of heterogeneity on the surface of habitats of the Palacky University (Integral Grant Agency, project no. PrF_2012_007)."

---

* Corresponding Author

6. References

[1] Muehrcke, P. C. et al. (2009). Map Use: Reading and Analysis. Redlands, ESRI, 6th edition, 528 pp.

[2] Voženílek, V. (2005). Cartography for GIS: Geovisualization and Map Communication. Univerzita Palackého v Olomouci, 142 pp.

[3] Golledge, R. G., Stimson, R. J. (1997). Spatial behavior: a geographic perspective. New York: Guilford Press, 620 pp.

[4] Board, C., Taylor, R. M. (1977). Perception and Maps: Human Factors in Map Design and Interpretation. In Transactions of the Institute of British Geographers, New Series, 2, 1, pp. 19-36.

[5] ISO 9241-11. Ergonomic requirements for office work with visual display terminals (VDTs) - Part 11: Guidance on usability. International Organization for Standardization, 1997. 22 pp.

[6] Duchowski, A. T. (2007). Eye tracking Methodology, Theory and Practice. Springer - Verlag London Limited, 2007, 321 pp.

[7] Nielsen, J. (2003). Usability 101: Introduction to Usability. Alertbox: Current Issues in Web Usability, [online]: < http://www.useit.com/alertbox/20030825.html>

[8] Hub, M., Víšek, O., Sedlák, P. (2011) Heuristic Evaluation of Geoweb: Case Study. Europan Computing Conference. Proceedings of the European Computing Conference (ECC '11) s. 142-146. ISBN: 978-960-474-297-4.

[9] Li, X., Coltekin, A. A., Kraak, M. J. (2010). Visual Exploration of Eye Movement Data Using the Space-Time-Cube. In Geographic Information Science, Lecture Notes in Computer Science, Vol. 6292/2010, pp. 295-309.

[10] Cutrell, E., Guan, Z. (2007). What are you looking for? An eye-tracking study of information usage in web search. In Proceedings of the SIGCHI conference on Human factors in computing systems, San Jose, California, USA, 10 pp.

[11] Alacam, O., Dalci, M. (2009). A Usability Study of WebMaps with Eye Tracking Tool. In Human-Computer Interaction, New Trends, Lecture Notes in Computer Science, Vol. 5610/2009, pp. 12-21.

[12] Sedlák, P., Hub, M., Komárková, J., Víšek, T. (2010) Nový přístup k testování a hodnocení kvality map. Geodetický a kartografický obzor. 56/98, 9, Praha, pp. 182-188.

[13] Dykes, J., Maceachren, A. M., Kraak, M. J. (2005). Exploring geovisualization. Elsevier Ltd., UK, 705 pp.

[14] Somersen, M. W. et al. (1994). Think aloud method: A practical guide to modelling cognitive processes. London, Academic Press. University of Amsterdam, Department of Social Science Informatics, 205 pp.

[15] Haklay Mordechai, Zafiri Antigoni (2008): Usability Engineering for GIS: Learning from a Screenshot. In The Cartographic Journal, Vol. 45, No. 2, pp. 87–97.

[16] Goldberg, J. H., Kotval, X. P. (1999). Computer interface evaluation using eye movements: methods and constructs. In International Journal of Industrial Ergonomics, Vol. 24, pp. 631-645.

[17] Gienko, G., Levine, E. (2005). Eye-tracking and augmented photogrammetric technologies. In Proceedings ASPRS 2005 Annual Conference, Baltimore, Maryland, 8 pp.

[18] Young, L. R. And Sheena, D. (1975) Eye-movement measurement techniques. American Psychologist 30, 315-330

[19] Mohamed, A. O. Da Silva, M. P., Courboulay, V. (2007). A History of eye gaze tracking. Computer Vision and Pattern Recognition, 16 pp., [online]: <http://hal.archives-ouvertes.fr/docs/00/21/59/67/PDF/ Rapport_interne_1.pdf>

[20] Holmqvist, K., Nyström, M., Andersson, R., Dewhurst, R., Halszka, J. & Van De Weijer, J. (2011). Eye tracking: A comprehensive guide to methods and measures. Oxford University Press, 560 pp.

[21] Glenstrup, A. J. (1995). Eye controlled media: Present and future state. Copenhagen, 90 pp. PhD Thesis. University of Copenhagen, Laboratory of Psychology.

[22] Dual Purkinje Eyetrackers. Faculteit Psychologie en Pedagogische Wetenschappen - KU Leuven [online]. [cit. 2012-04-11]. Dostupné z: http://ppw.kuleuven.be/english/lep/resources/purkinje

[23] Biedert, R., Buscher, G., Dengel, A. (2009). The EyeBook - Using Eye Tracking to Enhance the Reading Experience. In: Informatik-Spektrum, Vol. 33, Nr. 3, 2010, pp. 272 - 281.

[24] Iwasaki, M., Inorama, H. (1986). Relation Between Superficial Capillaries and Foveal Structures in the Human Retina (with nomenclature of fovea terms). In Investigative Ophthalmology & Visual Science, vol. 27, pp. 1698-1705, 1986, IOVS.org

[25] Pernice, K., Nielsen, J. (2009). Eye Tracking Web Usability: How to Conduct and Evaluate Usability Studies Using Eyetracking. Nielsen Norman Group, USA, 163 pp.

[26] Komogortsev, O. And Khan J. (2009). Predictive compression for real time multimedia communication using eye movement analysis. ACM Transactions on Multimedia Computing, Communications and Applications 6

[27] Salvucci, D. D., Goldberg, J. H. (2000). Identifying fixations and saccades in eye tracking protocols. In Proceedings of the Eye Tracking Research and Applications Symposium, New York: ACM Press, pp. 71-78.

[28] Just M. A.,Carpenter, P. A. (1976). Eye fixations and cognitive processes. Cognitive Psychology 8: 441–80

[29] Steinke, T. R. (1987). Eye movement studies in cartography and related fields. In Cartographica: The International Journal for Geographic Information and Geovisualization, Vol. 24, No. 2, pp. 40 - 73.

[30] Jones, R. E. et al. (1950). Eye movements of aircraft pilots during instrument landing approaches. In Aeronautical Engineering Review, Vol. 9, No. 2. (1950), pp. 24-29.

[31] Coltekin, A. et al. (2009). Evaluating the Effectiveness of Interactive Map Interface Designs: A Case Study Integrating Usability Metrics with Eye-movement Analysis. In Cartography and Geographic Information Science, Vol. 36, No. 1, pp. 5-17.

[32] Coltekin, A. et al. (2010). Exploring the efficiency of users' visual analytics strategies based on sequence analysis of eye movement recordings. In International Journal of Geographical Information Science, Vol. 24, No. 10, October 2010, pp. 1559–1575.

[33] Opach, T., Nossum, A. S. (2011). Evaluating the usability of cartographic animations with eye-movement analysis. In Proceedings of the 25th International Cartography ConferenceICC, Paris, 11 pp., ISBN 978-1-907075-05-6

[34] Fuhrmann, S., Komogortsev, O., Tamir, D. (2009). Investigating Hologram-Based Route Planning, Transactions in GIS, 2009, pp. 177–196

[35] Popelka, S., Brychtová, A. (2011): Metody virtuální rekonstrukce zaniklé pevnosti Olomouc. Historická geografie 37/2 Historický ústav AVČR, Praha, pp. 213-230.

[36] ICA Commision on Use and User Issues in Cartography and Geo-information Processing and Dissemination (2012).
http://www.univie.ac.at/icacomuse/index.php?title=Main_Page

[37] ICA Commission on Cognitive Visualization (2012).
https://www.geo.uzh.ch/microsite/icacogvis/ mission.html

[38] Raiha, K. J. et al. (2005). Static Visualization of Temporal Eye-Tracking Data. In Human -computer Interaction - INTERACT 2005, Lecture Notes in Computer Science, Vol. 3585/2005, pp. 946-949.

[39] Jarodzka, H., Holmqvist, K., & Nyström, M. (2010). A vector-based, multidimensional scanpath similarity measure. In C. Morimoto & H. Instance (Eds.), Proceedings of the 2010 Symposium on Eye Tracking Research & Applications ETRA '10 (pp. 211-218). New York, NY: ACM.

[40] Privitera, C. M., Stark, L. W. (2000). Algorithms for Defining Visual Regions-of-Interest: Comparison with Eye Fixations. IEEE Transactions on Pattern Analysis and Machine Intelligence (PAMI) 22, 9, 970–982.

[41] Duchowski, A. T., Driver, J., Jolaoso, S. Tan, W., Ramey, B. N., Robbins, A. (2010) Scanpath comparison revisited. In Proceedings of the 2010 Symposium on Eye-Tracking Research Applications (ETRA '10). ACM, New York, NY, USA, 219-226.

[42] Voßkühler, A., NORDMEIER, V., KUCHINKE, L., JACOBS, A.M. (2008). OGAMA - OpenGazeAndMouseAnalyzer: Open source software designed to analyze eye and mouse movements in slideshow study designs. Behavior Research Methods, 40(4), 1150-1162.

[43] Voženílek, V., Dobešová, Z. (2005). Metadescription - first step in an atlas production. In Co zwie sie koncepcja mapy?, Wroclaw, 154 pp.

[44] Fisher, R. A. (1925). Statistical Methods for Research Workers. Macmillan Pub Co. 362 pp.

[45] Hägerstrand, T. (1970). What about people in Regional Science? In Papers of the Regional Science Association, Vol. 24, pp. 6–21.

[46] Kraak, M.-J., Koussoulakou, A., 2004, A Visualization Environment for the Space-Time-Cube. Developments in Spatial Data Handling In Developments in Spatial Data Handling, pp. 189-200.

[47] Andrienko, N., Andrienko, G. And Gatalsky, P., (2003). Visual data exploration using Space-Time-Cube, In ICC 2003 Proceedings, Durban, south Africa, pp. 1981 – 1983.

[48] Nossum, A. S., Opach, T. (2011). Innovative analysis methods for eye-tracking data from dynamic, interactive and multi-component maps and interfaces, In Proceedings of the 25th International Cartography ConferenceICC, Paris, ISBN 978-1-907075-05-6

[49] Montello, D. R. (2002). Cognitive map-design research in the twentieth century: theoretical and empirical approaches. In Cartography and Geographic Information Science, 29, pp. 283-304.

[50] Slocum, T. A., Blok, C., Jiang, B., Koussoulakou, A., Montello, D. R., Fuhrman, S., Hedley, N. R. (2001). Cognitive and usability issues in geovisualisation. In Cartography and Geographic Information Science, 28, pp. 61-75.

[51] Downs, R. M., Stea, D. (1977). Maps in Minds. Reflection on cognitive mapping, New York, Harper & Row.

[52] Pravda, J. (2006). Metódy mapového vyjadrovania – klasifikácia a ukážky. Slovenská akadémia vied, Geografický ústav Bratislava, 128 pp.

Chapter 6

GPS Positioning of Some Objectives Which are Situated at Great Distances from the Roads by Means of a "Mobile Slide Monitor – MSM"

Axente Stoica, Dan Savastru and Marina Tautan

Additional information is available at the end of the chapter

1. Introduction

This chapter presents a Mobile Slide Monitor (MSM) which can be used for fast geographic positioning of some objectives or of some of their components that are situated at great distances from the roadways, (buildings, terrain markings) or are inaccessible in a direct mode, (dams, bridges, heaps of debris). The system assures an accurate geo-referencing of the off-road objective characteristics, an important problem for the infrastructure management at the level of public works administration.

The main applications of this MSM equipment are to estimate and alert, in due time, the occurrence of great proportion accidents caused by breaking down of civil constructions (buildings, bridges, tunnels, dams). These accidents are due to natural causes such as landslides and floods, in the areas with a high risk, or due to some human interventions such as the erroneous emplacement of some new constructions, the erroneous designing or even due to the oldness of some constructions.

Moreover, the bridges are part of a country's transportation infrastructure and are typically assessed and maintained by the authorities responsible for the appropriate transportation sector (road or rail). Nowadays, the deterioration of bridge structures is a serious problem due to issues related to modern society; reliance on the automobile, the increased bridge traffic, the environmental pollution, and the use of potentially corrosive substances (e.g. cleaning and de-icing). Bridge monitoring is necessary to ensure the safety of those who either use, or are affected by the structure itself, and the maintenance of the sector is usually part of the legislature governing.

Therefore the main objective of the present chapter is to describe a mobile laboratory for the monitoring of constructions deformations in the incipient phases, deformations that are due to terrain sliding from natural or human causes. As opposed to the systems which achieve in static regime the stability monitoring of the constructions by using some precision optical systems or GPS equipments with differential regime functioning but to which the follower receiver is attached in a fixed montage on the surveyed construction, the proposed mobile monitoring system permits that the measurements be rapidly performed, at a preset time interval, with a reduced cost on a multitude of objectives and with a minimum delay between the moment of some defection apparition and the moment of its identification and alarming.

2. Importance and relevance of the technical content

Monitoring of situations and territories with hydro-geological risk represent an institutional task of the Public Administrations. Therefore, in some areas, it becomes necessary to achieve systems for real-time survey, which are able to record the alarm signs of a potential risk for the population. An early-warning system provides, also, the foundation for an effective risk mitigation plan, given the uncertainties related to the mathematical prediction of the natural phenomena and the strong public demand for protection against natural hazards.

The World Bank promotes a proactive and strategic approach to managing natural hazard risk, by taking into account a comprehensive framework, based on the following five pillars:

Risk assessment - includes application of the hazard, exposure, vulnerability, and loss analyses and provides projections of the average annual expected loss and the probable maximum loss from a single catastrophic event;

Emergency preparedness. Citizens and government agencies need to be prepared for breakdowns în essential services, to develop plans for contingencies, and to implement the plans. They should be encouraged to make resources available for facilities and equipment, they need to provide emergency personnel, they need training, sponsor exercises, and get information available for the public;

Investments in risk mitigation. This may include inexpensive investments in increasing institutional capacity, strengthen enforcement of building codes, provide training, and involve communities, including mapping, monitoring and warning systems. As investments in physical infrastructure (flood protection, landslides prevention and retrofitting of housing and/or public buildings for seismic resistance) are very expensive, the selection of the most suitable of them should be carried out by applying cost-benefit or cost-effectiveness analysis;

Institutional capacity building. The efficiency and effectiveness of a comprehensive hazard risk management system depends on the knowledge, awareness, and capacity of the stakeholders involved. For that purpose, the following aspects are recommended:

• to create decentralized emergency management systems;
• to ensure community involvement and participation;
• to develop an efficient legal framework, and
• to provide training, education, and knowledge sharing.

It is also important to integrate hazard risk management into the economic development process. Emergency planning and risk mitigation need to be an integral part of the rural and urban development process, with the participation of all the stakeholders.

Develop a catastrophe risk financing strategy. Countries need to develop and introduce targeted risk financing strategies for dealing with catastrophic events that can have a severe impact on their economies. The strategy would address the funding gap caused by the need to recover economic losses and meet social obligations and other responsibilities, following a catastrophic event. Developing a risk financing strategy is particularly important for countries exposed to catastrophic earthquakes.

Therefore, the Regional Public Administrations from a country, which has more or less accentuated risk for the natural disasters, may have at their disposal the possibility to verify, at pre-established time intervals, the real state of the geological formations or of the building, which is suspected to be in danger.

The main application of the MSM equipment presented here, is the estimation and the alerting in due time regarding the risk of great proportion accidents, by break down of civil constructions due to some natural causes, such as landslides and floods, in areas with high risk of accidents.

The final result will be the achievement of a Geographic Informational System (GIS), which will have to integrate all the information and the all types of data, which are needed for the natural disasters management, from the prognosis to the post-factum measurements. Moreover, besides the hazard maps which must be elaborated for the all regions of the respective country, the local authorities must, also, draw up risk maps which refer to the most exposed areas to the natural calamities.

3. State of the art

Usually, the measurement of superficial displacement is the simplest way to observe the history of a landslide and to analyze the kinematics of the movement, so the investigation of the terrains sliding movements permits, also, the detection of possible precursor elements of the mass movements.

3.1. Existing fixed mapping equipments

In the past, a various surveying techniques were used to detect the superficial movements of unstable area. For examples, tapes and wire devices were used to measure changes in distance, between terrain points or crack walls. Levels, theodolites, Electronic Distance Measurement (EDM), and total station measurements provide both the coordinates and changes of target, control points and landslide features. In addition, aerial or terrestrial photogrammetry provides point coordinates, contour maps and cross-section of the landslides.

3.1.1. Leica Smart Station (Scott A., 2006)

A classical example of such optical measurement equipment for observing different targets with displacement probability is the "Leica Smart Station", a Total Station with integrated GPS offered on the market by the "Leica Geosystems AG" company. The introduction of SydNET, a network of Continuously Operating Reference Station (CORS), allows surveyors to perform Differential GPS without having to purchase a reference receiver. For distances of up to tens of kilometres away from the network reference stations, centimetre accuracy can be achieved, with the RTK-GPS Network.

This equipment involves the optical observation of the proposed object from different static locations of the operator, locations which are precisely determined by means of the highly performing GPS receiver. In these conditions, the determination of the geographic coordinates of a single distant objective involves multiple complex operations and in consequence, can be considered time consuming.

3.1.2. SEPA's system (Caporali A., 2008)

Another totally different technical solution for this problem, solution which aims to reduce the length of the measurement times involved by the use of the optical total stations, is represented by the "Fixed Satellitary Monitoring System of the Territory and Civil Infrastructures" (or SAMOS for short) achieved by SEPA company ("Sistemi Elettronici Per Automazione S.p.A.") from Torino (Italy).

The SEPA's system represents a solution for cost effective applications targeting the real time monitoring and diagnosis of ground deformation; for instance, landslides and the subsidence, or the infrastructure deformations affecting buildings, bridges, viaducts and dams, or even both simultaneously.

Based on measurements from a GPS L1-only carrier phase employing commercial receivers and using the basic principles of interferometric surveying, SAMOS provides continuous real-time monitoring of the area of interest, reporting the millimetric displacement of each sensor relative to a reference sensor.

Measurements are taken at a rate of 1 Hz and the processed results are updated using the same frequency. The system performance is equivalent, on short baselines up to a few km in length.

This Satellitary Monitoring System, in fact like other this kind of systems, is composed of two subsystems, namely:

-   a number of Field Sensors, (Fig. 2), deployed to collect the satellite data and which are fix mounted on different parts of the objective of interest (bridge, dam, building). These field sensors are continuously relayed by means of a radio connection to
-   a Base Station, (Fig. 3), for real-time processing of the data collected from the field sensors.

For its protection this Base Station is introduced in a Waterproof box (Fig. 4).

GPS Positioning of Some Objectives Which are Situated at Great Distances from the Roads by Means of a "Mobile Slide Monitor – MSM"

123

Figure 1. Fixed Satellitary Monitoring System of the Territory and Civil Infrastructures achieved by "SEPA Sistemi Elettronici Per Automazione S.p.A." from Torino (Italy).

Figure 2. GNSS receiver

The Field Sensor uses a single frequency GPS receiver to measure the carrier phase on the L1 GPS signal and an RF Modem for point-to-point communication with the base station via radio link. A microcontroller supervises the communication and the exchange of data between the GPS receiver and the radio modem, and in addition, it supplies diagnostic data related to the sensor itself.

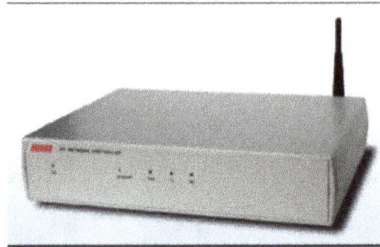

Figure 3. Base station

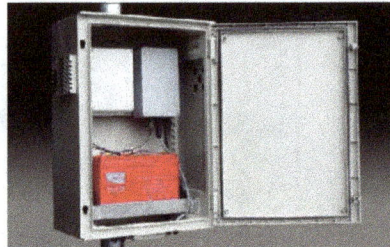

Figure 4. Waterproof box of the Base Station

It is also important to mention that, for this SAMOS system, it is necessary that, on the surveyed objectives, to be assured permanent electrical energy supplies (which can be constituted from photovoltaic panels or, if this is available, from the mains power supply of the area). In both cases, a backup battery is included. Moreover, we must notice that the receivers and the system GPS components remains in field in the majority of cases, without any surveillance from the operators.

The Base Station includes the network controller, used to receive the GPS data collected via radio links from sensors deployed in the field, and a computer running the software to process the data and display the results in real-time. The raw measurement data as well as the results are stored in a database for possible further processing, if required, or simply archiving. (Caporali A., 2008)

3.1.3. Conclusions referring to the fixed monitoring systems

In addition to collecting the measurement data, the base station can retrieve status information from the sensors, such as the accumulator supply voltage, RF link signal strength and temperature.

The Graphic User Interface (GUI) shows the real-time status of the network (satellites available and being tracked, nominal antenna locations and network geometry, status of each sensor) and the results of the data analysis as northing, easting and vertical displacements of each antenna with respect to the reference antenna, using diagrams and tables. The GUI can be tailored to the customer's requirements, and it will alert the operator when preset thresholds are exceeded.

The main disadvantage of the SEPA types fixed systems is that this equipment must be mounted in a fixed position on every point of interest, meaning that for an objective or an area suspected to present landslides, a great number of equipments of this kind is necessary, and, in consequence, high total costs.

Taking account of the newest improvements of the GNSS systems regarding the real-time positioning accuracy, (A. Stoica et al., 2008), the authors of this chapter present the achievement of a mobile equipment for the monitoring of the field stability as opposed to the most used in present, static optical total stations, "Leica" or "SEPA" type systems, which are based on GPS receivers with differential regime functioning but which are fixed mounted on the surveyed objectives.

3.2. Existing mobile mapping equipments

But one must specify that also for this mobile alternative, which is proposed here for the monitoring of some objective position, there are various mobile equipments for mapping and, respectively, for monitoring, introduced on international level, equipments which are described in the following subsections.

3.2.1. "GPSVision™" achieved by the "LAMBDA TECH International" company (Guangping He, 1996)

The Mobile Mapping Equipment "GPSVision™" (fig. 5) achieved by the "LAMBDA TECH International" company from Fort Wayne (USA), which is equipped with a positioning module composed of a GPS receiver with double frequency, an Inertial Navigation System (INS) and a linear Distance of running Measuring Instrument (DMI), in combination with four digital video cameras of high resolution. The digital video cameras are mounted above the vehicle and they can be oriented forward, to each side or backward in correspondence with the application needs, so that due to the fact that the video cameras pairs see, at a certain moment, the same field area from different positions, by using some triangulation algorithms, it is possible to calculate the locations relative to the lab vehicle of the sighted targets.

The main characteristics of the system are:

- The images are taken according to an operator-defined distance interval to provide full coverage of the route and its surroundings. By applying a sophisticated photogrammetric triangulation technology, any point that appears in any set of two images can be located in a global coordinate system during digitization with Lambda Tech's Feature Extraction software;
- Stereo imaging allows for determining absolute positions of features such as signs in latitude and longitude to sub-meter accuracy and it also allows very accurate relative measurements of all visible roadside attributes, such as the width, height and offset of a

sign. Stereo imagery allows for multiple views of the same object with 3D capabilities and the ability to recreate image views where the original cameras never took a picture;

- GPSVision™ specified absolute accuracy for terrestrial data positions is **one meter or less** depending on the distance bewtween the feature to be extracted and the camera lens. Depending on the image spacing this accuracy can be increased. The GPSVision™ system was designed to deliver sub-meter RMS positions when visible features are within the camera field of view of both image pairs and **no farther than 30 meters** in front of the camera lenses;

- From a photogrammetric perspective, GPSVision™ is a fix-based stereovision system with known position and attitude provided by the GPS/INS component. Just as a person uses two eyes to determine the distance of an object, every infrastructure feature that is "seen" by the cameras can be triangulated into a three-dimensional coordinate and then transferred into a global coordinate system (e.g., latitude, longitude, height);

- GPSVision™ Feature Extraction software is executable on Microsoft Windows operating systems. It is driven by an external rule base and is language neutral. The user interacts with the software by pointing at features of interest seen in the stereo image pairs with the mouse or stylus. Then, the software triangulates the relative position of the selected feature and transfers it into the global coordinate system and positioned to within one meter or less of their actual location;

- From an application perspective, the GPSVision™ system is used to collect digital images along highways, state roads, residential streets, alleys, and railroads while traveling at posted speed limits. These geo-referenced digital images are used for video log applications but most importantly, the software is used to position visible physical features, such as poles, curb lines, traffic signs, manholes, pedestals and building locations. In addition, the GPS/INS positioning component creates base maps of the route network for Geographic Information Systems (GIS) base map and Computer Aided Drafting and Design (CADD) applications.

Figure 5. GPSVision™ - with four video cameras achieved by the "LAMBDA TECH INTERNATIONAL" Company (USA) for objectives that are closed to the road arteries

GPS Positioning of Some Objectives Which are Situated at Great Distances from the Roads by Means of a
"Mobile Slide Monitor – MSM"

127

3.2.2. "GPSVan™" achieved by the "Mapping Center" from the Ohio State University (Brzezinska et al., 2004)

A similar example of a mobile system for mapping and data collection, which can map railroads, thoroughfares and transport infrastructures (as for example, roads, circulation signs and bridges) during its displacement at posted road speeds is represented by "GPSVan™" system achieved by the Mapping Center from the Ohio State University and which is also composed of two main components: a positioning module and an imagery module. This imagery module includes, also, in this case, a stereo metric system with two video cameras which record the stereo images of the roads during the displacement on the respective arteries of the lab vehicle. Each video frame is time marked with the GPS signal and the geodesic coordinates (latitude, longitude and ellipsoidal height, respectively) are attributed to each image.

From the above presented aspects, it results that both the GPSVision™ and GPSVan™ equipments represent, in their essence, a fix-based stereovision system with known position and attitude, provided by the GPS/INS component and, respectively, by the GPS/DRS component. This fixed base is represented by the distance, on the hood width of the lab vehicle, between the optical distances of the two video cameras, mounting distance which in the case of both GPSVision™ and GPSVan™ equipments is approximately of 1.2 meters. As a consequence of the fixed base reduced value, the difference between the angles of sighting directions of the two video cameras is under the minimum value it can be measured by the optical system when the sighted targets are situated at a distance, greater then approximately 40 meters, in respect to lab vehicle. This limitation of the observing distance, at a quite reduced value, represents in its essence the main disadvantage of the stereo metric systems, which are based on the use of a pair of video cameras, mounted on the same lab vehicle.

4. Originality of the proposed Mobile Slide Monitor (MSM)

The static regime systems, previously described, achieve the stability monitoring of the constructions by using precision optical systems or GPS equipment with differential regime functioning, but to which the follower receiver is attached in a fixed montage on the surveyed construction. As opposed to these systems the monitoring system presented here has the advantage that it permits measurements to be rapidly performed, at preset time interval, with a reduced cost on a multitude of objectives and with a minimum delay between the moment of some defection occurence and the moment of its identification and alarming.

Taking account of the above-mentioned characteristics of the bi-cameral stereo-metric systems, the INOE 2000 Institute from Bucharest elaborated the Invention Patent RO 126294 A2/2009 whose main objective was to increase up to U200 – 300 metersU, the distance up to which the sighted targets from the terrain can be positioned.

Conceptually this Mobile Slide Monitor - MSM involves:

- The acquisition of **successive images** achieved from a **moving vehicle** by means of a single **CMOS video camera** mounted above this vehicle (Fig.6);

Figure 6. Moving vehicle and the CMOS video Camera

- **The determination** with a sub-metric accuracy of the **vehicle position,** by means of a GPS device at the time of taking the picture (Fig. 7);

Figure 7. Multi-Frequency GPS receiver

- The use of an **innovative mathematical** algorithm based on a triangulation method for the **geographic position** computing of every object which appears in two different images.

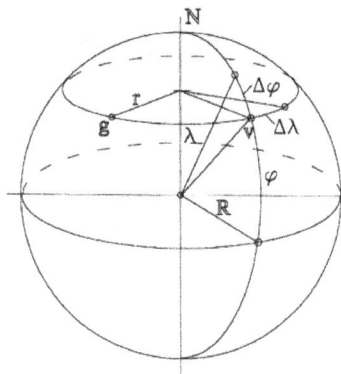

Figure 8. Geographic positioning of the sighted object.

4.1. The functioning and the use of the Mobile Slide Monitor

The general assembly of the proposed Mobile Slide Monitor is presented in Figure 9.

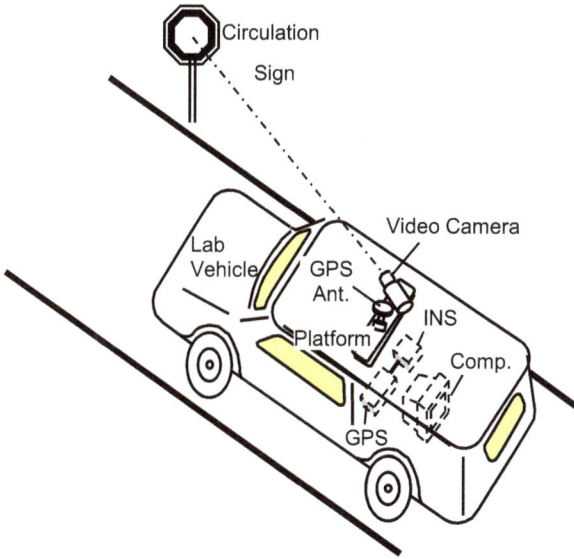

Figure 9. The equipment assembly during a normal functioning on a tested route

More precisely, the mobile positioning system achieved in conformity with the invention patent proposes itself to obviate the limits which affect the functioning of the above-mentioned equipments by introducing the following series of combined measures:

- In order to increase up to 200 – 300 meters, the distance up to which the sighted targets from the terrain can be positioned, the proposed MSM equipment resorts to the use of a single digital camera of high resolution in a fixed montage on a lab vehicle. This way, the measurement of the applied stereo-metric method is based not on the width of the lab vehicle, which uses the bi-cameral method, but on the distance of 20 – 30 meters between the positions resulted from the lab vehicle displacement and from which the camera sights the same objective (fig. 10). This single video camera has a telemetric type objective and a reduced angle of view;

On the basis of the notations entered in the figure 10, it is possible to compute the geographic position coordinates of the sighted target: λ_T, ϕ_T, h_T - longitude, latitude and height, respectively:

- Ψ_{T1} and Ψ_{T2}, azimuth angles of the sighted target in rapport with two positions of the Lab Vehicle, angles which are determined from the two target preloaded images with the Video Camera;
- λ_1, ϕ_1, h_1 and λ_2, ϕ_2, h_2, the vehicle geographic coordinates determined for the two positions of the Lab Vehicle by means of its GPS receiver;
- Ψ_1, Θ_1, Φ_1 and Ψ_2, Θ_2, Φ_2, the vehicle orientation angles determined for the two positions of the Lab Vehicle by means of its Inertial Navigation System (INS);
- M.B. – Measuring Base, namely the distance between the two positions, Pos.1 and Pos.2 of the Lab Vehicle,

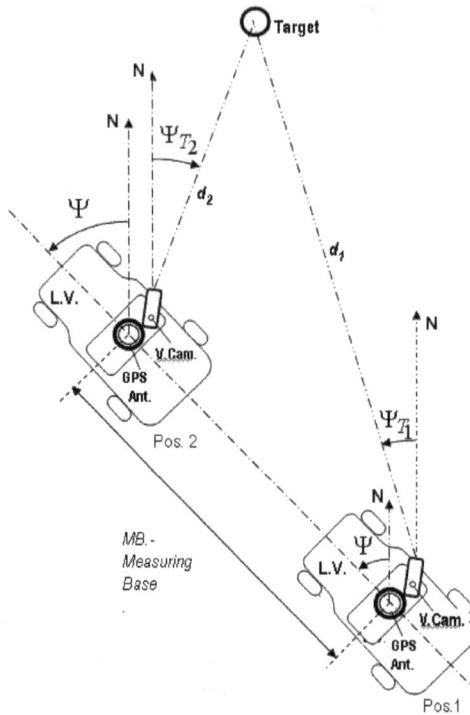

Figure 10. Fig.10 The Sighting of the same target from two different positions between which the Measuring Base - MB is forming.

The geographic positioning of a far away target (200 – 300 meters) involves in a first phase the use of the video camera which is triggered by the PPS (Pulse Per Second) signal furnished by the GPS receiver for taking photo images of the respective target from two different positions of the lab vehicle (about 20 – 30 meters) (Fig.11).

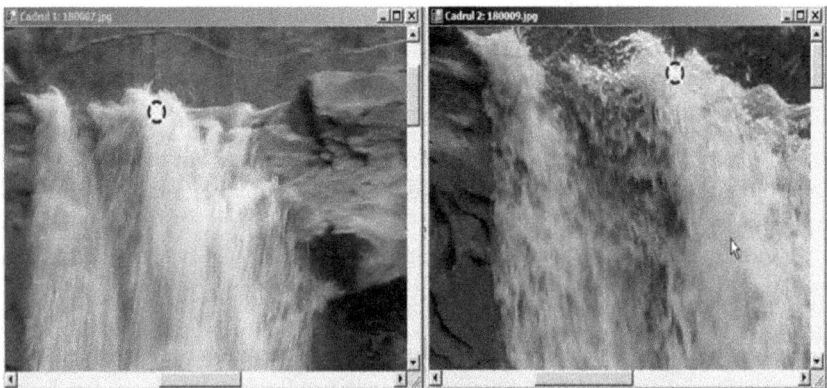

Figure 11. The sighting of a far away objective from two positions of the mobile monitoring system.

Afterwards, in a post processing regime, the pixel coordinates of the point, representing the target in each of the two images are determined, and, thus, it is possible to establish in each case the target angular position, relative to the video camera axis (Fig. 12).

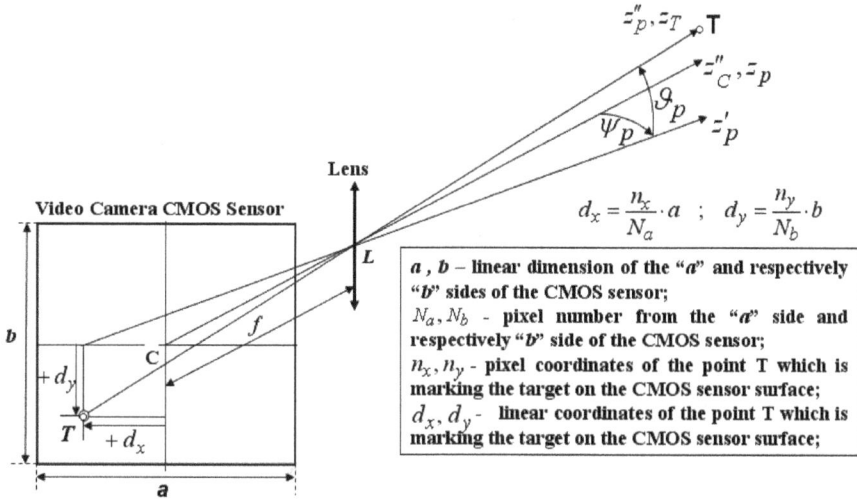

$$d_x = \frac{n_x}{N_a} \cdot a \quad ; \quad d_y = \frac{n_y}{N_b} \cdot b$$

a, b – linear dimension of the "a" and respectively "b" sides of the CMOS sensor;
N_a, N_b - pixel number from the "a" side and respectively "b" side of the CMOS sensor;
n_x, n_y - pixel coordinates of the point T which is marking the target on the CMOS sensor surface;
d_x, d_y - linear coordinates of the point T which is marking the target on the CMOS sensor surface;

Figure 12. The definition of the deviation angles ψ_p and θ_p of the target direction relative to the central axis of the viewing field of the video camera

- To assure the positioning of the sighted objectives in the frame of the global terrestrial system of coordinates and their registering in files of GIS (Geographic Information System) type, at the operator returning at the computing centre, from the obtained images, series of two images are selected. In these series of two images the same sighted objective is evidenced in a corresponding mode, which will allow the selection of this objective in an electronic modality and the determination of the pixel coordinates which achieves the objective displaying on the monitor screen. These coordinates together with other data which accompany the two selected images, permit to compute the geographic and elevation coordinates of the sighted objective with the use of using triangulation proceedings as well as methods to report to the spherical system of the terrestrial coordinates.

So, as it is presented in figure 12, in the post processing regime, the pixel coordinates of the point representing the target in each of the two images are determined, and, on this basis, it is possible to establish in each case the target angular position relative to the video camera axis.

- To achieve, in real time, the precise positioning of the lab vehicle, with errors that can be included between few millimetres and some centimetres, it resorts to the use of a GPS positioning system with multiple frequencies and with differential RTK (Real Time Kinematic) regime functioning capability. This means that it has the possibility to be connected through the Internet network, to a reference GPS station, at which it is subscribed and from which it can take the differential corrections. In relation to this, it is mentioned that the National Romanian Service ROMPOS – DGNSS provides corrections for the real time kinematic applications with a positioning accuracy evaluated at the interval between 3.0 and 0.5 meters for the receiver with a single frequency. The ROMPOS – RTK service delivers corrections for the real time kinematic applications with a positioning accuracy value situated between 0.5 and 2.0 cm for the receivers with two frequencies (Stoica, 2008);

- To increase the time marking accuracy of the obtained video images, the video cameras triggering is achieved from exterior by the PPS signal (Pulse Per Second) received from the GPS satellite system, which, also, contains in its message, besides of the positioning data, the data regarding the Universal Time – UTC;

4.2. Tightly coupled GPS/INS

- In order to continuously maintain the achieved positioning precision, the GPS system is tight coupled with an IMU (Inertial Measuring Unit) unit, the data of these systems being distributed through a filtering element of Kalman type. By integrating GPS and INS, the accurate GPS position is used to update the INS, and the latter then produces the high rate of accurate position and attitude data of mobile mapping system. The INS is needed for continuously measure the camera location and orientation. Combining GPS, INS and Distance Measurement Indicator (DMI) data is a very efficient and accurate method to determine the position (lat/long/height), azimuth, pitch and roll angles of the system cameras. The measurements of the INS come from two sensor triads, an accelerometer block and a gyro block. They are defined as three components of the specific force vector and three components of the body rotation rate. Integrated with GPS data, the system geometry data are calculated using the Kalman method. (Moafipoor S. et al., 2004)

The integrated GPS/INS solution produces continuous, smooth position and orientation of the system even when the GPS signals are lost due to obstructions such as bridges, trees, tunnels, mountains, high-rise buildings or limited and sporadic satellite coverage.

Taking into account of the functioning details described here, the figure 13 presents the complete set of devices which compose our Mobile Slide Monitor.

4.3. SPAN (Synchronized Position Attitude Navigation) technology:
Bidirectional INS/GPS coupling to obtain the objectives positioning with an improved accuracy and continuity

Inertial Measuring Systems – INS are used on land, at sea and in the air as well as in space to determine the dynamic properties and trajectory of a moving object. They are, also, used for navigation, guidance and control or stabilisation of objects.

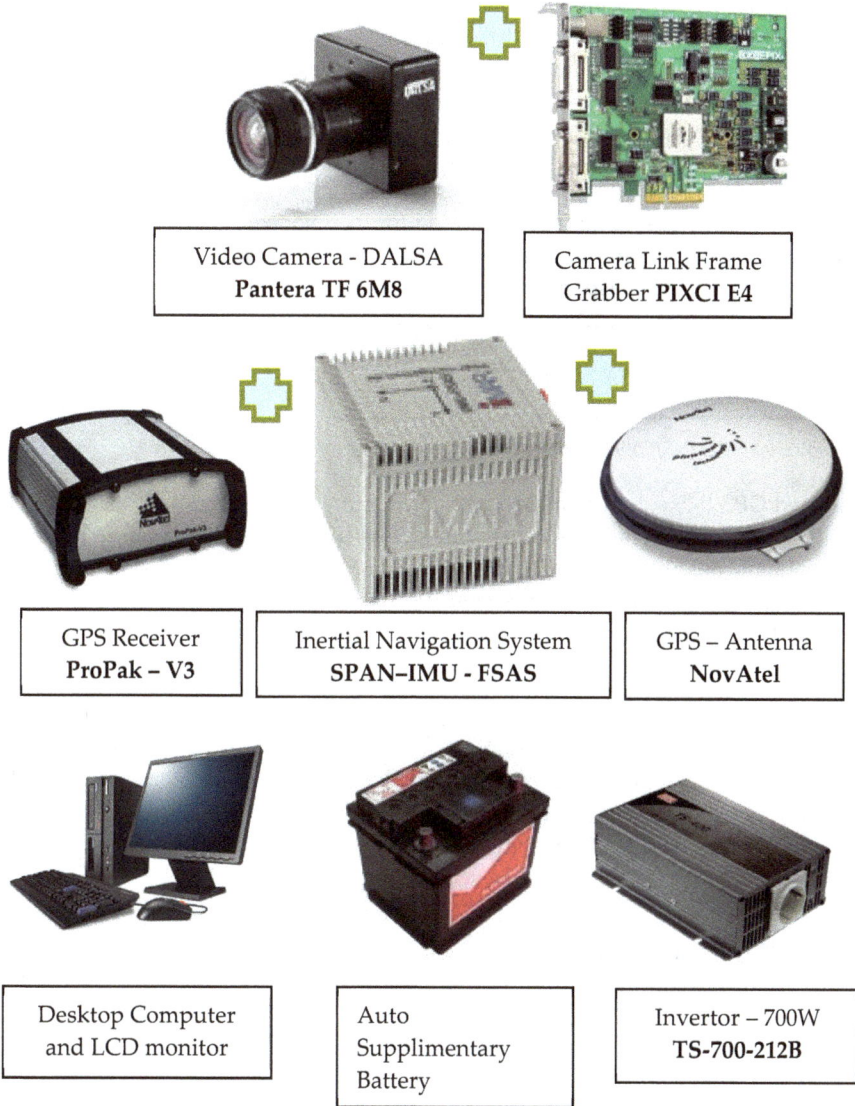

Figure 13. The components of the Mobile Slide Monitor

In general, an INS system uses forces and rotations measured by an IMU (Inertial Measuring Unit) to calculate position, velocity and attitude. Forces are measured by accelerometers in three perpendicular axes within the IMU and the gyros measure angular rotation rates around those axes. Over short periods of time, inertial navigation gives very accurate acceleration, velocity and attitude output. The INS must have prior knowledge of its initial position, initial velocity, initial attitude, Earth rotation rate and gravity field. Since the IMU measures changes in orientation and acceleration, the INS determines changes in position and attitude, but the initial values for these parameters must be provided from an external source. Once these parameters are known, an INS is capable provide an autonomous solution with no external inputs. However, because of errors in the IMU measurements that accumulate over time, an inertial-only solution degrades in time unless external updates such as position, velocity or attitude are supplied.

The GPS receiver provides auxiliary information for the INS, and it is reciprocally aided by feedback from the INS to improve signal tracking. The feedback from the INS to the GPS engine is the deeply coupled aspect of the system (Fig. 14).

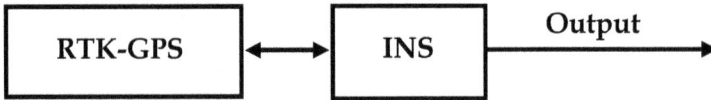

Figure 14. Bidirectional INS/GPS Coupling

The combined GPS/INS solution of the SPAN (Synchronized Position Attitude Navigation) integrates the raw inertial measurements with all available GPS information to provide the optimum solution possible in any situation. By using the high accuracy GPS solution, the IMU errors can be modelled and mitigated. Conversely, the continuity and the relative accuracy of the INS solution enable faster GPS signal reacquisition and RTK solution convergence.

GPS signal reacquisition is dramatically improved when running SPAN. This is a key performance feature in restricted coverage environments, such as urban canyons, where the user may have only a few seconds of satellite visibility before another blockage occurs. With SPAN technology, the user will be able to get GPS measurements in that small window of visibility. That means the INS will have shorter periods of free navigation and smaller errors, since the GPS is available more often for aiding (Kennedy and Rossi, 2005).

4.3.1. The IMU unit montage conditions

It is necessary to mount the IMU unit in a fixed location where the distance from the IMU to the GPS antenna phase center is constant. Also, the use must ensure that the orientation, with respect to the vehicle and antenna, is also constant. For the attitude output to be meaningful, the IMU should be mounted such that the positive Z-axis marked on the IMU enclosure points up and the Y-axis points forward through the front of the vehicle, in the direction of track and X pointing to right. (IMAR-iTraceRT-F200, 2008).

The body coordinate system is defined as given in figure 15:

4.3.2. GPS antenna montage conditions

Mount the GPS antenna close to the IMU housing. It is recommended to mount the antenna in top of the IMU if the system is mounted on a car, truck, ship or aircraft. In the

GPS Positioning of Some Objectives Which are Situated at Great Distances from the Roads by Means of a "Mobile Slide Monitor – MSM"

135

cases in which the geographic positioning of some objectives which are at some distance from the surveying vehicle, an optimal variant in the case of an aircraft can be constituted by IMU unit mounting above the video camera in the mode presented in the following schema:

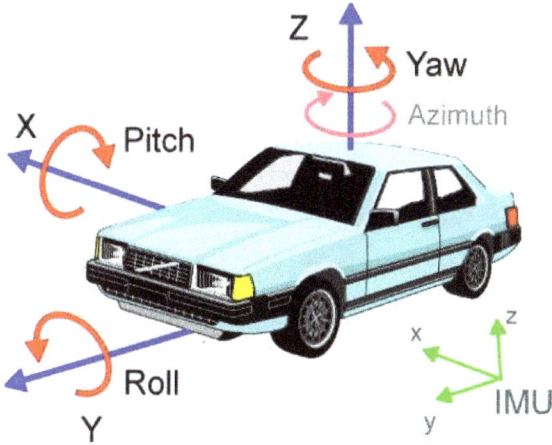

Figure 15. The reciprocal orientation of the IMU unit and the carrying vehicle. (IMAR - iTraceRT-F200, 2008)

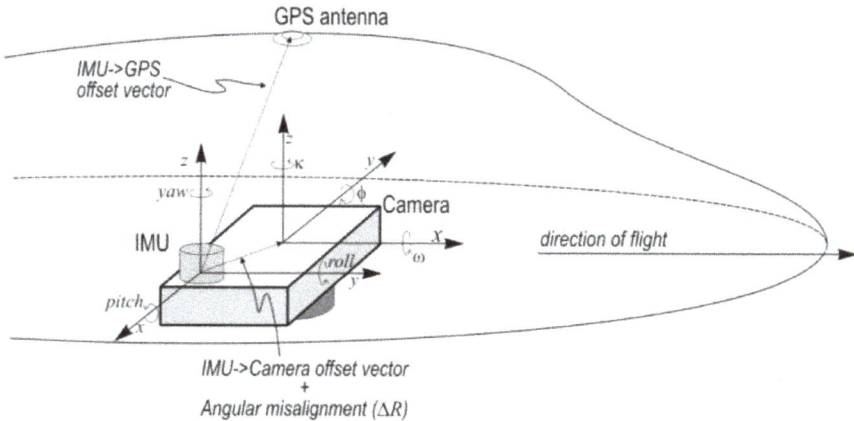

Figure 16. The mounting variant of the IMU unit in a surveying aircraft. (Novatel, Inertial Explorer, 2009)

The lever arm (offset) between IMU and GPS antenna has to be measured in the IMU coordinate system and with an accuracy better than 1 centimeter. Even slight deviations in the measurement of the lever arm may lead to significant position errors and will degrade the total system performance (Fig.17).

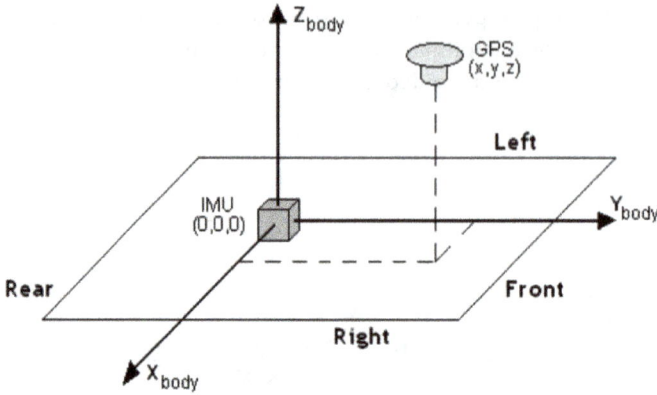

Figure 17. Fig.17 Body Frame Definition for Lever Arm Offset (Novatel Inertial Explorer 8.20, 2009)
The IMU is the local origin of the system and the measurements are defined as follows:
X: The measured lateral distance in the vehicle body frame from the IMU to the GNSS antenna;
Y: The measured distance along the longitudinal axis of the vehicle from the IMU to the GNSS antenna.
Z: The measured height change from the IMU to the GNSS antenna (Fig.18).

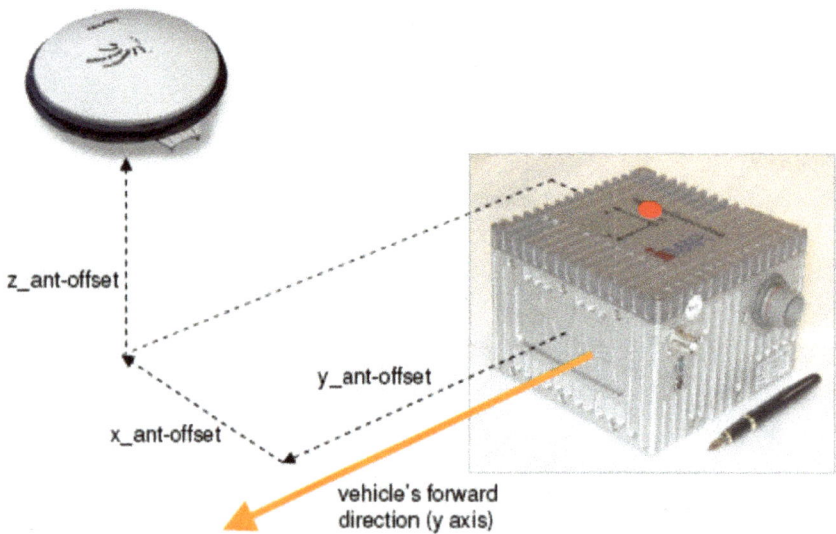

Figure 18. The definition of the offsets between the IMU unit and the GPS antenna
(IMAR iTrace-F200, 2008)

4.3.3. Complex positioning equipment for the slide monitoring

In the same time it is very important to mention that the necessary accuracies of some centimetres can be obtained due to the capability of a device such as Novatel GPS which is to function in a differential RTK (Real Time Kinematic) regime. (Kennedy S. et al., 2007). This type of operation is obtained by connecting the Novatel GPS receiver via Internet to a network of fixed reference base stations as it is the **ROMPOS network** in Romania, which is able to transmit to its customers differential correction data.

4.4. Determination of the offset values between the GPS positioning point of the Inertial Measuring Unit and the central reference point "L" of the video camera

4.4.1. The offset linear components in relation to the reference system (xv, yv, zv) of the lab vehicle

- The geographic coordinates (longitude, latitude) for the vehicle positioning are provided by the GPS system, more precisely, for the IMU point in which the Inertial Measuring Unit is placed (Fig.19);

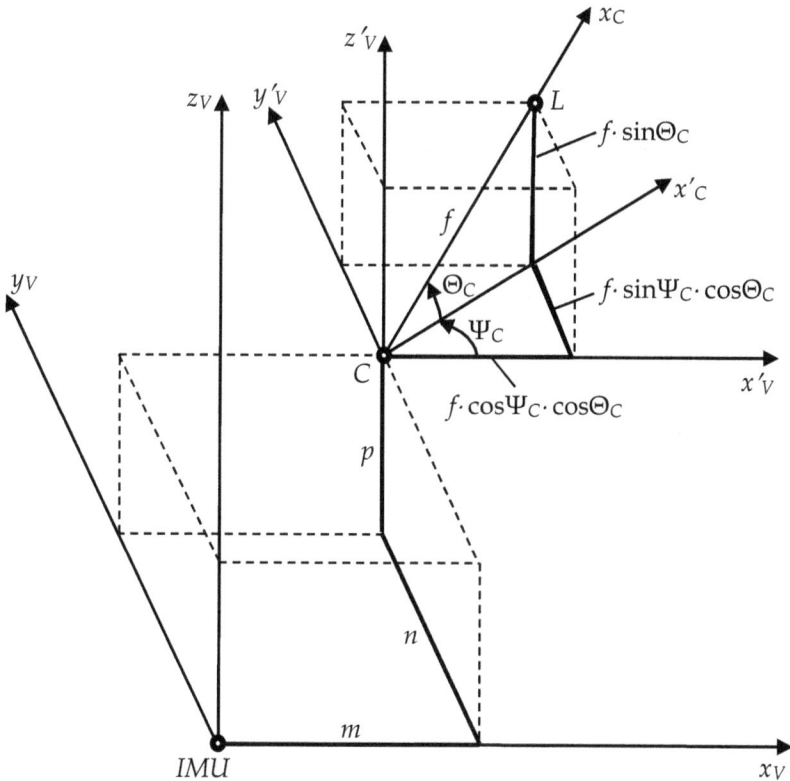

Figure 19. The relative positioning between the Inertial Measuring Unit (IMU) and the Video Camera coordinate systems

-   The reference point relatively to which the positioning measurements of the sighted target are achieved, is constituted by the central point L of the video camera lens, which is placed at the focal distance f from the video camera matrix sensor;
-   It arises the problem to establish the geographic coordinates for the reference point L on the basis of the same type of coordinates of the point IMU;
-   The distance between the points IMU and L presents the following components:
a. On the x'_V axis of the vehicle coordinate system:

$$D_{x'_V} = m + f \cdot \cos \Psi_C \cdot \cos \Theta_C; \tag{1}$$

b. On the y'_V axis of the vehicle coordinate system:

$$D_{y'_V} = n + f \cdot \sin \Psi_C \cdot \cos \Theta_C; \tag{2}$$

c. On the z'_V axis of the vehicle coordinate system:

$$D_{z'_V} = p + f \cdot \sin \Theta_C. \tag{3}$$

In the above-mentioned relations, by m, n and p were noted the components, on the axis x_V, y_V, z_V, of the distance between the point IMU and the central point C of the video camera sensor.

By Ψ_C and Θ_C were noted the video camera montage (fixed) angles in relation with the vehicle coordinate system.

$$D_{x_V}, D_{y_V}, D_{z_V}$$

-   In order to compute the offset components, in relation with the reference system x_V, y_V, z_V, with IMU point as origin, the following group (2) of relations is used:

$$D_{x_V} = D_{x'_V} \cdot \left(\cos \Psi \cdot \cos \Phi - \sin \Theta \cdot \sin \Phi \cdot \sin \Psi \right) - D_{y'_V} \cdot$$
$$\cos \Theta \cdot \sin \Psi + D_{z'_V} \cdot \left(\cos \Phi \cdot \sin \Psi \cdot \sin \Theta + \sin \Phi \cdot \cos \Psi \right)$$

$$D_{y_V} = D_{x'_V} \cdot \left(\cos \Phi \cdot \sin \Psi + \sin \Theta \cdot \sin \Phi \cdot \cos \Psi \right) + D_{y'_V} \cdot$$
$$\cos \Theta \cdot \cos \Psi + D_{z'_V} \cdot \left(\sin \Phi \cdot \sin \Psi - \sin \Theta \cdot \cos \Phi \cdot \cos \Psi \right)$$

$$D_{z_V} = -D_{x'_V} \cdot \cos \Theta \cdot \sin \Phi + D_{y'_V} \cdot \sin \Theta + D_{z'_V} \cdot \cos \Theta \cdot \cos \Phi$$

In these relations by Ψ, Θ and Φ, were noted the rotational angles of the lab vehicle, angles which were measured by the Inertial Unite IMU in relation with the reference system (x, y and z) of the current location.

4.4.2. Transformation of the linear offset distances, D_{x_v} and D_{y_v} , in circular arcs corresponding to the angular coordinate segments: $\Delta\lambda$ and respectively $\Delta\varphi$

$$\Delta\lambda\big[\min.\big] = \big(360 \times 60 \times D_{x_v}\big)\big/\big(2 \cdot \pi \cdot R \cdot \cos\varphi\big)\,\big[meters\big] \tag{5}$$

- for the distances on the longitude λ direction ;

$$\Delta\phi\big[\min.\big] = \big(360 \times 60 \times D_{y_v}\big)\big/\big(2 \cdot \pi \cdot R\big)\,\big[meters\big] \tag{6}$$

- for the distances on the latitude φ direction.

It is also adopted the notation:

$$D_z = \Delta h. \tag{7}$$

On this basis, the geographical coordinates, λ_L, φ_L, h_L, of the reference point L which is constituted from the video camera objective centre, are obtained as follows:

$$\lambda_L = \lambda + \Delta\lambda; \quad \varphi_L = \varphi + \Delta\varphi; \quad h_L = h + \Delta h. \tag{8}$$

where λ, φ, h, represent the geographical coordinates supplied by the IMU for the point in which this is situated.

5. The computing relations group with which it is achieved the determination of the target T geographic position in the horizontal plane of the referential ellipsoid

At the computing of the linear distances, on the longitude and, respectively, on the latitude direction, between the video camera successive positions and, respectively, between the camera positions and the sighted target, it, also, takes account from the fact that this mono-cameral stereo-fotogrammetric system, permits the sighting of some objectives which are situated at distances of up to 200 – 300 meters from the lab vehicles. This way, it is possible to adopt the hypothesis consisting in the approximation of the terrestrial globe with an equivalent sphere with a radius R = 6.367.472 km., as it is presented in the Fig. 20.

The geographic position of the target T in the horizontal plane of the referential ellipsoid is achieved, by combining the determinations of the absolute angular coordinates, $\Psi_{T_1}, \Theta_{T_1}, \Phi_{T_1}$ and respectively, $\Psi_{T_2}, \Theta_{T_2}, \Phi_{T_2}$, of the sighted target, for two different positions, L_1 and, respectively, L_2, of the video camera, positions which are obtained as a result of the lab vehicle displacement with a distance in limits of which the target T is maintained in the video camera viewing field.

In the positioning scheme presented in Figure 20, the following notations were introduced:

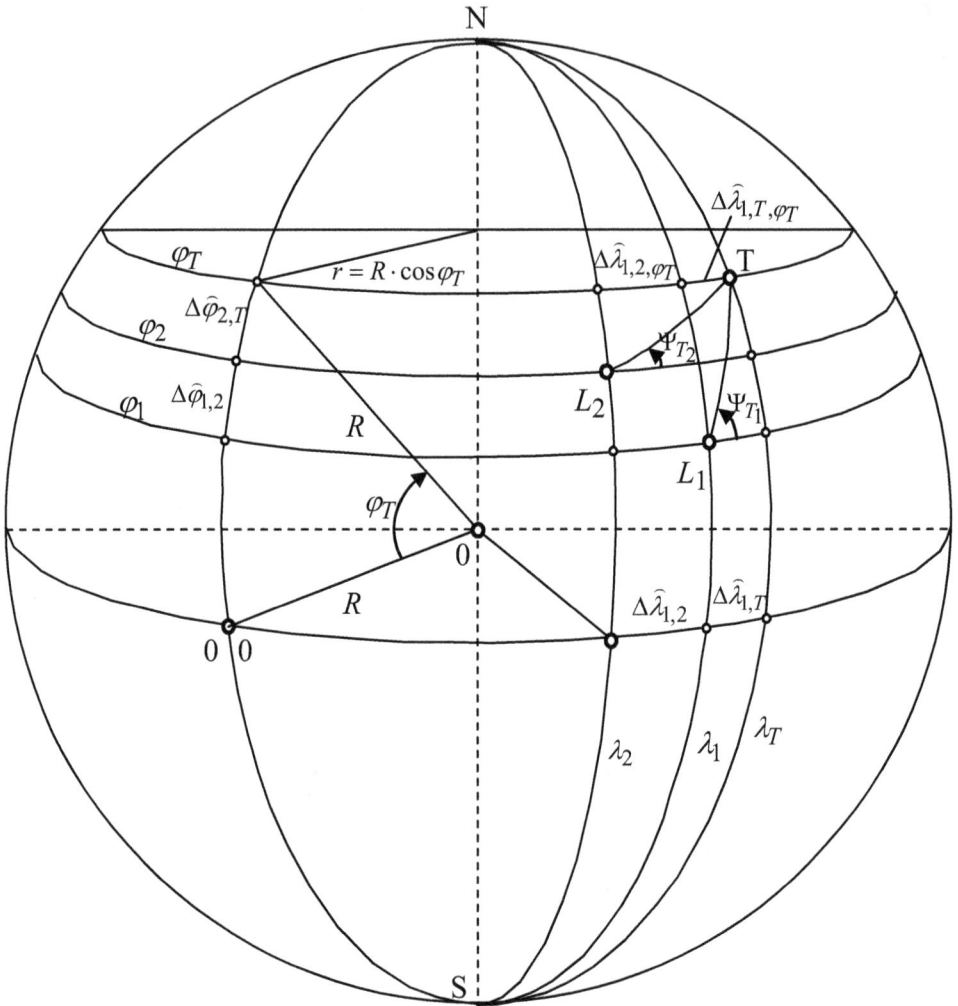

Figure 20. The Lab Vehicle and the sighted target positioning on the equivalent sphere of the terrestrial globe.

λ_1, φ_1 and λ_2, φ_2 - the geographic coordinates of longitude and, respectively, of latitude, which are deduced for two sighting successive positions, L_1 and respectively L_2, of the video camera from the determinations performed by the GPS-INS group, by introducing the offset corrections, which correspond to the distances between the emplacement location of this group in the lab vehicle and the camera objective:

λ_T, φ_T - the target geographic coordinates of longitude and, respectively, latitude;

$\Delta\lambda_{1,T} = \lambda_T - \lambda_1$; $\Delta\lambda_{1,2} = \lambda_1 - \lambda_2$; - the angular differences of longitude;

GPS Positioning of Some Objectives Which are Situated at Great Distances from the Roads by Means of a
"Mobile Slide Monitor – MSM"

141

$\Delta\varphi_{1,2} = \varphi_2 - \varphi_1$; $\Delta\varphi_{2,T} = \varphi_T - \varphi_2$; - the angular differences of latitude, between the target and, respectively, the two positions of the lab vehicle.

Between two circular arcs on latitude, $\Delta\hat{\lambda}$ and $\Delta\hat{\lambda}_{\varphi_T}$, which are delimited by two meridian circles and which are situated, the first at the latitude 0, and the other at the latitude φ_T , the following relation exists:

$$\Delta\hat{\lambda}_{\varphi_T} = \Delta\hat{\lambda} \cdot \cos\varphi_T \tag{9}$$

Also on the basis of the scheme from the figure 20 which presents the positioning mode of a target on an equivalent sphere of the terrestrial globe, the linear distances can be calculated on the basis of angular coordinate differences by means of an equation set, with the following form:

$$a\left[meters\right] = \frac{\Delta\hat{\lambda}\left[min.\right]}{360\times60} \cdot 2\pi\cdot r\left[meters\right] - \text{ for the distances on the longitude } \lambda \text{ direction;} \tag{10}$$

$$b\left[meters\right] = \frac{\Delta\hat{\varphi}\left[min.\right]}{360\times60} \cdot 2\pi\cdot R\left[meters\right] - \text{ for the distances on the latitude } \varphi\text{direction;} \tag{11}$$

where: $r = R \cdot \cos\varphi$.

For the establishment, on this basis, of the computing relations for the geographic position absolute coordinates of the target T, it resorts to the positioning scheme presented in figure 21, taking account of the fact that due to the relative reduced dimensions of the sighting field, its spherical curved surface is approximated by in plan projection of this field.

By this, in plane projection of the sighting field, the circle arcs are replaced by linear segments, as follows:

$$a_1 = \frac{\Delta\hat{\lambda}_{1,T}}{360\times60} \cdot 2\pi\cdot R\cdot\cos\varphi_T = \frac{\Delta\hat{\lambda}_{1,T}}{360x60} \cdot 2\pi\cdot R\cdot\cos\left(\varphi_2 + \Delta\varphi_{2,T}\right); \tag{12}$$

$$a_2 = \frac{\Delta\hat{\lambda}_{1,2}}{360\times60} \cdot 2\pi\cdot R\cdot\cos\varphi_T = \frac{\Delta\hat{\lambda}_{1,2}}{360\times60} \cdot 2\pi\cdot R\cdot\cos\left(\varphi_2 + \Delta\varphi_{2,T}\right); \tag{13}$$

$$b_1 = \frac{\Delta\hat{\varphi}_{1,2}}{360\times60} \cdot 2\pi\cdot R; \tag{14}$$

$$b_2 = \frac{\Delta\hat{\varphi}_{2,T}}{360\times60} \cdot 2\pi\cdot R; \tag{15}$$

These result in the following expressions for the azimuth angles:

$$\tan\Psi_{T_1} = \frac{b_1 + b_2}{a_1} = \frac{\Delta\varphi_{1,2} + \Delta\varphi_{2,T}}{\Delta\hat{\lambda}_{1,T}\cdot\cos\left(\varphi_2 + \Delta\varphi_{2,T}\right)}; \tag{16}$$

$$\tan \Psi_{T_2} = \frac{b_2}{a_1 + a_2} = \frac{\Delta\varphi_{2,T}}{\left(\Delta\lambda_{1,2} + \Delta\lambda_{1,T}\right) \cdot \cos\left(\varphi_2 + \Delta\varphi_{2,T}\right)}. \tag{17}$$

From the first expression, we obtain:

$$\Delta\lambda_{1,T} = \frac{\Delta\varphi_{1,2} + \Delta\varphi_{2,T}}{\tan \Psi_{T_1} \cdot \cos\left(\varphi_2 + \Delta\varphi_{2,T}\right)}; \tag{18}$$

and from the second expression it results that:

$$\Delta\lambda_{1,T} = \frac{\Delta\varphi_{2,T} - \Delta\lambda_{1,2} \cdot \cos\left(\varphi_2 + \Delta\varphi_{2,T}\right) \cdot \tan \Psi_{T_2}}{\cos\left(\varphi_2 + \Delta\varphi_{2,T}\right) \cdot \tan \Psi_{T_2}}; \tag{19}$$

So by elimianatig the $\Delta\lambda_{1,T}$ parameter, we obtain:

$$\frac{\Delta\varphi_{1,2} + \Delta\varphi_{2,T}}{\tan \Psi_{T_1}} = \frac{\Delta\varphi_{2,T} - \Delta\lambda_{1,2} \cdot \cos\left(\varphi_2 + \Delta\varphi_{2,T}\right) \cdot \tan \Psi_{T_2}}{\tan \Psi_{T_2}} \text{and :}$$

$$\left(\Delta\varphi_{1,2} + \Delta\varphi_{2,T}\right) \cdot \tan \Psi_{T_2} = \Delta\varphi_{2,T} \cdot \tan \Psi_{T_1} - \Delta\lambda_{1,2} \cdot \cos\left(\varphi_2 + \Delta\varphi_{2,T}\right) \cdot \tan \Psi_{T_1} \cdot \tan \Psi_{T_2};$$

On this basis, the following implicit computing relation of the latitude angular difference $\Delta\varphi_{2,T}$ is obtained:

$$\Delta\varphi_{2,T} = \frac{\tan \Psi_{T_2}}{\tan \Psi_{T_1} - \tan \Psi_{T_2}} \cdot \left[\Delta\varphi_{1,2} + \Delta\lambda_{1,2} \cdot \cos\left(\varphi_2 + \Delta\varphi_{2,T}\right) \cdot \tan \Psi_{T_1}\right] \tag{20}$$

and in continuation:

$$\varphi_T = \varphi_2 + \Delta\varphi_{2,T} \tag{21}$$

With the determinated in this mode value of the angular difference $\Delta\varphi_{2,T}$, can be calculated in this phase and the value $\Delta\lambda_{1,T}$ of the longitude angular difference by means of one of the two explicit relations (7.1) or (7.2).

In similar mode:

$$\lambda_T = \lambda_1 + \Delta\lambda_{1,T} \tag{22}$$

After obtaining, in the presented mode, of the target geographic coordinates, λ_T and φ_T , in continuation it is possible to calculate and the linear distances: a_1 , a_2 and b_1 , b_2 , on the longitude and respectively latitude directions, between the video camera successive positions and, respectively, between these positions and the sighted target, with the relations:

GPS Positioning of Some Objectives Which are Situated at Great Distances from the Roads by Means of a "Mobile Slide Monitor – MSM"

143

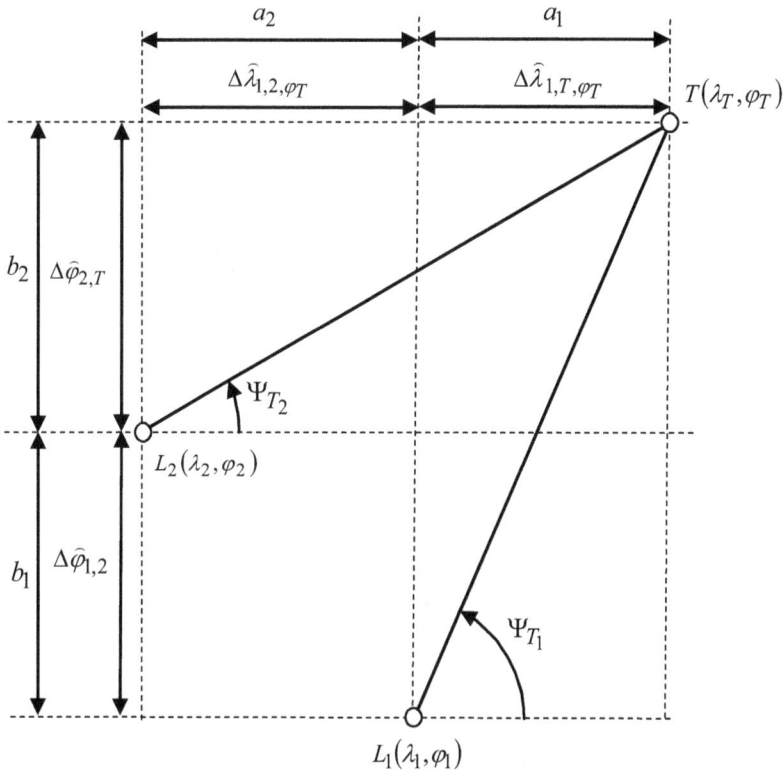

Figure 21. In plane projection of the target sighting field

$$a_1 = \frac{\Delta\lambda_{1T}\left[\min.\right]}{360\times 60}\cdot 2\pi\cdot R\cdot\cos\varphi_T; \quad a_2 = \frac{\Delta\lambda_{1,2}\left[\min.\right]}{360\times 60}\cdot 2\pi\cdot R\cdot\cos\varphi_T;$$

$$b_1 = \frac{\Delta\varphi_{1,2}\left[\min.\right]}{360\times 60}\cdot 2\pi\cdot R; \quad b_2 = \frac{\Delta\varphi_{2,T}\left[\min.\right]}{360\times 60}\cdot 2\pi\cdot R. \tag{23}$$

On this basis, it is, also, possible to calculate the direct distances between the target **T** and the positions, L_1 and L_2, of the two video cameras by means of the relations:

$$L_1T = \sqrt{a_1^2 + \left(b_1 + b_2\right)^2} \text{ and } L_2T = \sqrt{\left(a_1 + a_2\right)^2 + b_2^2} \tag{24}$$

Moreover, in conformity with the schema presented in figure 22, the height h_T of the target **T** in the horizontal plane O.P. of the reference ellipsoid can also be calculated with one of the relations:

$$h_T = h_{L_1} + L_1T\cdot\sin\Theta_{T_1} \text{ or}: h_T = h_{L_2} + L_2T\cdot\sin\Theta_{T_2} \tag{25}$$

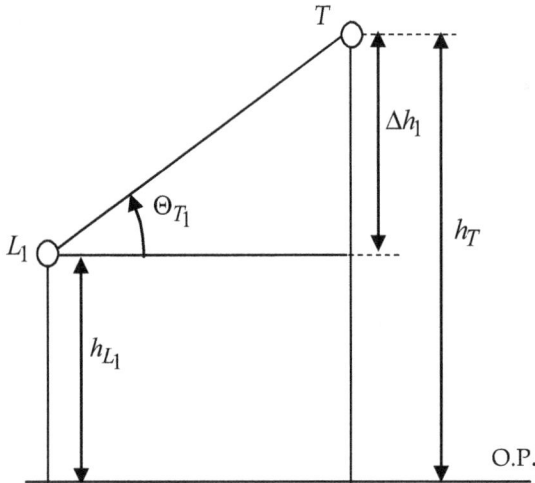

Figure 22. The diagram for the target height h_T computing.

6. Inertial Explorer post-processing software for the final perfection of the MSM accuracy data

Waypoint Products Group's **Inertial Explorer** post-processing software suite integrates rate data from six degrees of freedom IMU sensor arrays with GNSS information processed with an integrated GNSS post-processor (same as GrafNav's). Inertial Explorer use strapdown accelerometer (Δv) and angular rate ($\Delta \theta$) information to produce high rate coordinate and attitude information from a wide variety of IMUs. (Kennedy S. NovAtel Inc., Canada & Hinueber E., iMAR GmbH, 2005)

Inertial Explorer implements either a loose coupling of the GNSS and inertial data or tightly coupled (TC) processing that uses GPS carrier phase to limit error during periods where satellite tracking is limited or variable (even if only 2 or 3 satellites are visible). It is important to time-tag the inertial measurements to the GPS time frame during the data collection process. Proper synchronization is vital. Otherwise, the IMU data will not process. In NovAtel's SPAN system, IMU data is automatically synchronized and the Inertial Explorer's GNSS decoder automatically extracts the IMU data.

7. Conclusions

In order to increase up to 200 – 300 meters, the distance up to which the sighted targets from the terrain can be positioned, it resorts to the use of a single digital camera of high resolution in a fixed montage on a lab vehicle, instead of two cameras which usually are used in the case of classical stereo photogrammetric systems and for which the measurement basis is limited by the montage distance between the two cameras on the lab vehicle, respectively by

GPS Positioning of Some Objectives Which are Situated at Great Distances from the Roads by Means of a "Mobile Slide Monitor – MSM"

145

the dimensional width of the respective vehicle, which is around of 1.2 meters. Due to this fact, the two viewing lines are not intersected on the sighted target, but only with great errors and therefore the bi-cameral stereo photogrammetric systems cannot precisely identify the positions of the objectives that are at distances of more than approximatively 30 meters from the equipment.

So by using a single video camera, the measurement basis of the applied stereo metric method is constituted from the distance interval of 20 – 30 meters, between two triggerings of the camera during the lab vehicle displacement, distance from which the camera sights the same objective.

On this basis, at the returning at the computing center, it follows that from the obtained images to be selected by two images in which the interesting objective is evidenced in a corresponding mode, which will permit the selection of this objective in an electronic modality and the determination of the pixel coordinates, which achieves the objective displaying on the monitor screen. These coordinates together with the other data which accompany the two selected images, will permit to compute, using triangulation proceedings as well as methods for reporting to the spherical system of the terrestrial coordinates, the geographic and elevation coordinates of the sighted objective.

The immediate result of this equipment functioning is represented by the obtaining of a series of digital documents structured in GIS format, documents which contain the data registered in field, with the possibility to update them anytime.

Taking account of its conception, the "Slide Monitor" equipment can be installed not only on a terrestrial vehicle, but also on any kind of boats for the surveillance from an aquatic medium of some isolated objectives disposed on an inaccessible border.

Author details

Axente Stoica, Dan Savastru and Marina Tautan
National Institute of R&D for Optoelectronics – INOE 2000, Romania

8. References

Brzezinska D., Ron Li, Haal N. & Toth C., (2004), GPSVan™ Project: "From Mobile Mapping to Telegeoinformatics: Paradigm Shift in Geospatial Data Acquisition, Processing and Management,"
Photogrammetric Engineering and Remote Sensing, vol. 70, No.2, February 2004, pp. 197-210.
American Society for Photogrammetry and Remote Sensing
Caporali A. (2008) . System and Method For Monitoring and Surveying Movements of the Terrain, Large Infrastructures and Civil Building Works In General, Based Upon the Signals Transmitted by the GPS Navigation Satellite System. Patent application number:

20080204315 (Este (Padova), IT) Assignees: HSEPA-Sistemi Elettronici Per Automazione S.p.A.

Guangping He (1996). Design of a Mobile Mapping System for GIS Data Collection Lambda Tech International, Inc. - Waukesha, WI-53188, USA International Archives of Photogrammetry and Remote Sensing. Vol. XXXI, Part B2. Vienna 1996.

Kennedy S., Hamilton J., NovAtel Inc., Canada & Hinueber E., iMAR GmbH, Germany (2005). Integration of Inertial Measurements with GNSS - NovAtel SPAN Architecture Symposium Gyro Technology, Stuttgart 09/2005

Kennedy S., Cosandier D., & Hamilton J. (2007). GPS/INS Integration in Real-time and Postprocessing with NovAtel's SPAN System, NovAtel Inc.,Canada IGNSS Symposium 2007, The University of New South Wales, Sydney, Australia

IMAR - iTrace-F200 – Interface Description –Sensors and Systeme for OEM applications,- Product Guide, Hwww.inertial-navigation.deH, (2008)

Moafipoor S., Brzezinska D., & Toth, C. (2004). Tightly coupled GPS/INS/CCD integration based on GPS carrier phase velocity update, *Proceedings of the ION 2004 National Technical Meeting*, 26-28 January, San Diego, California.

NovAtel – Inertial Explorer 8.20 , WayPoint – A Novatel Precise Positioning Product, User Guide , www.novatel.com , (2009)

Scott A. (2006). Precision & Accuracy of Leica SmartStation using Network RTK GPS "SydNET"
University of New South Wales - *Undergraduate Thesis – October 2006*

Stoica A., Savastru D. & Tautan M. (2008). The use of High Precision Global Navigation Satellitary Systems (GNSS) for monitoring deformation of buildings at risk for landslides, in flooding areas. "Journal of Optoelectronics and Advanced Materials", 10 (6), 2008, Romania.

Stoica A., Savastru D., Tautan M., Miclos S. & Tenciu D., (2009). Patent: Mobile System and Mapping proceeding of some objectives which are situated at great distances from the roads. Patent RO 126294 A2/ 24.07.2009, Romania

Unexpected 16th Century Finding to Have Disappeared Just After Its Printing – Anthony Jenkinson's Map of Russia, 1562

Krystyna Szykuła

Additional information is available at the end of the chapter

1. Introduction

Nowadays it is rather not common to find 16th century map to be unknown for five centuries. In case of Jenkinson's map of Russia (1562), it was well known by historians of cartography, however, only from their renditions. It occurs possible thanks to the outstanding Flemish 16th century cartographer and editor Abraham Ortelius. One day he simply decided to collect the maps of his times to create an atlas. In this way the first atlas in a quite new editorial form came into existence, equally becoming a rich historical source for the scientists of different fields. Ortelius collected the maps of different regions of the world made by different excellent cartographers of his times. One of the maps that gained his interest was the map being a result of the first English travels to explore the way to China and India by water. English traders travelled along the north-east passage. At the same time Ortelius has just learned that the map of Moscovia has been printed in London by a king's printer, Reginald Wolf. Twenty-five copies of the map in question have been sent to him by Nicolaus Reinoldus to Antwerp, what the latter mentions in his letter. One of the copies has been assigned to be remade in the form of rendition and to fit the size by Ortelius' new Atlas *"Theatrum Orbis Terrarum"*, first edition (1570).

2. The map itself, its author, his travels and the differences among the genuine copy and its renditions. The importance of its recovery

The map of Moscovia – today's western Russia, by Anthony Jenkinson with the 1562 date is hand-colored copper-cut and it measures 101,7 x 81,7 cm including 6 cm decorative border. It is considered by historians as a wall map and has been made in the north orientation.

Doubtful in this respect is only Finland Gulf (*"Sinus Finlandicus"* in the map), which is situated north-south instead of rather east-west.

In the north part of the map there is today's north-western coast of Russia. Far eastern part of the coast in geographical sense reaches the Ural mountains and the lower course of the Ob river. The river discharges into so-called North Sea (*"Mare Septentrionale"*) – today's Kara Sea. The source of the Ob river in the map is in mysterious Chinese Lake (*"Kitaia Lacvs"*). On the opposite side of the lake, i.e. from the south, the river continues its course, however not as the Ob but the Sur river, which bifurcates in its upper course in the *"Shamarghan"* and *"Baida"*regions. The tributary of the Sur is *"Amow"*river.

Figure 1. Jenkinson's genuine copy, 1562 (size of the map see in text). From Wrocław University Library cartographic collection

An eastern part of the map, i.e. east of Ob and Sur rivers, covers quite a wide space of this part of the map. However, geographically this is a rather poor fragment. Namely, there are only names of the following regions, going from the north: *"Casackia"*, *"Samoyeda"* and *"Molgomzaia"*, *"Baida"* and *"Colmac"*. In the south part of the map we can see the name *"Persia"* and in the lower right-hand corner there are *"Mhoghol"* and *"Kirges"*. To the north there is *"Taskent"*, where the Sur river takes its source. In the most south-eastern end of the map there is a city called *"Audeghen"*. In south-western part is the Black Sea but it has been almost wholly covered by the cartouche with the dedication for the sponsor of the map. Above this cartouche we can read the names of *"Lithuania"* and *"Livonia"*. In the western part is the above mentioned Finland Gulf and White Sea, named not as it is today but as the gulf of the *"North Sea"*(*"Mare Septentrionale"* on the map)

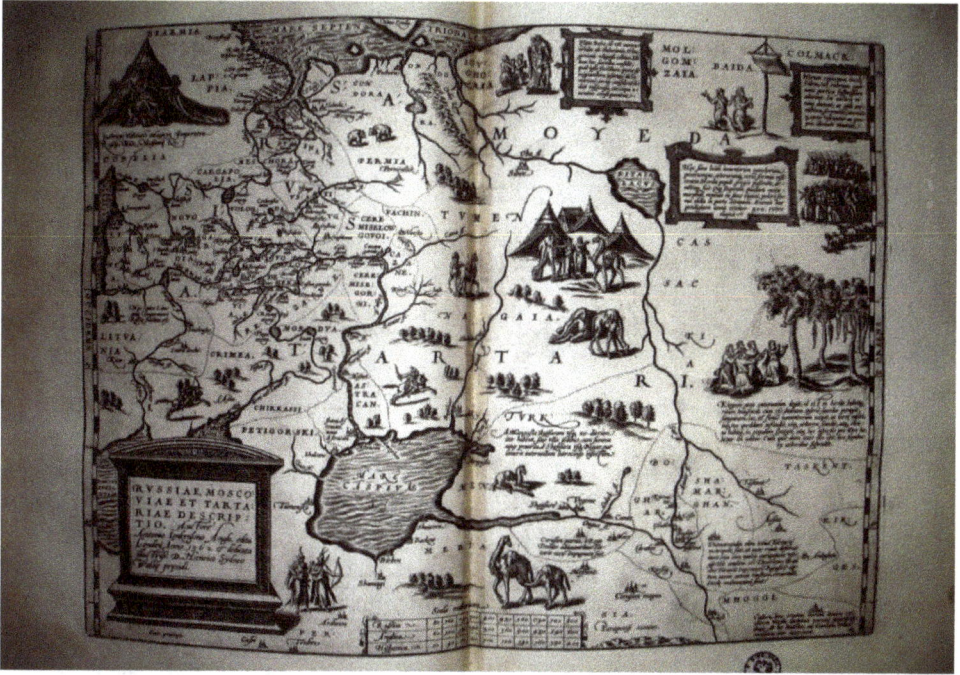

Figure 2. Ortelius' rendition, one of the edition in his *"Theatrum orbis Terrarum"* (here much enlarged to Jenkinson's original copy above – size of the map see in text). From Wrocław University Library cartographic collection

Characteristic feature of the genuine copy of the map in question is its unusually rich decorativeness. It is undoubtedly worth wider discussion. The variety of the content of the map we can study is especially interesting. There are ethnographic, religious, military and historical elements. Finally, the map can be examined in respect of its rich fauna, too. These features make the map an outstanding document of the epoch of the territory portrayed here. The decorative border with metal design is an additional element which makes the map even more interesting. We can find the special value of the map in the numerous texts distributed all over it. These decorative elements are of historical value, too. Mentioned texts are boxed in different cartouches.

The author of the map in question – Anthony Jenkinson (1525/29/30?-1611) is one of the first English travelers, and simultaneously a member of the founded in the years 1552-1553 and chartered in 1555 Muscovy Company. The Company, the society of the English merchants has been later called the Russia Company, English Trading Company, Company of Merchant Adventures or differently. Its purpose was to penetrate and to explore the north-east sea passage to reach China and India.

Sebastian Cabot (ca. 1482 - 1557), Robert Thorn (1492-1532) and John Dee (1527-1608) were the first who became the impellers of this enterprise. As already mentioned, Jenkinson was not the first voyager who travelled to Russia. Earlier there were two brothers – Stephen Borough (1525 - 1584) and William Borough (1536- 1599) who undertook the task to pass the way to mouth of the Ob river. Richard Chancellor (1520-1556) and Sir Hugh Willoughby (1516-1554), the explorers of the northern part of Russia were the next travelers, but unfortunately they both perished in their voyages.

Anthony Jenkinson was the trustworthy agent of the Queen Elizabeth I (1533-1603, dominated 1558 - 1603). He began his journeys in 1557. First of them were in 1557-1560. The next: 1561-1563 (to London 28 Sept. 1564), 1566 - 1567 and the latest 1571 - 1572. His map is dated back to the 1562, however, when we take into account the examinations by Samuel H. Baron's (1989)[1] and Krystyna Szykuła's, (2000)[2], the map in question has been probably published between 1567 and 1569.

An exciting moment when the genuine copy of the map has been rediscovered was finally that could be compared with the existing renditions, i.e. made by Abraham Ortelius (1527 - 1598) and Gerard de Jode (1509 - 1591). It was especially exciting because of the different representations by Ortelius and de Jode which differ with one another. Namely de Jode's picture displays only, even not in whole, the left-hand part of the original. Before the genuine copy was recognized, scholars had discussed who of these two cartographers has been right. Finally today we know that Ortelius' representation was correct in respect to the territorial range.

The size of the three maps – the prototype and its renditions was the essential distinction to be seen at first glance, because of the quite other way of situating the title and dedication cartouche. The existence of the dedication has been only mentioned by Ortelius, but we can learn about its content as recently as from the genuine copy. The same concerns the content of boxed texts, decorative elements and borders – much more of them and changed in style in the genuine copy of the map. An appearance of the quite new creators in the main title cartouche was of the most importance issue. Namely, both of the creators of renditions placed Jenkinson as an author in their cartouches, but only Ortelius additionally included the name of Henry Sidney (1529-1586), the sponsor of the genuine map to whom the above mentioned dedication has been devoted by the editor Clement Adams (1519? – 1587). The second unknown yet to us was an engraver Nicolaus Reinoldus (Nicolas Reynold). We already know him thanks to his letter mentioned in the Introduction, unfortunately undated, which has been estimated to be written about 1573[3]. The name of the printer of the

---

[1] The date has been established by Professor regarding the journeys of Thomas Southam and John Spark, quoted in their account, in early English: *"The way discouered by water by vs Thomas Southam, and Iohn Sparke, from the towne of Colmogro, vnto the citie of Nouogrode in Russia, conteining many particulers of the way, and distance of miles as hereafter foloweth. Anno 1566"* (Hakluyt, 1589, p. 390)

[2] The date has been established by the author according to *"A book of heraldry"* at Cambridge University Library, call number Kk. I. 26 (Szykula, K., 2000). One of the coats of arms have been conferred to Henry Sidney in 1566 and visually it is undoubtedly connected with the composed under the dedication in the genuine map

[3] In the light of the recovered map the date is now not possible to be correct, because of the date of the Ortelius' first edition *"Theatrum Orbis Terrarum"* (1570), where the rendition has been included. The possible date must be previous to the first edition, then about 1569 (?). The date 1573 had been estimated before the genuine map has been recovered and placed in Hessels J.H., 1887 (the mentioned printed collection of the Ortelius' correspondence)

map has been recovered, too, to be Reginald Wolf(ius) - Dutchman, settled in England since 1530, d. 1573. He was the member of the Muscovy Company, as well, in the following years: 1559, 1564, 1567 and 1572. In the letter we can find one more interesting person, Jan de (van) Schille (1533 - 1586), who was an Antwerp painter and engraver and could be also engaged in creating the genuine map, maybe even responsible for the decorative part of it. In the letter he is the person who was allowed by Wolfius to keep one of the 25 copies of the map. The original copy of the letter quoted in Hessels's Ortelius' correspondence, too (Hessels J.H., 1887, letter number 43) had been indicated by Peter Barber – then the Head of the British Library Manuscript Department (Barber, P., 1989). Unfortunately, the letter is still the only document in which genuine Jenkinson's map had been mentioned, and even not quite directly. We can only presume that mentioned 25 copies were not a full size of its edition, and therefore ask where is the rest (if there were any at all) of the 24 copies which are missing, if we take into account only those mentioned in the letter.

As far as the above mentioned differences in arrangement and the content of cartouches are concerned – in Ortelius' rendition the title cartouche is placed in the left hand bottom corner, but on the genuine map in the upper left corner. Close to the title cartouche in genuine copy there is another very important one in which all the regions belonging to Moscovia at that time are mentioned. De Jode's title is also placed in the left upper corner but only with the name of Jenkinson (without information on Henry Sidney). The latter is distinguished only by name in Ortelius' title cartouche. Then we can read the comprehensive dedication to this noble man, as the sponsor of the map, only in its genuine copy.

As far as the dimensions of the three maps are concerned they go are as follows: the genuine copy – 101,7 x 81,7 cm.[4], Ortelius' rendition – 44 x 35,3 cm[5], and de Jode's – 26,3 x 32,6 cm[6]. Quite a long in size horizontally is one more rendition, by brothers Jan and Lucas Deutecum (Doetecum, too) – 104 x 50 cm[7], which the author kindly received from Dr Aleksy K. Zajcev.

3. How and where the Jenkinson's map of "Moscovia" has been found?

This coincidence took place in the author's domestic city, Wrocław (Poland). It was during a visit in the cathedral library, that the head of the library informed her about one lady, who brought him a 16th century map. It was obviously extremely exciting news for the author. The owner of the map, the lady who was a teacher in one of the Wrocław high schools, decided to sell it. That is why she brought the map to the author (then the head of the Wrocław University Library Department of the Cartographic Collection). An expertise has

[4]"*Nova absolutaque Russiae, Moscoviae et Tartariae, descriptio. Authore' Antonio Jenkinsono Anglo, Clemente Adamo èdita, et a Nicolao Reinoldo Londinensi aeri insculpta. Anno salutis, 1562*"

[5]"*Russiae, Moscoviae et Tartariae descriptio. Auctore Antonio Ienkensono Anglo, edita Londini Anno. 1562 et dedicata illustriβ: D. Henrico Sÿdneo Walliae praesidi*"

[6]"*Moscoviae Maximi amplissimi que Ducatus Chorographica descriptio. Authore Anthonio Iankinsono Anglo*"

[7]"*„Regionum Septentrionalium, Moscoviam, Rutenos, Tartaros, eorumque hordas comprehendentium, ex Antonij Jenkensonij et Sigismundi liberi Baronis ab Herberstein itinerariis, nova descriptio. 1569; Joa. & Lucas Duetecum tgraefscap vä Holland Anno 1569 geprint i Hollant in des Gravenhage. 1569; Joa. & Lucas Duetecum tgraefscap vä Holland Anno 1569 geprint i Hollant in des Gravenhage.*

shown that just found genuine map is the one used by Ortelius and de Jode as the basic picture to their most popular renditions. Additionally in the famous *History of Cartography* (Bagrow, L. & R.A. Skelton, 1964, p. 172) , there is only one sentence on the genuine map – *"the map not survived and it is known only from the copies in the atlases of Ortelius and de Jode"*. Then the conclusion was quite clear – that is the only copy of the map in the world! What is yet more interesting, before bringing the map to the University Library the owner has been showing it in some eminent libraries but nobody showed any interest in this map. Because of the great interest of the author finally it had been purchased by the Wrocław University Library.

According to information of the owner, the present map has been used by her for years as a didactic (teaching) aid on her history lessons. Because of its big size it has been folded twice and finally brought in a plastic bag, simply because the lady was not aware of its value. To the question of how the map ended up in a teacher's hands, she answered that it was a gift from her pupil, who found it after the World War II, probably in some cellar or attic!

Figure 3. The content of the dedication by the editor Clement Adams to Henry Sidney – the sponsor of the Jenkinson map. Below Sidney's coat of arms

4. How could Jenkinson's map be examined by scholars if the genuine copy of the map has been lost immediately after its printing? (About the development of the examination by numerous scholars)

Undoubtedly we have to remember that it happened thanks to above mentioned famous cartographer Abraham Ortelius as the creator of the first Atlas *"Theatrum Orbis terrarum"* to which the Jenkinson's map had been incorporated, too. Although he did not use it in an original form, however, thanks to him the knowledge on the map and its author has survived. On the other hand, to include the map in its original big size was even technically not possible. He was then obliged to reduce its format, as well as to limit its decorative elements and the number of boxed texts. However, he never limited the geographical content of the maps he reworked. He also always put the original authors in them, in spite of his own authorship as the new author of rendition. The engraver of Ortelius' rendition was Franciscus (Frans) Hogenberg.

Apart from the Ortelius' and de Jode's renditions, another one had been made by above mentioned Deutecum brothers with the date 1569[8]. As well as rather mysterious is the Ortelius' rendition reworked by Antonio Possevino (1533/34 --1611), in 1587 edition of his book (reproduction of the map in : Szykula, K., 2000, p. 79). One more rendition, has been published by B. Langens much later than the first edition of Ortelius' Atlas in Amsterdam (1598)[9].

The first historian of cartography who revealed his interest in Jenkinson's map was Richard Hakluyt (1552 – 1616). It resulted in the comprehensive edition of the book entitled *"The Principal Navigations, Voyages and Discoveries of the English Nations"*(1589), where he included Jenkinson's accounts from his diary about his journeys. Among them he placed the description on the journey by the first Russian ambassador in England during the reign of Mary Tudor. The interest in the period of journeys and Jenkinson's map has reappeared and raised in 19th century. It was the topic which became the fundament of the work by Edward Delmar Morgan and Charles Henry Coote and resulted in their book published in 1886 – *"Early Voyages and Travels to Russia and Persia, by Anthony Jenkinson and Other Englishmen"*(Morgan, E.D. & Coote, H., 1886). At the end of the 19th century we can notice a great interest by Russian scholars, as well. The leading one at that time was the scholar of Russian history of cartography Veniamin Aleksandrovich Kordt (1899), who has published absolutely fundamental work on the early maps of his country in which he included their reproductions, too. At the beginning of the 20th century another Russian historian of cartography H. von Michow (1906, pp. 22-25) showed the same interest, and in the interwar period also Leo Bagrow (1928) – the Russian emigrant settled in Germany and then Sweden. Again, after the World War II, we can observe an interest in our subject. There are articles by Dutch historian of cartography Johannes Keuning (1956), and by mentioned Leo Bagrow

---

[8] The map is known as well as "Dashkov map" because the only genuine copy survived in Dashkov collection in Petrograd (St. Petersburg)
[9] Sketch map published by Petrus Kaerius entitled *"Ruffia"* According to Wikipedia, the map is available in New York Public Library.

(1962). In the same year the book by Margaret B.G. Morton had been published, but rather from the Jenkinson's private life point of view. Then in turn we have works by: Rybakov, A. B. (1974), Sager, P. (1974), Bagrow L. (1975) and Oakeshott W. (1984) and finally by still uncertain to the original finding S.H. Baron, (1989) together with his several works connected with the epoch in question. So, these are all of the works issued before the genuine map has been found (1987), i.e. the period when none of the mentioned authors was aware of the original picture of the Jenkinson's map, i.e. the period when they had only two main existing renditions as a proof of the map's existence to their disposal.

In the year 1987 starts a new epoch for the Jenkinson's map. The second step of the author taken was to announce the subject of the rediscovery to Organizers of the next International Conference on History of Cartography, then it was to be in Amsterdam. Just after the announcement of the author's abstract of the subject, Canadian editor of *"Cartographica"* Edward Dahl showed an interest in this exciting news. The editor needed the confirmation that the announcement on the genuine map in an Abstract is trustworthy. It was because of the next paper which professor Baron prepared for printing, still about the rendition of the Jenkinson map. Hence, professor Baron had to rework just before delivering the article (Baron S.H., 1989) to the editor and write in the footnote this sensational news, however as has been said, carefully informing on the new discovery. The presentation during the conference brought unexpected effect. Englishmen who noticed the genuine map reproduction in the poster session called: sensational, incredible, unbelievable. The first post conference publications on the discovery were: short article in conference book (Szykuła, K., 1989); conference account (Scott, V.G., 1989), and the same author short information with small reproduction of the rediscovered map (Scott, V.G., 1990), finally additional information (Barber P., 1989). At the same time in accounts by Eckhard Jäger (1989) and R. W. Karrow (1989) the copy in question had been announced, too.

In the meantime, there has been established a friendly scientific correspondence cooperation between Professor Baron and the author which resulted in the Professor's first article about the genuine copy (1993). He considered the relations between original and its renditions, and tried to establish the real dating of the genuine copy.

5. The picture of the Jenkinson's map. The description and analysis of the richness of the content

5.1. Historical background of the map

From the historical point of view Jenkinson's map is the 16th century document of the epoch during the reign of Elizabeth I (1533–1603) in England and of Ivan IV Terrible (1530-1584) in Russia. To be more clear why the map had been depicted we have to go back to the epoch of Edward VI (1537-1553) and Mary Tudor (1516-1558). The date of the death of Edward is at the same time the date of establishing the Muscovy Company. The date of the death of Mary is in turn the period when our Jenkinson had been travelling on his first voyage to Russia (he left London on May 1557).

Figure 4. Ivan IV Terrible and Anthony Jenkinson (from the genuine map)

Important dates for the mentioned period were following events as attachment of Khanate of Kazan (1552) and Astrakhan Khanate (1556), as well as subordination by tsar the Nogai Orda and Khanate of Sibir. At the time the Russian neighborhood played an important role – Poland and Lithuania, for instance Ivan's suffering the defeat in the war with Livonia (1557/1558 – 1570). To have an access to the Baltic coast was the main reason of the battle at that time.

This short introduction to historical epoch of our map of Russia let us take a look at the map of this point of view. Historical content is reflected in numerous texts on the map, as well as, for instance in silhouettes of numerous khans which are placed in the right-hand part of the map in its south-eastern fragment of the territory. In the left-hand part of the map there is only one figure of khan -*"Ismail Sophi"* near the Ardevil town *("Ardabil"* in the map). This city is situated close to the other important city, Tabris *("Tenbres")* and in the western direction there is yet one more important city Kazvin *("Caſby"* in the map).

Opposite to the left-hand side of the map, in the right-hand side, are five figures of khans. There are (going from the south): *"Kvrcot chan"*- to the north of *"montes paraponise"* , "Alie chan"– in *"Kirges"* region, *"Blag chan"*- in *"Boghar"* region, *"Azim chan"*- in *"Turkmen"* region and *"Aphis chan"*- in *"Taskent"* region, to the north of *"Taſkent"* city.

As far as the content on regions included is concerned the best idea is to quote full list mentioned in the bottom of the title cartouche, i.e. left hand corner of the map. This goes as follows:

"Johannes Baſilius Dei gratia, Magnus Imperator totius Ruβiae, Magnus Dux Vladimiriae, Moſcoviae, Nouogardiae, Imperator Aſtrachaniae atque Liuoniae, magnus Dux Plaſrouiae,

Smolenciae, Tueriae, Iogoriae, Permiae, Viatiae, Bolgoriae, etc. Imperator et magnus Dux Nouogardiae Niuociorum, Chernigouiae, Rezaniae, Volotiae, Erzeuiae, Bieliae, Jaroslauiae, Belozeriae, Vdoriae, Obdoriae, Condiniae, et aliarum multarum regionum Imperator atque …(?)totius Septentrionis dominus".

As the historical description we can also consider the following one[10]: *"Haec pars Lituaniae, hic descripta, Imperatori Ruſſiae ſubdita"* (transl. "this part of Lithuania, here depicted, is subordinated to tsar"). Similar description concerns the subordination of the part of Livonia.

One of the descriptions is a historical one only at its beginning: *"Permiani et Condoriani, aliquando Ethnici fuerunt, at nunc a Ruβorum Caeſare perdomiti,…"* (transl."Permians and Condorians were in past times one nation, but now they have been conquered by Russian tsar").

The subsequent text of historical content is: *"Crimae ſunt Mahumetiſtae, quibus cum Moſcouitis aſſiduum bellum intercedit"* (transl. "Crymens are Mahometans, who still proceed the war against Moscow"), *"Cazane Regnum Tartariae fuit Anno 1551 expugnatum ac Imperatori Ruβiae subiectum"* (transl. "Cazane kingdom was taken away from Tartaria and subjected to Russia in 1551"), *"Aſtracan Tartarorum regnum fuit anno 1554 ſubactum, ac imperio Ruſſiae adiectum"* (transl. "Tartar region Astrakhan was conquered and attached to Russian Empire in 1554").

Here again there is only the beginning of the text which can be interesting from the historical point of view: *"Medi, Perſaeque Mahumetani ſunt, aβidueque cum Turcis Tartariſque pugna confligunt…"*(transl. "Meds and Persians are Mahometans, they are still fighting with Turkish Tartars").

Interesting is the description close to Caspian east coast region: *"Turcomannorum imperium inter quinque fratres eſt partitum, quorum qui primas tenet Azim Chan nominatus eſt. Reliquero, Sultani appellatur. Quique ſolum oppida uel potius caſtra subiectione et imperio suo tennet"* (transl. "Turkmens Empire is divided between five brothers, one of them, the leading one, is called Azim Khan, the next are designated Sultans. Only five towns or rather camps are subordinated to them").

The next text is to be continuation of the previous one and it goes: *"Horum Vrgence Principem locum tenet. Incolae Mahumeticam ſectam agnoſcunt, uiuuntque iuxta Nagaiorum conſuetudinem, ac cum Perſarum Principe (uulgo Sophi nuncupato) continenter belligerantur"*(transl. "The period and place of the Urgench Duchy. Habitants practice the Mahometan religion, and live according to Nagai customs and with the duke of Persians (so called Sophi), they permanently wage the war").

---

[10] Every one of the texts quoted here is transliterated according to the originals. The two exceptions to the rule the author used are in the cases when the end of the word was for instance "rū". Than the form "rum" was used or instead of "e'" in text author used the end "ae".

For the translated texts of genuine Jenkinson's map from Latin to Polish the author wishes to thank Dr Wojciech Mrozowicz from Wrocław University Institute of History

Figure 5. Northwards of the Caspian Sea in the genuine Jenkinson's map

Some other interesting descriptions are as follows: *"Vrbs Coraſon a Rege Perſico adiuuantibus Tartaris anno 1558 expugnata fuit"* (transl. "The city Corason had been captured in 1558 by the Persian king, who was supported by Tartars"), *"Shamarcandia olim totius Tartariae metropolis fuit, at nunc ruinis deformis iacet, una cum multis antiquitatis ueſtigijs. Hic conditus eſt = Tamerlanns ille, qui olim Turcarum Imperatorem Baiazitem captiui aureis catenis uinctum circumtulit. Incolae mahumetani ſunt"* (transl. "Samarcanda was a capitol of the entire Tartaria, now is in ruinous state, however with numerous ancient relics. It was founded by Tamerlan, who in early times conducted emperor of Turks Bajazyd, who was chained in golden chains. The habitants are Mahometans"), *"Boghar urbs ampliſſima, aliquando Perſis fuit subdita. Ciues Mahumeticam hereſim amplexantur, Perſicaeque loquuntur. Frequentia hic ſunt commercia, tum ex Cataya, India, Perſia, alijsque orbis tractibus"* (transl. "Boghar is the most extentsive city, in early times it was subjected to Persians. The habitants are heretic Mahometans and they speak Persian. There are often trade fairs, /merchants/ come from China, India, Persia and from other districts of different countries"), *"Rex hic aduerſus Caſsachios aſſidua bella mouet, quae gens nuper prope exterminata fuerat"* (transl. "This King is still fighting against Khasack tribes. Once they have been close to be chase away"), *"Princeps hic cum Indis plurimae habet certamina, qui ad auſtrum illi finitimi ſunt"* (transl. "This duke is still fighting with Hinduss, who are his neighbours from the south"), *"Caſcarae princeps Mahumetanus est, ac cum Kirgijs bella mouet"* (transl. "Duke of Caskhara is Mahometan and he is fighting against Kirghiz").

Conclusion: these above mentioned descriptions placed in the map give us quite a rich material on history of the territory in question.

Figure 6. Right hand lower part of the genuine map

5.2. Genuine map from the geographical point of view with a short introduction of the history of cartography of Russia

To examine the subject of the Jenkinson's map from the geographical point of view, it is worth devoting some place to history of cartography of Russia. Obviously, this field is best known by native scholars. One of them is for instance Professor Alexey V. Postnikov. Before we take into account his publication from 2000, we should go back to the ancient times. Here should be mentioned for example Hecateus of Miletus (c. 550 – 480 BC), Herodot from Halicarnassus (c. 484 – 425 BC), Dicearchus of Messana (c. 350 – 285 BC) or Eratostenes (c. 276 – 196 BC) and of course many others. On the map of the world by Dicaearchus it is the most amazing because already depicted in south-north extension of the Caspian Sea, and as we know this error was presented on early maps up until the beginning of the 18th century, for instance on J.B. Homann's map in 1720 (*"Generalis Totius Impeii Moscoviti"*). As well as on the Dicaearchus map, there is already a symbol of the cartographic net in form of two perpendicular lines – the meridian which crosses Rhodos Island and parallel, so called diaphragma, which start from Pillars of Hercules (Strait of Gibraltar), and it runs to the Himalaya Mountains. The proper shape of Caspian Sea was undoubtedly known by ancient people, what we can learn from the article by Leo Bagrow[11].

On the maps of ancient geographers and cartographers untill the times of Claudius Ptolemaeus (100 – c.168) one may observe the development of geographical knowledge.

---

[11] Bagrow, L., *Italians on the Caspian*, Imago Mundi 13, pp. 2-10

Some information were repeated together with the development from one to the next generation. This happened with the presentation of the Oxus or Ougus river[12] which was so depicted until the first quarter of the 18th century. As an example can be shown the map by Christfried Kircher of 1734 or J.B. Homann's map of *Kilania* (and different other dates of its editions)[13].

As far as the domestic Russian cartography is concerned, it is necessary to come back to the above mentioned Alexey V. Postnikov's article. We read there that the first document of Russian domestic cartography was so-called *"Nikon's latopis"*. The earliest Russian maps initially were composed for small fragments of areas, for instance a vicinities of rivers, meadows, then strongholds, and finally cities. The last were created mainly for military needs. Road maps in turn were created for mission needs to be used by monks. The maps of northern sea lands were made because of sailors' and fishermen's needs. However, despite of existence of much information in Russian transmissions in maps and drawings, practically they had not survived. There are, however, many maps, which have been made by foreign cartographers, who in their diaries or accounts were writing about the politeness Russian natives showed towards the foreigners. They were particularly very helpful in every aspect in terrain. It is even possible that they served some sketch maps of a small parts of a given area like those experienced traders and voyagers of the Muscovy Company. Professor Postnikov writes about the Polish cartographer G. Maintsky, who, according to the Professor, was an author of the world map of 1100, where he already marked Russia as a country situated northerly of the Danube. The next cartographic document where the Russia territory is marked is the famous Ebstorf map of the end of 13th century. On the map Professor Postnikov notices fourteen times the different names connected with the region of Russia.

As far as the territory of South Asia is concerned, we should not forget about the voyages by 13th century latest half traveller Marco Polo (1254 – 1324). Together with his brother he passed the so-called Silk Road (south of the Black Sea and Caspian Sea) and reached China, and, like Jenkinson, later on, Marco Polo passed the same dangerous Bokhara, as well as was a guest on the court of Great Khan of Persia.

From the year 1459 comes the world map by Fra Mauro, however, there is a quite detailed fragment of territory of Russia which is therefore why it should be quoted here (Borodaev & Kontev, A. V., 2007). Very good picture of the part of the region in the book illustrates the fragment of the map (p. 20). We find the description on this map on the next page. On subsequent maps appear more detailed pictures of Sarmatia[14] – for instance in 1513 Strasburg edition of *"Geography"* by Ptolemaeus.

---

[12] The name Oxus is the Arabic name of Ougus (more read later on)
[13] This map has been kindly indicated by Professor Alexander Podosinov
[14] Sarmatia is the region situated between Vistula river and Caspian Sea. This name became famous thanks to well-known Polish classic writer Henryk Sienkiewicz. In 16th century there was another Polish writer – Matthaeus of Miechov (1457 - 1523, so called Miechowita in Polish), who described this region in his "Tractatus de duabus

16th Century is a golden age of very comprehensive geographical works where the maps became quite often illustrations of the texts. The most famous is so-called *"Cosmography"* by Sebastian Münster (1488 – 1552). In its first edition of 1544 in Basilea we can find the map of Moscovia, as well. However, the real cartography of Russia begins from Dmitry Gerasimov (c.1465 – c.1535), who was the Russian ambassador. He passed his observations on Moscovia to Paolo Jovio[15] (1483 – 1552). Next, Battista Agnese (1500 – 1564) published the Jovius map in his Venice edition of 1554 (reproduction of the map in Szykula's article, 2000).

The next important map of Russia was prepared by Baron Sigismund Herberstein (1486 – 1566). It is the result of his travels to Russia. He was an Austrian diplomat and the messenger of the emperor Maximilian I (1459 – 1519). The map has been made in wood by Augustin Hirschfogel in 1546, and published in 1549 in the book by Herberstein *"Rerum Moscoviticarum commentarii"* and it is the first comprehensive report on the Moscovia State.

Dates 1537, 1542, 1555 and 1570 are the years of subsequent editions of the map by Anton Wied. This one has been made on basis of an information by I(van?) V(asilevich?) Liacky.

Now it is time when the Jenkinson's map should be already described from the geographical point of view. The left-hand half of the map which has been already mentioned in this respect is the richer one than the right. The latter is not only poorer in those physiographical elements but generally speaking in most degree erroneous in its representation. On the other hand it is richer in decorative components. As was already mentioned, the Ortelius rendition is the most faithful to the original map, especially at its left hand part. Vaughan, Earnest Vancourt (1912) was the one who very accurately analyzed Ortelius' rendition.

Obviously many other historians were engaged in analyzing the map, too, but still before the original has been found. After the discovery of the genuine copy of Jenkinson's map we can find the first descriptions on the relation between genuine copy and the Ortelius' rendition: Szykula, K., (1989) in the conference book and Baron S.H., (1993) more comprehensive description. Next in: Szykula K. (1995), Szykula K. (2000).

Now to attempt to analyze the genuine copy in this respect, it is worth remembering some common opinion which was expressed by many scholars, that the north-western part of the map has been made by Jenkinson on the basis of the manuscript map by William Borough (1558) – reproduction in Szykula's, K. (2000), south-western part on the basis of Anton Wied's, but the most erroneous east part by Anthony Jenkinson himself. Obviously, it would be nothing strange that Jenkinson should use the existing maps of his predecessors, but on the other hand such an opinion is to some degree rather unjust. At first, because he himself personally first overcame the roads so far inside Russia and as a first Englishman reached Buchara region at the time - previously Bokhara region have been reached only by Marco Polo. Before the map has been depicted, he has been exploring the western region of

---

Sarmatiis" (1517) where he paid attention to the error made by Ptolemaeus regarding the Ripheans and Hyperboreans mountains in lower Moscovia state. However, later on Herberstein in his work (1549) identified them as Ural mountains.

[15] See Rybakov B.A., *Novoodkrytaja karta Moskovii 1525 g.* Otečestvennye archivy, 4. pp. 3-8

Moscovia three times. This shows his great and rather correct knowledge of the way in question, to Persia, too. Some proof makes, too, his detailed diary in which we read about his numerous measurements along the way, by log (distances) and astrolabe (latitudes). He also gives many concrete data on estimated value of latitudes as well as distances in miles or in number of days. The north orientation of his map is the next proof of the modern attitude to cartography, however, known already from the Ptolemaeus atlas, but more innovative to Wied's map, which is, according to its author, an eastern one, but south-eastern because of some elements on the map.

Figure 7. White Sea region on the genuine map

The visual eastern border of the left half of the map reaches the lower course of the Ob river, *"Tiumen"*region, east coast of Caspian Sea, crosses the *"Ougus"*river, and in the southern part of the territory reaches the north end of the Persia and Hindukush mountains, where river Ougus takes its source. As far as the descriptions of the rivers here is concerned, the author asks the reader to be understanding if she will not always follow the principle to describe them from the upper course as it is usually practiced. To continue the subject and going from the south-eastern part there is the Black Sea but it is covered by the cartouche containing the comprehensive dedication to Henry Sidney, as the sponsor of the map. We can only notice the north part of the Sea, i.e. Azov Sea (*"Palus Meotis"* in the map). In the western part of the Sea is the Dnieper river (*"Biriſtines uel neper"* in the map). The river goes from the north reaching its source in the non-existing *"volock"* lake.

Into the Azov Sea flows Don river (*"don uel tanais fl."*). Then it turns to the north-eastern direction, where from the west it leaves the *"Crimea"* region. Further the river passes

"Mordva" and *"Reza"* regions crossing two lakes – smaller one *"Iuan ozera"* and the larger one *"ploglar ozera"*. Finally it reaches the smallest of them *"rezanskoy ozera"*.

Between the two mentioned seas runs two mountain ranges marked with small hillocks, which are almost for sure Kaukaz mountains. The name of this region is marked *"Chirkassi Petigorski"*. Below is Kura river *("Cirus fl.")*. To finish the description of the left lower part of the map, it is worth mentioning the cities of the region, but they will be included in the *"Dictionary"* of further *"Monography "*. In this chapter there should be yet considered Kaspian Sea *("Mare Caspium")* with the estuary of Volga river. It changes its names *("Volga fl"* or *"Volga Rha uel Edel fl."* or *"volgha fl.")*. The another one is Ural river mouth *("bogthiar"* & *"Yaik fl.")*. However, coming back to the course of proper Volga going from its estuary in Caspian Sea - its first half of the course is better shown than in other early maps, however we can say that there is not one Volga river! The proper course of Volga river goes from the Caspian Sea under the name *"volga fl.* and *"Volga fl."".* Then from Kazan *("Cazane gorode")* it turns by 90⁰ angle to the west to Nizhny Novgorod *("Nifhougorod")*, next it takes the name *"Volga, Rha, uel Edel, fl."*. Between the city Balachna *("Balaghna")* and Kastroma *("Caftrome")* the river takes back the name of *"volga fl."* and just behind Yaroslav *("yearaslaue")* it bifurcate into two Volga rivers. One of them flows to the south *("volga, rha, uel edel, fl.")* and finishes its course in mysterious *"volock lacus"* going subsequently through Yaroslav city *("yearoslaue")* "finishing" its proper course to Tver *("Tuer")* and then to *"volock lacus"*as *"volga, rha, vel edel, fl."*. The other one *("volgha fl")* reaches White See *("biatla ozera")*. In the mentioned *"volock lacus"* are also sources of so-called western Dvina and Dnieper river. *"Duina fl"* flows through Vitebsk city *("Widepfky")*. To-day it flows to the Riga Gulf and its source it takes in the Valdai hills (Latvia).Coming back to the Caspian Sea from Volga toward the east is the river Ural, which today flows through Nagaya region *("Nagaia")*, and in a wider sense Tartaria. In this quarter of the map – in the east side we have yet today Kara-Bogaz Gulf, in the shape of widely extent and long mouth of the Ougus river (Amu-Daria). In this part of the map we have many different regions. Going from the south there are: *"Petigorski"*, *"Astracan"*, *"Chirkassi"*, Crimea", Mordva", "Rezane", "Volodemer", "Novogardia", "Ceremise Gorni", " "Ceremise Lowgovoi", "Casane", "Vstimia", "Vologda", "Cargapolia", "Meschora", "Dvina", "Permia", "Condora", and "Obdora". Coming back to the north-western border of the quarter of the map there is east part of Livonia *("Livonia")* and Lithuania *("Litvania")*. Neva river flows into Finland Gulf *("Sinus Finlandicus")*. In the map Neva takes name of *"volgha fl."*As far as the west-northern part of the western half of the map is concerned, there is one more Volga river but this one are today's Neva and Svir rivers. There are equally two well known lakes Ladoga and Onega, but in Jenkinson's map they change places and sizes with each other. Namely, Onega Lake *("Ourfhock lacus")* is larger than Ladoga Lake *("Vladifcoy lacus")*. Further to the north is situated White Lake *("biatla ozera")*, which is on the way of Onega river *("Onega fl.")* and the river finally flows correctly to the bay of White Sea, today's Onega Bay. In the north-east fragment of the western half of the map there are yet three great rivers – North Dvina *("duina")*, Mezen river *("mezena fl")* and Petshora river *("Pechora fl.")*. The fourth river Ob makes some kind of the border between west and east half of the Jenkinson map. All of the rivers mentioned here have its own tributaries in the map, but not every one of them is named. One more river in the western half of the map is Kama river *("Cama fl.")* and Samara river *("ʃamar fl.")* as the easternmost tributaries of the Volga river.As has been said, the right hand half of the map is quite a controversial one. It has to be remembered it had been taken into consideration by Ortelius, too, but neglected in de Jode's rendition. In an eastern part of the western Jenkinson's map territory we have only the lower course of the Ob. In the map it flows to North Sea if we translate *"Mare Septentrionale"*, but its source is in the Chinese Lake *("Kitaia Lacvs")*. From the opposite side of the lake the river called by Jenkinson *"ʃur"* flows into the lake. Its source in the map is in Tashkent as far as to Tashkent *("Taskent")* and probably it is to-day Sir-Darya river with its tributary Zeravshan *("Amov fl.")* Today Zeravshan is the tributary of Amu-Darya river. In the map *"Amow"* river flows around three relatively big cities: *"Coʃin"*,*"ghudowa"*and Bokhara *("Boghar")*.The Ougus river with its tributary *"ardock"* is situated in the southern part of the map and the source of this river is in Hindukush mountains *("montes paraponiʃi")* at the south-eastern end of the map. There is the name *"Mhoghol"* and *"Kirges"*, too, in the right lower corner of the fragment. In this part of the map, to begin from the west, are marked following cities: *"Shayʃare"* and *"Vrgencz"*, and in bifurcation of two rivers *"ardock"* and *"Ougus"* is the city *"Cante"*. Further on the river Ougus there is *"Caracoll"* and to the south *"kwʃhy"*. Close to the *"montes paraponise"* there is a city *"balgh"*. In the east we have a city *"Audeghen"*, and a little higher to the west there are: *"Samarcandia"* and *"Arʃow"*. Close to the upper course of the *"ur"* river is *"Taʃkent"*, and close to the river *"Amow"* – *"kyrmina"*. In this right hand fragment of the map there are also the names of the regions: *"Taskent"*, *"Marghan"*, *"Tvrkmen"*, *"Boghar"*, *"Kiata"*, and as the biggest one *"Tartaria"*. Further:*"Cassac"*, Molgomzaia"*, *"Baida"*, *"Colmack"* and again the biggest ones *"Samoieda"* and *"Tvmen"*(to the west of *"Kitaia Lacvs"*) and finally below *"Tumen"*- *"Nagaia"* region. At the end of this subchapter worth mentioning is one more physiographic element - clusters of forests in Jenkinson's map. Here they are represented very picturesquely by smaller or wider clusters of the forests in the form of the trees which are not distinguished as the deciduous or coniferous forests, but overlapped one on another. Fauna, the topic which somebody would include in the physiographic elements, here is to be reviewed together with the ethnographic content of the map.

5.3. Mathematical basis of the map. Experimental attempt to the method of examining the differences of the old maps and their cartographical nets by means of the nets of triangles

The distinctive feature of early maps are different scales on the same map. It is because the cartographers compiled different maps which were made by using a different scales. Hence, we can observe numerous errors in several parts of the maps. According to numerous measurements of the distances made by the author the average scale of Jenkinson's map occurred to be about 1: 5 083 871 (Szykuła, 1995). Whereas, the scale which has been counted on the basis of three scales on the map, i.e. in Russian miles, English and Spanish miles , according to professor Baron has been assessed to be between 1:6 000 000 to 1: 7 800 000 (Baron, 1993).

The other reason of the distortions in the early maps was obviously the imperfection of the measuring instruments used then. In 16th century there were used so-called logs for measuring the distances. Jenkinson used it for instance during his journey by Volga river, whereas for measuring the latitude he used so-called astrolabe, the most popular device in 16th century which enabled him to measure the position of the sun or stars. This instrument was still modernized and used until the 18th century. As far as the designation of the longitude is concerned, it was still a very difficult task at that time. Therefore Jenkinson in his map marked only latitudes on the frames of the map. The map has no cartographical net. It was the reason why the author used her own method in the form of a rectangular net to show the differences between the examined maps and in the same way to try to determine its projection. Simply because well-known distortion net is possible to be used only if the cartographical net is used. The method the author first time has shown in 1994 during the implementation of the grant Project given by so-called Polish "State Committee for Scientific Research" (in Polish KBN). Then, the author invented the method and shown in graphic form in two poster sessions on Zurich (1994) and Vienna Conference (Szykuła, K., 1995). Next in a sketch description in the publication from the national Conference in Pobierowo (Szykula, K., 2008). However, its final graphic result the author is going to present in her *"Monograph"* – then to be evidently proven by specialists of geodesy. Now short description should be presented to the readers to let them see what is a basic principle of this method.

The nets of triangles have been depicted on every one of the maps, including contemporary maps, which have been chosen to be compared with Jenkinson's and other 16th century maps of different authors. The number of triangles could be assumed by the author himself. In case of Jenkinson's map it has been used ten triangles. The points of cities, mouth of rivers and so on have been used as a vertexes of the triangles. The triangle nets to be compared were fixed in two position – first in natural position (according to the lower frame of the map), i.e. the given orientation and the second one according to one of the side of the triangle which has been chosen by the author, but fixed vertically for every one of the maps the same. In this way it was possible to observe how the whole given figure turned out and how subsequent triangles changed its angles, surfaces and sides of the triangles. Sometimes they changed not only their shape, but also vanished at all or they overlapped. Every figure with precisely measured angles of the triangles had been then introduced to the computer in a special Program. The results on the printed triangle nets were noticeable at the first glance. It was especially interesting when we turned the figures according to the same vertically fixed side of the given triangle. Then we could define closer an orientation of the map, as well as to compare the triangle nets depicted on contemporary maps, too.

Summing up the benefits of the method we can shortly mention them as follows: 1/ possibility to identify the non-existing projection, 2/ to notice the differences in the localization of the subsequent points as cities, mouth of rivers and so on, 3/ to define an orientation of the map and 4/ to find most uncorrected places in maps of our interest according to subsequent triangles.

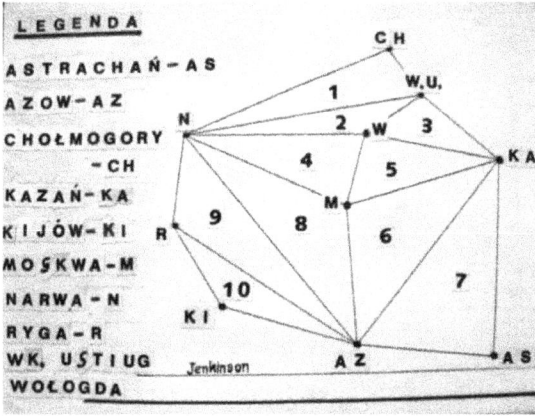

Figure 8. The triangle net from the Jenkinson genuine map - an initial (basic) figure to be compared with every other maps (the sketch and legend in Polish because it has been made in Polish Project then)

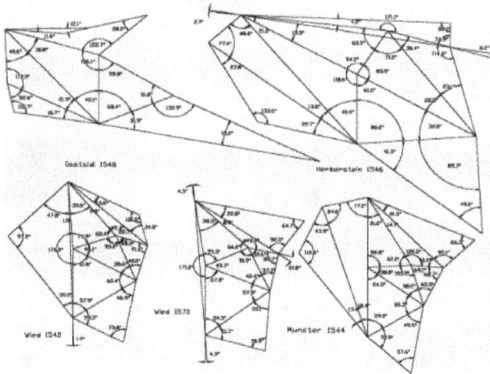

Figure 9. Triangle nets of the same region from maps of Russia by different authors (author of the whole conception Krystyna Szykuła, the triangle nets introduced to a computer program by Mariusz Ożarowski)

Going back to our map, the above mentioned method clearly has shown the innovativeness of our genuine map and the similarities between renditions and the genuine copy of the Jenkinson map. Amazing similarity the author has found, too, if a figure of the one of contemporary maps of the territory in question has been taken. Then it evidently occurred that possibly Jenkinson used a similar or even the same projection which has been used in the contemporary map.

From the geometrical point of view Jenkinson's map is undoubtedly quite innovative. The same conclusion concerns an orientation of the map when we compare two figures - taken for instance from Jenkinson's and Wied's map.

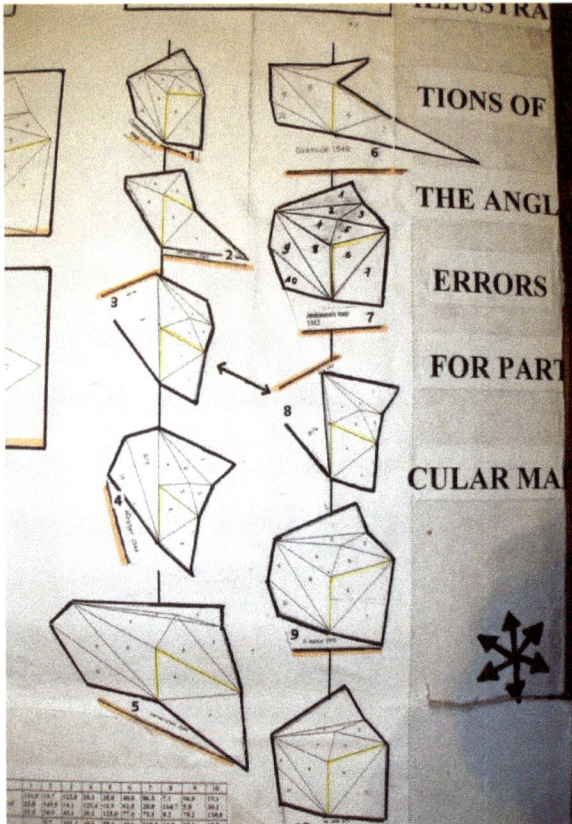

Figure 10. Triangle nets depicted on maps of different authors being compared with Jenkinson's map. They are fixed vertically according to one common axis Moscow-Azov. The fragment of the poster presentation in mentioned Zurich and Vienna Conferences

5.4. Ethnographic content on the Jenkinson's map, including its fauna

As far as the ethnographic elements in Jenkinson's map are concerned, it is exceptionally rich, although this is rather typical for the 16th century maps. However, it can be an outstanding source of information for the specialists of different fields[16]. In this respect the genuine copy significantly dominates over its renditions. Although fauna is usually joined with physiography, however, in this article the author decided to join it with ethnography because the human being from the earliest times has lived with animals to use them as the means of transport and, unfortunately, as the foodstuffs, too. As the means of transport we can see numerous camels and as the means of food we can see hunters with their trophies.

---

[16] During the Polish Project 1994/1995 the author found interesting work in Antwerp Plantin Museum by B. Vuylsteke, entitled "Het Theatrum orbis terrarium van Abraham Ortelius (1595). Een studie van decoratieve elementen en de gehistreerde voorstellingen (unpublished Ph.D. thesis), Louvain 1984.

We can also notice the connections between animals and people in so-called numerous genre scenes. There are camps of Cossacks, Tartars, Samoyeds depending on the given region in the map. From the human figures there are mostly warriors of different kinds depending on the region they are depicting. The warriors are equipped with a bows. Especially in the left part of the map we can notice many warriors who are shooting with the bows. In the east in turn we have already mentioned procession of the figures of sitting khans. There are also some dog-teams or deer-teams in the north of the map territory and camel-teams in the middle of the map – north of the Caspian Sea. In this fragment of the map we can see herds of sheep and tarpans, too. In the south-eastern part there are even panthers in Tashkent region. In the east are hordes of Tartars living in characteristic tents. Worth to draw attention to are two-wheeled carts, to which camels are hitched. Numerous horses are used to horse riding, furnished with bows or lances.

There are three religious scenes. First of them in the upper right corner close to camp in *"Colmack"* region (the Khanate of Sibir at that time), is the scene where the group of a few believers worships the sun. The comment on the map to this scene goes as follows: *"Molgomzaiani, Baidai, Colmachij, Ethinici ſunt, ſolem, uel rubrum pannum, de pertica ſuſpenſum adorant. In castris uitam ducunt, ac omnium animantium, erpenitum, uermiamque, ac proprio idiomate utuntur."* (transl. "Molgomzaians, Baidais, Colmachias, they are tribes, who worship the sun or a coat which is erected up on the perch. They lived in strongholds and use all the creatures as snakes and every of worms, as well as they use their own way of speaking or language")

The next object is worshipped by Samoyeds. It is so-called Golden Woman (Zlata Baba in Russian). The figure of the woman is sitting on some kind of pedestal and holding a small child. In Jenkinson's map it is situated in *"Obdora"* region, between the mouth of Ob toward the sea and the unnamed mountain range. Undoubtedly it is the North Ural. Quite a long description on the scene in Latin is placed on the opposite side of Ob river where we read: *"Zlata Baba id eſt (aurea uetula) ſedet, puerum ad genua tenens, qui nepos dicitur, ſtatua haec, ab Obdorianis, et Iogorianis, religioſe colitur. Qui laudatiſſimas, et maximi precij pelles Zebellinas Idole huic offerunt, una cum reliquis ferarum pellibus. Ceruos etiam ſacrificio mactant, quorum ſanguine, os, oculos, ac reliqua ſimulachri membra ungunt. Inteſtina uero etiam cruda deuorant, sacrificij autem tempore, ſacerdos Idolum consulit, quid ipſis faciendum, quoue ſit migrandum, ipſumque (dictu mirum) certa conſulentibus dat reſponsa, certique euentus conſequuntur."* (transl.: "Zlata Baba (Golden Woman), is seated and holds a boy-known as the ancestor at her knees. Obdorians and Iogorians worship the statue of Golden Woman and offer her their most valuable animal skins. They sacrifice deer to her, smearing the mouth, eyes, and other parts of the goddess with the animals' blood. They eat the entrails raw. During the sacrificial ceremony, their priest asks the goddess for advice, and strange to believe - receives credible answers, and certain incidents follow", see fig. 6)

Above mentioned pedestal with "Zlata Baba" has been used not only by Jenkinson but by Herberstein, Mercator and Wied in their maps as well. However, in every of these maps they differ from each other in subsequent representations. Accompanying descriptions on every of these representations on these maps are different as well.

Third interesting religious scene is the picture in Kirgiz region as follows: *"Kirgeſi gens eſt, quae cateruatim degit, id est in Hordis, aßidueque cum mhogholis gerit, habetque ritum iſtiuſmodi. Ipsorum untiſtes aut ſacrificus, quo tempore rem diuinam peragit, ſanguine, lucte et fimo iumentorum acceptis, ac terrae mixtis, ac in uas quiddam infusis, una cum hoc arborem ſcandit, atque hinc diu uel populum concionatus, in ſtultam plebeculum ſpergit. Populus uero in terram pronus, adorabunduſque, aspersiunculam hanc pro deo colit: firmeque credit, nihil eſſe perinde ſalutare ac terram, pecus, armentaque et cum quis inter eos diem obit, loco ſepulture arboribus ſuſpendit."* (transl. "Kirgiz are tribes who live in teams, i.e. hordes, they still are at war with Mongols and practice their own ritual: their priests when they are serving the God make an offering. They take the blood, milk and cattle excrements, mix them together with a soil and pour to the dish. Then, the priest takes this mixture and climbs up a tree and next has a long teaching talk he spreads the mixture blessing the people. For the people who are bowing to the ground it means to be worshipping the idol (God?). He probably considers that it is more important to bless from up than from the ground every animal or people. As far as the dead people are concerned they are hanging here on the trees instead of being buried.

We can classify the signatures of the cities rather to decorative elements, however they were popular in 16th century maps. The examples can be picturesque signatures of *"Vſtiuge"* (Veliki Ustiug), *"Tourſhock"* (Toržok) or *"Cazane gorode"* (Kazan)

5.5. Toponymy on the Jenkinson's map; the language and ortography

Toponymy and the orthography on the Jenkinson's map has been already considered by the author in her article (Szykula, K., 2010), published in a special jubilee volume devoted to the 50th work anniversary of Professor of the Stettin University Olga Molchanova = Molczanowa or Molčanova (*Ad Fontes*, 2010).

In the first part of the paper the author shared with the readers her remarks, doubts and difficulties, which accompanied her during the creation of the dictionary being in preparation. The cause of the difficulties were the differences in the orthography on every of the examined maps. We sometimes meet interesting phenomenon on some of the maps where the represented region is not a native for given cartographer. This happens for instance in case of Mercator's and de Jode's map when they marked the same cities many times because of the different versions of their names they met on several maps. In Mercator's map it concerns for instance Polish city Bydgoszcz which takes following names to be in fact the same city. There are: *"Bromberg"*, *"Bidgostia"*, *"Bizgelaw"* and *"Biltgotz"*. In de Jode's rendition of Jenkinson's map there are in turn: *"Bobroueſko"*, *"Bobranſko" and "Bobrouensko"* (to-day Bobruysk in Russian).

In the mentioned article the author was comparing the toponymy on different 16th century maps in their relation to the Jenkinson's map. Examined in this respect were Wied's map, by brother's Doetecum (so called Daškov[17] = Dashkow map), Sigismund Herberstein and

---

[17] Pavel Jakovlevich Dashkov (1849-1910) was the bibliographer and collector of the historical documents to St. Petersbourg in which the Doetecum map has survived and now is in the State Historical Museum in Moscow.

Giacomo Gastaldi (1500 - 1566) – map of 1551 (in the map itself the 1550 date has been written).

When we are examining toponymy we take into account such physiographic elements as: mountains, rivers, lakes, gulfs or bays, islands, peninsulas but as well as names of cities, ethnic or administrative names and every other elements that bear the geographical names in the maps.

The other phenomenon is when the names differ sometimes only in their orthography, but in some cases there are quite unrecognizable changes to the names. To mention some of them there are for instance: *"Kinieshma"* in temporary map, *"Kmysma"* in the Jenkinson's map, *"Kmÿshma"* in Ortelius' rendition and *"Kmijshma"* in de Jode's rendition. If we compare the genuine Jenkinson's map with Doetecum map there is for instance *"Choghloma"* in Jenkinson's and *"Czohloma"* in Doetecum map.

The case is very interesting if we compare Jenkinson's map with Wied's map. The latter used so-called Cyrillic script, not only for the names but for the quite comprehensive text situated in the bottom of the map as well.

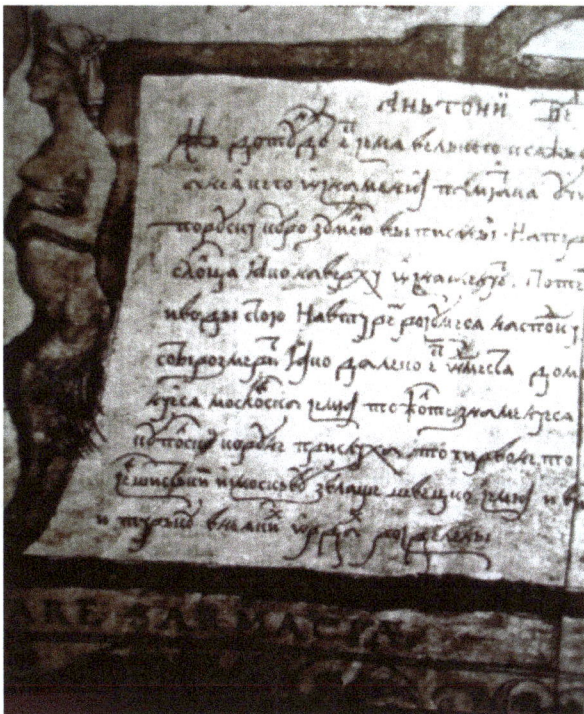

Figure 11. Fragment of the cartouche from the Wied's map of Russia

The names in question have double forms – in Latin and Cyrillic script. The next question is how the two editions of the Wied's map differ from each other. In his map of Moscovia published in 1542 there is the city *"Wollozeck"* , but in the edition from 1570 – there is the version *"Wolloſek"*. Interesting as well as is the name of then north sea in Jenkinson's and Wied's maps. Jenkinson called it *"Mare Septentrionale"*, whereas Wied (1542) – *"Mare Sarmacia"*.

Jenkinson's map in relation to Herbestein's and Doetecum map gave interesting results in case of the name "Kiev": in Jenkinson's the name of the city is *"Kiou"*, in Herberstein – *"Kiow"*, and in Doetecum – *"Kioff"*. In Wied's map we have even a typical Polish letter "ę" in the sentence *"Dux Moſcovię tranſfert"*. To summarize, the mentioned subject has been here only touched. In *"Monograph"*, however, it will be obviously extended. As far as the Borough's manuscript map is concerned we can find the similarities, but not in every respect. There are some cities or other geographical names which significantly differ geographically. We then may raise the question if Jenkinson used Borough's map at all as a model in this small north-western part of Russia.

6. Intriguing geographical elements in relation to the 16ᵗʰ century geographical picture of Russia

Taking into account the most intriguing geographical elements in the Jenkinson's map we have to mention: northern Volga river, lakes: *"volock lacvs"* and *"Kitaia Lacvs"*, as well as the rivers *"Amow fl."* and *"Ougus fl."*

Every of the mentioned elements has been already considered in Polish articles published by the author (Szykuła K., 1995 & 2000) and in one published in English (2005) – available in the Internet, too. Unfortunately, the latest has been limited by the editor and therefore it is there not in its original form and without most of the figures.

The problem of north Volga river has been already described in the subchapter on the physiographic elements. To remind it, the error is a consequence of the incorrect name of the Volga river, because instead of *"Volgha fl."* it should be there today's Neva, Swir and Volkhov rivers. Quite a rich history have already the so-called *"volock lacvs"*. We can find some information in very useful 18ᵗʰ century dictionary of geographical names *"Historisch=Politisch=Geographischer Atlas, 1774-1750"*. This is the German version of the French Lexicon by Bruzen de la Martiniere. Under the head *"Wolochs, Volock"* we read there (in transl.): *"the city in Russia, see "Wolocz"*, and there: *"a small city in State Russia in the province Rzeva, on the border of the Dutch of Moscovia, not far from the Fronowo Lake, on the outer edge "Wolkonsky Forest"* (Volkonski les in Russian). The description could be correct in respect of the place of the lake in question in the Wied's map, however, this author carefully left this lake without geographical name. Herberstein placed the Fronowo Lake at the source of Dvina as *"Dwina Lacq."* We find another explanation in the history of cartography by well known Polish historian of cartography Stanisław Alexandrowicz in his book on the history of cartography of Lithuania (1989, p. 57, footnote 25). He considers there that the incorrect information has been found in Polish historiographer's Jan Długosz *"Chorographia"*, and he quotes *"Annales seu cronicae"*, lib. I, p. 99. There the author of the work writes about a big lake or marsh which lays 30 miles from Smolensk towards Novgorod, where three rivers have their mouth: so called western Dvina, Volga and Dnieper. This information has been used again by famous Polish cartographer Bernard Wapowski. He placed the lake in question in his not surviving map of the Northern Sarmatia. Then, the information has been taken over again by Wied and placed in his map. Finally, the lake's name has been retaken by Jenkinson. According to Professor Alexandrowicz (1989) the name of the lake comes from the city *"Wyšnij Woloček"* (Wyshni Wolochek). In turn in above mentioned map from 1525 (Gerasimow-Jovius-Agnese map) this lake is called *"Palus magna"*, and there meets Volgha, Dnieper, Dvina, and additionally to Jenkinson's representation – the Neva river. Professor Samuel H. Baron (1993, p. 58, footnote 10) gives some other conception. He claims that this mistake comes from Gerasimov and then from Münster's map. This mistake could be also explained by the translation of the word *"volock"*– in Russian language it means "the carriage across the river". Then, it could be understood as the lake, especially that the terrain on which the lake is situated is full of marshes. The lake could be also identified with mentioned here Fronovo Lake, which is also mentioned by Jovius (Baron, S.H., 1986) and was confirmed above by the quotation from the Bruzen de la Martiniere Lexicon on the lake.

The next problematical element worth to be considered here is *"Kitaia Lacvs"* which has been already discussed several times in many articles, and which representation has been depicted on numerous maps. The number of conceptions, too, was presented. However, in spite of so many theories, which based on quite a real research results, it is difficult to resist an impression that both in shape and in its relation with the river Sur (Sir-Darya today), the lake can be automatically associated with the Aral Sea. Especially that it does not exist at all in its proper place in the Jenkinson's genuine map.

Figure 12. "Kitaia Lacvs" and Ob river in Jenkinson's map

Quite different but interesting is the attitude of the Russian scholars to this open question. In *"Zapiski"* by Herberstein (Gerberstein, S., 1988, footnote 546, 547), there are following theories presented by different scholars. As far as the Zajsan lake is concerned it has been considered by A.F. Middendorf as an *"Upsa=Ubsa Nur"* or *"Uvs nuur"* by G. Genning and M. P. Alekseev, as the Aral sea by G. Michow , L.S. Berg, and K.M. Ber, and as the Teleckoye lake by A.Ch. Lerberg and D. N. Anucin. To the mentioned conception refer another Russian scholar, Vadim F. Starkov (1994), who writes about the participation of Gerasimov in creating the theories, like that of above mentioned professor Baron. Aleksey K. Zaytsev focuses on *"Teleckoye Lake"*. He explains that Teleckoye Lake is placed on the way of the Ob river, or rather its tributary called Bija, which joins with the Kotunia river and they create the Ob river. The note we can consider as the real one if we take into account mentioned here *"Historisch=Politisch=Geographische Atlas"* (Bd. 1, 1744,; Bd.4, 1746,), where under the entry we read (in transl. from German): *"Kitaius lacus"*, so is in Latin named a great lake in the Kingdom of Altin, see "Altin" and then under the entry *"Altin"* we read – so is named (by somebody) the lake, which is situated in the eastern part of the kingdom of the same name. The completion of this information we find in another geographical dictionary (Šchekatov, 1808, č. 6, v. X, columns 164-165), where under the entry of the "lake" we read that the lake bears as well the name *"Altyn"* or *"Altaj"*[18] and is situated in Tomsk Gubernya, and Kuznieck district. Having so many conceptions, it is necessary to consider their reliability. Then, we should look at lakes' geographical placement. Every one of the lakes is situated in the mountain region, close to the Altai mountains. When we look at the Jenkinson's map, east of Kitaia Lacvs, he depicted high mountain range. As has been already said, Teleckoe Lake is situated on the way of Bija river – the tributary of the Ob river, whereas the Zaisan Lake on the Irtysh river the tributary of the Ob river, too. Only Upsa is not connected with the Ob river. The latter is in most degree in the shape of *"Kitaia Lacvs"*, because two other have, using the geographical vocabulary, the shape of finger lakes. There are, however, many indications against these theories on the three lakes. Mainly because they are situated far away to the east from the territory represented in the Jenkinson's map, i.e. above 1500 km. from the Aral Sea. The Altai mountains are situated to the south-eastern direction, but not in northern Siberia. On the other hand, the Teleckoye lake could be accepted to be *"Kitaia Lacvs"* because of the reasonable argument which has been already mentioned above and quoted from the geographical Lexicon by Bruzen de la Martiniere. Simultaneously we know that neither Jenkinson nor Herberstein entered so far in the Asia interior. They had been not able to get in so high ranges of mountains and because of wild tribes living east of Bokhara region. Then, to be remembered, there is yet the possibility that the Lake was retaken from Gerasimov, what has been already mentioned in Starkov's work (1994) and confirmed by S.H. Baron (see above).

Then, maybe we should return to the Aral Sea conception. As it was already mentioned about its shape and the course of river Sur (Syr-Darya) there is *"Amow"* tributary of *"Sur"* river, too. It seems very unlikely if Jenkinson, who travelled across the territory between the Mare Caspium and Tashkent including Bukhara region, did not notice such a great lake as the Aral Sea. We are obviously not talking about today's drying up lake.

---

[18] Special work has been devoted to this "country" by two authors: Borodaev, V.B. & Kontev, A.V., Istoričeskij atlas Altajskogo kraja. Kartografičeskie materialy po istorii Verchnego Priob'ja i Priirtyš'ja (ot antičnosti do načala XXI veka). Vtoroe izdanie, ispravlennoe i dopolnennoe". Barnaul, Azbuka 2007 (in the Bibliography here, English title, too).

However, coming back to the previous thought, that is *"Kitaia Lacus"*, at the times thanks to brothers Boroughs' voyages (Baron, S.H., 1989 & Mayers, K., 2005) there was known only the lower course of the Ob river, and yet before the Jenkinson's map has been made. Then it is very probable, that Jenkinson, who did not know the further course of the Ob river, could add up information acquired by the mentioned brothers with his own information collected during his second travel to Russia, when he reached Bokhara. Hence, he made similar mistake as Wied on his map - to join two pieces of news together - because he did not know the territory between the lower Ob river and the Ougus river. That is why there is also Sur extended to the south and the Aral Sea shifted too much to the north that was Wied's some kind of idea or trick done by eliminating the unknown territory.

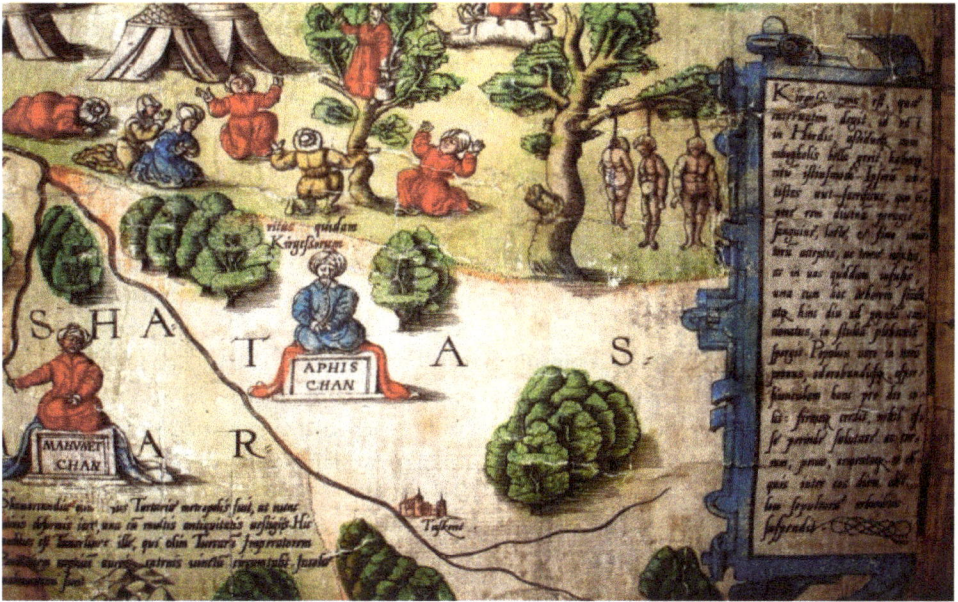

Figure 13. Funeral ceremony by Kirghizes

On the other hand there are still many unclear points in the presented "discussion". We should raise the question, why under the entry *"Arall"* in the quoted dictionary by Bruzen de la Martiniere there is only a short description on the settlement of the same name. *"Arall"* as a sea is only briefly mentioned there, however in the same dictionary is that the view is so great as the sea! Another argument which indicates that the Kitaia Lake is the Aral Sea is the opinion expressed by zoologist J. Bartmańska[19], who claims that these kinds of animals as camels, tarpans and sheep should be presented on the latitudes $45^0 - 52^0$ but not on the $60^0 - 65^0$. However, some of them are situated in a correct place, too, for instance in northern direction of the Caspian Sea in the genuine map. We know that the latitude of the Aral Sea is about 45^0 and there is Kyzyl-Kum desert but not the region of the high mountains as it is

---

19 According to the manuscript expertise made during the Polish grant (1994/95) in the form of the order.

shown in the Jenkinson's map. Whereas the existence of the camels and sheep on the same latitude as *"Kitaia Lacvs"* and between the mountains steppes indicate in turn the Altai mountains. Then, as we can see, solving the problem is not so easy and probably it will be never solved. Some final conclusion could be supported, too, by the description of the name *"Kitaya"* in Wikipedia under the entry *"Kitay-gorod"*.

There is one more interesting geographically questionable element in Jenkinson's map. This is the Amu-Darya river which as Ougus flows into the Caspian Sea. The problem have been already interestingly and accurately described by many authors (Menn G.F.C., 1839, Alenicin V.D., 1879, Barthold W., 1910). As have been here already mentioned the representation of the mouth of Ougus river can still be found in the 18th century maps. The history of the mouth of Amu-Darya from the oldest times had been very accurately described by G.F.C. Menn in Latin[20]. There are some interesting testimonies which we can find in Herodot's work in following sentence (translated: "the biggest river in this region is *"Arakses"* or *"Oksos"* and that one of the branch flows into the Caspian Sea"). What is interesting, he also calls the river as "marshes of Aral". Alenicin in turn asks himself a question – which way Jenkinson has gone when, as he writes in his diary, he met so-called "priasna" water, i.e. sweet water. The author comes to conclusion that at some point Jenkinson had to confuse the directions of the world. Alenicin realized that if Jenkinson met a sweet water it could be neither the Aral Sea nor the Caspian Sea – it must have been probably Sary-Kamysh – a big lake situated between the Caspian Sea and the Aral Sea, because it goes from Mangyshlak peninsula across the Ust'- Urt'. On the 12 page of his book Alenicin writes about yet another conception. He there claims that in 1878 came into the Caspian Sea a branch of the river which probably reached it. Interesting description on Amow river (a tributary of the Sur river on Jenkinson's map) we read in already quoted geographical dictionary (transl.: *"Amou or Amu it is the river of Asia, which by our contemporary geographers is named Amu. Because "Ab" word in Persian means water or river. Arabs call it Gibon, but accurately Balkh, river Balkh, because it flows across city of the same name. The old called it Oxus and Bactus. This river flows out from the Imaus mountains and directs its course from the east to the west. As a matter of fact, when it comes close to the Khovarezm country, it runs in meander way, and seems to flow to its source direction, however it comes back again, and flows into Caspian Sea in the west"*). It is worth here to quote before mentioned Menn (G.F.C., 1839). In his book in the chapter I entitled *"Oxi fluminis vetustae navigationis in mare Caspium documenta"* on 5th page we read (transl.:"...Oxus flows into the Caspian Sea across Scythia..." W. Barthold (1910, s. 68), who discusses the Jenkinson's map writes (in transl.: "south branch flows to the lengthened gulf of the Caspian Sea, i.e. Sary-Kamysh" (see Tolstov 1962, pp. 261-267) where the author writes that the river Amu-daria at the beginning of the 10th century had flown into the Caspian Sea. This Russian archeologist Tolstov (1953, p. 62) recalls Jenkinson's opinions from his diary, which are opposite to his map's picture. Namely, Jenkinson claims that the water in his times did not flow into the Caspian Sea as it was in the early times. So

---

[20] For translation of the fragment of this interesting work I would here like to express my gratitude to Aleksandra Krajczyk, the teacher of Latin at the University of Wrocław.

we may presume, that the picture of the Ougus river in his map has been simply retaken from Ptolemaeus' map of this territory. The history of this early course of Amu-Darya comprehensively describes S.P. Tolstov in his another book (1962, pp. 17 – 26), where he gives the bibliography of this subject.

Every one of these theories could be taken into account in the light of the interesting text quoted in Internet from the famous book by Ryszard Kapuściński. In this description we are told a beautiful story of the Uzboj river, which was examinated by yet another Russian archeologist A.H. Jusupov. The story has been told to Kapuścinski by a local called "Raszyd"(Rashyd). The author writes: (transl."Raszyd has shown me on the map the course of the Uzboy river. The river Uzboy has taken its waters from the Amu-Darya river, the river has flown across the desert Kara-Kum into the Caspian Sea. It was a beautiful river – long as the Seine river. This river died, as he said, and from the time of its death the war begun.")

7. Coclusion

In the article, the author tried to show a kind of sketch of the further full *Monograph* on the Jenkinson's map. The first conclusion which is easily noticeable is that the subject in question constitutes an inexhaustible material for investigation from different points of view. What one could notice if we enter more deeply into some of the described questions is that every one of them opens new paths to be further investigated. After all, we have here following questions: the history of geographical discoveries in Asia, relations between the two continents, Europe and Russia in relation to Asia, and next the subsequent regions – their history, ethnography, especially connected with Cossacks and Tartars, the history of links between England and Russia[21], history of Persia, including the history of consecutive khans. In this subject it is difficult to neglect the Russian relations to the western frontier states or historical regions which, especially at that times, were Lithuania and Poland. Characteristic and equally very interesting is that the stories of several regions and states connects with one another. Thanks to this phenomenon new topics still emerge and it is difficult to resist them. New and equally very interesting riddles still arise. The sources which needed to be compared were very exciting for the author. Such was for instance the story of the course of the Amu-Darya river and the Aral Sea in connection with the "*Kitaia Lacus*" as well. Very interesting is the interdisciplinary character of the subject in question, too. For instance, some evidence is the participation of the archeologists in the examination of the bed of the Amu-Darya river, its significance and results. To continue the idea presented by the author it could be also worth using the infra-red pictures to confirm one more argument for the theory of the old river-bed of Amu-Darya. We can say the same about the other questions considered here. The long-lasting investigation of the author and the rich literature create valuable occasion for scientific contacts with other authors in the world. It was a very valuable exchange with mentioned here Professor S.H. Baron, Dr. K.

---

[21] This investigation entitled „*Cartographical links between England and Russia in mid of 16th century*" has been carried out in the British Academy Project by the author together with Magdalena Peszko in 2011.

Mayers and Dr. Osipov Igor A. (2008 and 2009). We can notice that the papers by many other scholars of the western Europe and of the USA already mentioned here that were issued brought benefits for the subject considered here from the time since the Jenkinson's map has been rediscovered (1987). The subsequent papers by the author during the period as well tried to enlarge the field of the investigation and still are bringing new reflections.

Therefore the author hopes that the full version of *Monograph* will bring much more valuable results and will do its good for the Jenkinson's map and its examination.

At the end, summing up the benefit of the rediscovered genuine copy of the map in question we should answer the question "what has its recovery brought?" There are as follows:

1. Quite a new image of the Jenkinson's map – both in general view and its size.
2. The confirmation of the territorial range which has been portrayed by Ortelius and territorial discrepancy of de Jode's rendition.
3. The differences in arrangement, number and the contents of cartouches and boxed texts placed in them in every of these maps.
4. The knowledge of the two new co-creators of the genuine copy of the map. They are the engraver Nicolaus Reinoldus and the editor Clement Adams, who both to the very moment of rediscovery were not connected with the map in question. The notice directed to the printer Reginald Wolf and the painter Jan de Schille should be not neglected as well.
5. The knowledge of the much richer content in relation to the original and renditions in all elements as: genre scenes, boxed texts, decorative borders and so on.
6. Jenkinson was the first who determined the distance and the direction to China by means of the description which he placed in the lower right-hand corner as follows (in translation from Latin to Polish and then to English): "*Thirty days of the travel in the eastern direction from Kashkhar begins the frontiers of the Chinese Empire (Cataye). From this frontier is three-month of travel to Cumbalcu*"(= Cambaluc – the early name of Beijing, now Peking). However, we also know that in the place of Siberia region "*Regnum Cathay*" Waldseemüller puts in his world map of 1513. Hence maybe "*Kitaia Lacvs*" taken his name.
7. We can presume, how it happened that the genuine Jenkinson's copy has been found in Wrocław city. Probably it was sent from Antwerp or from Ortelius himself to one of the learned officials in the city who were in close relation at that time. The author could drawn this conclusion thanks to the examinations by Curator of maps Joost Depuydt at Antwerp FelixArchief who carried out the investigation on Ortelius' activity. However, this is not yet proved by appropriate document because of the huge amount of correspondence to be read and dispersed in different places of the world, among others in Wrocław University Library or Harry Ranson Humanities Research Center (from the paper kindly sent to the author by Professor Christian Coppens)
8. Unfortunately, an information on one more copy of the genuine Jenkinson's map which was said to be found in State Library in Leningrad could be sensational if trustworthy. It appears to be untruthfull, and has been given by Leo Bagrow in his article in Imago Mundi 5, 1965, p. 62 . Reprint edition of 1925 I.M.

All reproduced photographs presented in this Chapter were made by Krystyna Szykuła.

Author details

Krystyna Szykuła
Wrocław University Library Department of the Cartographical Collection (retired), Wrocław, Poland

Acknowledgement

For supporting of every works connected with the library collections, author of the Chapter wish to thank to her Director Grażyna Piotrowicz and her Successors in the Department of Cartographic Collections of Wrocław University Library – Anna Osowska and to Dariusz Przybytek – the Head of the Department.

8. References

A book of Heraldry, (1566). (manuscript) Cambrigde University Library. Kk.I.26., found during the Polish KBN Project 1994/1995

Alenicin, V.D. (1879). Neskol'ko zamiečanij o putešestvii Dženkinsona v' Chivu v' 1559 godu, V. Bezobrazov i K º, S. Peter'Burg.

Alexandrowicz, S. (1989). *Rozwój kartografii Wielkiego Księstwa Litewskiego od XV do połowy XVIII wieku*. Wyd. 2   popr., Uniwersytet im. Adama Mickiewicza, ISBN 83-232-0239-7, ISSN 0554-8217, Poznań.

Bagrow, L. (1928). *Anthony Jenkinson*, [w:] *A. Ortelius Catalogus Cartographorum*. 1.Teil: Ergänzungsheft 199 zu Petermanns Mitteilungen, Justus Perthes, Gotha, pp. 120 – 121.

Bagrow, L. (1962). *At the source of the cartography of Russia*, Imago Mundi, 16, pp. 33-48.

Bagrow, L. (1964). *History of cartography*. Revised and enlarged by R.A. Skelton, Harvard University Press, Cambridge, Mass., ISBN 0913750336.

Bagrow, L. (1965). *A page from the history of distribution of maps*. Imago Mundi, 5, pp. 53-62.

Bagrow, L. (1965). *Italians on the Caspian*, Imago Mundi, 13, pp. 2-10.

Bagrow, L. (1975). *A history of cartography of Russia up to 1600*. Ed. by H.W. Castner, Walker Press, ISBN 0969051409, Wolfe Island, Ontario.

Barber, P. (1989). *Additional information on Jenkinson map*, The Map Collector, 48, p. 39.

Baron, S.H. (1986). *Herberstein and the english „discovery" of Muscovy*, Terrae Incognitae, 18, pp. 43 – 54.

Baron, S.H. (1989). William Borough and the Jenkinson map of Russia (1562), Cartographica 26, 2, pp. 72 – 85.

Baron, S.H. (1993). The lost Jenkinson map of Russia (1562). Recovered, redated and retitled, Terrae Incognitae, 25, pp. 53 – 66.

Baron, S.H. (1993). B.A.Rybakov on the Jenkinson map of 1562, [w:] New perspectives on muscovite history. Selected papers from the Fourth World Congress for Soviet and East European Studies. Harrogate 1990. Ed. by L. Hughes, St. Martin Press, London, pp. 3 – 13.

Barthold, W. (1910). Nachrichten über den Aral-See und den unteren Lauf des Amu-darja von den ältesten Zeiten bis zum XVII. Jahrhundert (transl. from Russ.) Otto Wigand, Leipzig.

Borodaev, V.B. & Kontev, A.V. (2007). Istoričeskij atlas Altajskogo kraja. Kartografičeskie materialy po istorii Verchnego Priob'ja i Priirtyš'ja (ot antičnosti do načala XXI veka). Vtoroe izdanie, ispravlennoe i dopolnennoe". ISBN 978-5-93957-198-2, Barnaul, Azbuka 2007 (English title, too): The Historical Atlas of the Altai region. Cartographic materials on the history of the Ob and the Irtysh rivers upper reaches (from antiquity to the early XXI century. Second edition, corrected and added.

Gerberštein, S. (1988). *Zapiski o Moskovii*. Vstup. st. A. L. Choroškevič, primeč. S.V. Dumina i dr., Izdatel'stvo Moskovskogo universiteta, [Moskva].

Hakluyt, R. (1589, Repr.1965). *The principall navigations voiages and discoveries of the english nation. Imprinted at London 1589*. A photolithographic facsimile with an introduction by D.B. Quinn and R.A. Skelton, University Press, Cambridge.

Hakluyt, R. 1907, *Voyages*. Vol.1. Introduction by John Masefield, J.M. Dent & Sons Ltd., London.

Herberstein, S. (1557) Rerum Moscoviticarum Commentarii Sigismundo Liberi authore, Antverpiae.

Hessels, J.H. (1887) *Abrahami Ortelii (geographi Antverpiensis) et virorum eruditorum ad eundem et ad Jacobum Colium Ortelianum (Abrahami Ortelii sororis filium) epistulae. Cum aliquot aliis epistulis et tractatibus quibusdam ab utroque collectis (1524-1628). Ex autographis mandante ecclesia Londino-Batava, Cantabrigiae Typis Academiae, [Cantabriga].*

Historisch=Politisch=Geographischer Atlas der gantzen Welt; Oder Grosses und vollständiges geographisch=und critisches Lexicon, (...) Aus des berühmten Königl. spanischen Geographi Mr. Bruzen la Martiniere (...) ins Deutsche übersetzt, 1744-1750, Johann Samuel Heinsius, Leipzig.

Jäger, E. (1989). Die XIII. Internationale Konferenz zur Geschichte der Kartographie in Amsterdam, Nordost-Archiv 22, 96, pp. 239 – 245.

Karrow, R. W. (1989). Amsterdam Conference on the History of Cartography, Mapline, 55, pp. 1 – 5.

Karrow, R. W. (1993). *Anthony Jenkinson.[in:] Mapmakers of the sixteenth century and their maps. Bio-bibliographies of the cartographers of Abraham Ortelius, 1570. Based on Leo Bagrow's A. Ortelii Catalogus Cartographorum, Orbis Press, Chicago, pp. 317-318.*

Keuning, J. (1956). *Map of Russia,* Imago Mundi, 13, pp. 172 – 175.

Kordt, V.A. (1899). Karta Rossii A. Dženkinsona 1562, [in:] Materialy po istorii russkoj kartografii. Vyp. I. Karty vsej Rossii i južnych' oblastej do poloviny XVII veka, S.V. Kulženko, Kiev, pp. 10 – 11.

Mayers, K. (2005). North-East Passage to Muscovy. Stephen Borough and the First Tudor Explorations. Sutton Publishing Limited. ISBN 0-7509-4069-7

Menn, G.F.C. Rhenani. (1839). Meletematum historicorum praemiis regiis ornatorum specimen duplex, apud Eduardum Weber, Bonnae.

Meurer, P.H. (1991). Fontes Cartographici Orteliani. Das „Theatrum Orbis Terrarum" von Abraham Ortelius und seine Kartenquellen, VCH Acta humaniora, ISBN: 3527177272, Weinheim, pp. 175 – 176.

Michow, H. von. (1906). Karte des Anton Jenkinson von 1562, [in:] Das erste Jahrhundert russischer Kartographie 1525-1631 und die Originalkarte des Anton Wied von 1542, Mitteilungen der Geographischen Gesellschaft in Hamburg 21, pp. 1 – 61.

Molchanova, O.T. (2010). *Ad Fontes: Księga jubileuszowa ofiarowana Profesor Oldze Mołczanowej,* pod redakcją naukową Ewy Komorowskiej i Patrycji Kamińskiej.: Volumina.pl Daniel Krzanowski ISBN 978-83-62355-19-8, Szczecin.

Morgan, E.D. & Coote, H. (1886). Early Voyages and Travels to Russia and Persia, by Anthony Jenkinson and other Englishmen.

Morton, M.B.G. (1962). *The Jenkinson story,* Scottish National Press by William Maclellan, Glasgow.

Oakeshott, W. (1984). *A Tudor explorer and his map of Russia,* The Times. Literatury Supplement, 22 June, p. 238.

Osipov, I.A. (2008). *Antonij Dženkinson i karta Rossii 1562 goda.* Wydawca I.A. Osipov, Syktyvkar 2008, 46 p.

Osipov, I.A. (2009). Opisanija Rossijskogo Gosudarstva XVI veka: vlijanie kartografičeskich materialov na narrativnye istočniki. The Editor I.A. Osipov. Syktyvkar.

Possevino A. (1587). Moscovia Antoni Possevini Societatis Iesu, Antverpiae.

Postnikov A.V. (2000). *Outline of the History of Russian Cartography. 1. Introduction (Internet)*

Retish, A.B. (1995). A foreign perception of Russia. An analysis of Anthony Jenkinson's map of Russia, Muscovy and Tartaria, The Portolan, 33, pp. 1 – 10.

Rybakov, B.A. (1974). *Čertež Moskovskich zemel' 1497 goda (karta A. Dženkinsona 1562 goda), [w:] Russkie karty Moskovii XV/načala XVI veka, Moskva.*

Rybakov, B.A. (1974). *Drevnejše russkie karty Moskovii,* Nauka i žizn', 4, pp. 33 – 38.

Rybakov, B.A. (1994). *Novoodkrytaja karta Moskovii 1525 g.,* Otečestvennye archivy, 4, pp. 3-8.

Sager, P. *(1974). Zu Anthony Jenkinson's Karte von Russia (1562), Nordost-Archiv 7, 32, pp. 1 – 7.*

Ščekatov, A. (1808). Slovar' geografičeskij Rossijskogo gosudarstva. Č.6, T.10, Moskva.

Scott, V.G. (1989). *Map of Russia revealed at conference,* The Map Collector, 48, pp. 38 -39.

Scott, V.G. (1990). *The Jenkinson map,* The Map Collector, 52, p. 29.

Starkov, V.F. (1994). *Opisanie karty 1525 g.,* Otečestvennye Archivy, 4, pp. 8 – 15.

Tolstov, S.P. (1962). Po drevnim del'tam Oksa i Jaksarta, Moskva.1953, *Śladami cywilizacji starożytnego Chorezmu* [transl. From Russian.], Czytelnik, Warszawa.

Vuylsteke, B. (1984). Het Theatrum orbis terrarum van Abraham Ortelius (1595). Een studie van decoratieve elementen en de gehistrieerde voorstelligen. Del 1-2, Leuven, manuscript PhD Thesis, Museum Plantin-Moretus, Antwerpen.

Vaughn, E.V. (1912). English Trading Expeditions into Asia under the Authority of the Muscovy Company (1557-1581), In:. Gerson, Armand J., Studies in the history of English commerce in the Tudor period. 1912, Part III.

Krystyna Szykuła's publications on the Jenkinson's map

(1989). *The Newly Jenkinson's Map of Russia of 1562.* W: 13th International Conference on the History of Cartography. Amsterdam and the Hague. June 26 to July 1, 1989. Abstracts. Amsterdam 1989, pp. 109 - 111, Size of the book A4 .

(1990). *Odnalezienie mapy Rosji Jenkinsona z 1562 roku.* Polski Przegląd Kartograficzny. 1990. T. 22, nr pp. 81 – 82.

(1995). *Jenkinson's Map of Russia of 1562. Further investigations.* W: 16th International Conference on the History of Cartography. Vienna (Austria), 11-16 September1995, Abstracts, p.105, size of the book A4.

(1995). *Odnaleziona mapa Rosji Jenkinsona z 1562 roku. Pierwsza próba analizy mapy.* Acta Universitatis Wratislaviensis nr 1678. Bibliothecalia Wratislaviensia. Wrocław 1995,T.2, pp. 7 – 31.

(2000). *Mapa Rosji Jenkinsona (1562) – kolejne podsumowanie wyników badań.* Czasopismo Geograficzne, Wrocław 2000, T. 71, Z. 1, pp. 67 – 97.

(2005). *The Jenkinson Map of Russia (1562) A Research Summary.* Polish Library Today. Foreign Collections in Polish Libraries. Biblioteka Narodowa. Warszawa 2005, Vol.6, pp. 57 - 69 (It is an abbreviation of the previous above mentioned paper in Polish in "Czasopismo geograficzne", and without important illustrations). Therefore it is better to read it in original form, i.e. in Polish!).

(2007). *Anthony Jenkinson's unique wall map of Russia (1562) and its influence on European cartography.* Abstract in: Formating Europe – Mapping a Continent, Proceedings of the 6th International BIMCC Conference at the Royal Library of Belgium Brussels 16 November 2007 pp. 13-16 in size A4.

(2008). *Anthony Jenkinson's unique wall map of Russia (1562) and its influence on European cartography, Belgeo 2008, pp. 325 - 340.*

(2008). *Projekt monografii i faksymilowego wydania mapy Jenkinsona z 1562 roku.* W: Dawna mapa źródłem wiedzy o świecie. Pod redakcją Stanisława Alexandrowicza i Radosława Skryckiego. Szczecin 2008, s. 149 - 160 (the paper is a final result of author's lecture given on the national conference in Pobierowo).

(2009). Problem nazewnictwa geograficznego i celowości opracowania słownika nazw geograficznych na bazie XVI-wiecznej mapy Rosji Anthony'ego Jenkinsona. W: Ad fontes. Księga jubileuszowa ofiarowana profesor Oldze T. Mołczanowej pod redakcją naukową Ewy Komorowskiej i Patrycji Kamińskiej. Szczecin 2010, pp. 353 – 371.(2012).Sensacyjny nabytek kartograficznych zbiorów Biblioteki Uniwersyteckiej we Wrocławiu – mapa zachodniej Rosji Anthony'ego Jenkinsona. 1562. Iluminacje. Kwartalnik o iluminowanych manuskryptach. 1/2012 (1), pp 33 – 36.

Chapter 8

Contribution of New Sensors to Cartography

Carla Bernadete Madureira Cruz and Rafael Silva de Barros

Additional information is available at the end of the chapter

1. Introduction

Different cartographic representations are considered fundamental for the Geographic Science, whose characteristics are related, almost always, with the need for representation and analysis of spatially distributed phenomena. Therefore, it is fundamental that the cartographic bases are available, meeting the needs regarding the accuracy and updating, in different scales.

Generation and maintenance of these bases are considered as highly complex, considering all the planialtimetric information that they should contain, mainly in the case of a large country as Brazil, whose territory extends for about 8.5 million km², with many areas of difficult access. Furthermore, there is a need for concern also with the thematic maps, whose origins depend on the cartographic bases themselves, which operate as spatial reference, besides other inputs originated normally from field surveys and remote sensing. The diversity of these maps is enormous, creating an immense variety of methods and specialists involved in their preparation.

With all this progress, cartography is presenting growing changes in the form of preparation and divulging of its products. The availability of new technologies, for acquisition, storage and dissemination of spatial data, made the use of maps and images increasingly popular, reaching not only government organizations and private companies, as habitual, but also individual users. On the other hand, this same technological progress has entailed a new context of methodological solutions, some of which presenting themselves as interesting alternatives for the process of maps generation, despite presenting also many future challenges. Changes to the way work is carried out are fundamental to attend viable solutions in ongoing applications, in the long-term, that also require a high level of quality, as those involving systematic cartography.

It is a fact that the consolidation of the Information Technology in practically all areas of knowledge, with the consequent cost reduction regarding software and hardware, was fundamental in this evolutionary process. Despite the intense and growing offer of technologies, the adsorption of the new methods and products, which were slowly converted from analog to digital, involves countless challenges, being slower than expected or desired. It should be highlighted also that the conversion requires the review of many concepts that need to be adapted to the new reality, as in the case of the sheets of charts in a certain scale in view of the challenge of preparation of continuous bases.

Therefore, an interesting new panorama is configured, which presents a diversity of uses with a growing demand for Geographic Information, associated, increasingly, with the use of Geographic Information Systems and products originated from Remote Sensing. The lack of control of individual actions, without the support of a detailed analysis on the real potentialities of the new products, can entail serious problems caused by inappropriate use of methods and/or inputs, making a wide discussion on the theme urgent, with establishment of a set of rules that supports such discussions.

Considering the images from Remote Sensing as growingly important inputs for cartography, it is emphasized that the diversity of new remote sensors, at different acquisition levels, and values of resolution and accuracy, allows to meet a variety of reference and thematic maps, in several scales.

2. Considerations on remote sensing in cartography

The use of remote sensing as important data source has been a reality for several decades, is ongoing and will probably be part of the everyday life of the population for many years to come - although most people are not even aware of it. The aerial level has supplied great volumes of data for cartography since the beginning of the 20th Century, whether in military or civilian applications, such as reference and thematic maps. In its evolution, it is also considered that the great revolution of the orbital phase has as main marker the launching of satellites focused on natural resources in the beginning of the 1970's.

In Brazil, practically all existent officially-based cartography was produced from aerial photos and, despite receiving satellite images from Landsat on the Cuiabá (MT) land station since 1973 in [1], only recently its use as input was adopted for the generation of reference cartographic bases.

The diffusion of micro-information technology also helped to increase the availability of free or low-cost images aimed at remote sensing image processing, contributing to a significant increase in the number and diversity of uses and applications. This growth occurred both in reference cartography as, mainly, in thematic cartography. Perhaps the climax of the dissemination occurred with the launching of Google Earth in 2005 in [2], making such products available for non-specialist users.

If on one hand such dissemination carted many benefits for most of the users (whether specialized or not), on the other hand it may contribute for the generation of products which do not present the expected, and often, required quality. Ignoring the phase of amazement which is common after launching products which in some way become object of consumerism, the products derived from remote sensing images may end up presenting a lower quality than desired, for, among other reasons, generating an expectation of intrinsic good quality. This may occur for being considered something modern, conveying the impression that the leading-edge technology is synonymous of - or is very near to present - guaranteed quality. In an opposite vision of the former, it is noted that there is a group of users who reveal suspicion or misbelief regarding the promised potentialities. Probably this occurs for different reasons, such as insufficient information, prejudice, or still, concerns with loss of jobs. Such resistances delay the renewal of methods and techniques, ending up to configure "pre-judgments" that end up making the performance of an analysis difficult, even in an emergency character, regarding their real efficacy. The results of our studies offer greater clarifications on products originated from different sensors, leading the reader to the need to break myths - pessimistic or optimistic - regarding the use of remote sensing, mainly at orbital level for cartographic ends.

It is important also to consider that during recent years a lot of investment was carried out in new sensors, aiming to meet different applications. It is in place to state that we are being overrun by growing offers of novelties associated to many promises. The universe of remote sensors aimed at terrestrial studies is therefore very wide, involving products with several and variable spatial, radiometric, spectral and temporal resolutions. Such diversity allows the production of mappings with different detail and accuracy levels, of greater or minor complexity of legend and the possibility to follow up dynamic phenomena.

Another relevant consideration is the fact that many sensors allow also the extraction of 3-D data, also with different resolutions and accuracy levels. One fundamental mark in this area is undoubtedly SRTM (Shuttle Radar Topography Mission), which despite not having been the pioneer, is the most encompassing and popular Digital Elevation Model (DEM) available worldwide. Consequently, there is a growing need to study the quality of the altimetry obtained from such sensors, and also an assessment of how much the implemented advances in the data acquisition process can influence the results, as for instance the along-track stereoscopy (images acquired during satellite orbit) and interferometry.

A differentiation between DEM and Digital Terrain Models (DTM) can be useful, because often this is one of the arguments against the use of orbital images for the generation of terrain representations. In a simplified way, it will be assumed that the DEMs represent the land surface added by any existent objects on it and which influence the value of the pixel reflectance. In other words if there are trees and constructions, the surface represented refers to the top of them. The DTMs on the other hand, represent the actual ground surface.

Despite so many investments in orbital sensors, the airborne surveys still offer greater accuracy for their products. It is emphasized also that there are some benefits of the active sensors in relation to passive sensors, as the case of the interferometer radar and LIDAR (LIght Detection And Ranging), which are not subject to atmospheric interferences and allow extraction of DEMs and DTMs - depending on the band for radar and the density of points for LIDAR.

In the case of new methodologies for generation of DEM/DTM to meet the expectations, reaching greater detail scales, an important step was taken in the sense of generating data with a good cost/benefit relation, mainly regarding time reduction, if compared with conventional methodologies. In such case the topographic features can be obtained for the most extensive areas of our country, meeting a demand of great part of the studies and activities related to topographic representation.

The interest in investigation on altimetry spawned by orbital products arises from the fact that planimetry has already been studied, both in meeting thematic and reference cartography. Besides the need of studies related to altimetry, this category has been raising growing uses and interest. For some examples of the use of altimetry in studies related to thematic cartography, see [3-7]. To exemplify applications related to reference cartography, see [8-12].

In the case of Brazil, cartography still presents serious problems related to the updating of the existent bases and availability of appropriate scales for the most different studies in [13], especially in scales with greater detail of systematic mapping, as for instance, 1:100,000 and 1:50,000 which are very useful in environmental and urban studies.

3. Accuracy of the planimetric and altimetric representations: Case studies

A proposal to minimize the problems presented - rejection of indiscriminate utilization - is the creation of a culture for assessment of the quality and validation of the generated products. This way, maybe, we come closer to know what is real. There is a norm called Cartographic Accuracy Standard (PEC, in Portuguese – [14]) to classify cartographic products. However its application has always been limited. PEC was created in the 1980's, when cartography was still in the analog phase. With the change to digital cartography, new approaches are required regarding assessment of planimetric, and especially altimetric accuracy. In [15] it is possible to see a discussion on the use of PEC in present days, calling attention for the need of changes and inclusion of statistical tests, not foreseen in the present norm.

PEC makes reference to the assessment of final products through the use of control points for checking how much the points located on a map deviate from their homologous located on the ground. For each quality class - which can also be understood as accuracy - a standard value is specified to be reached by at least 90% of the points, as well as a limit value for the Mean Square Error - understood as synonymous of standard deviation. Said simply, this norm applies both for planimetric as altimetric assessments.

Notwithstanding all our concerns regarding the quantity and quality of information that can be extracted from remote sensing products, the volume of products originated from new sensors grows every day and the demand for investigation regarding accuracy of its geometry, considering what is established in the PEC, is still very large. As suggested above, probably this is explained by some factors such as insufficient knowledge of such need or the lack of norms or standards for other types of assessments that may be used as reference.

ESPAÇO Remote Sensing and Environmental Studies Laboratory of the Department of Geography of the Federal University of Rio de Janeiro (UFRJ) has a research line aimed at assessment regarding the geometry of several products originated from remote sensing since

2004. As these products are increasingly used in the production of cartographic data, a brief summary of the results obtained in the case studies carried out by members of the research group during that period is presented. To facilitate the search, the results were grouped in three different Classes of scales: 1) scales over 1:25,000; 2) 1:25,000; 3) 1:50,000 or less.

All assessments of the planimetric and/or altimetric accuracies used as reference what is provided in the PEC; which establishes that 90% of the points tested should present errors below certain tolerance values and the total set of points cannot surpass a standard deviation limit. These thresholds are determined for each quality class, which varies from A (higher accuracy) to C (less accuracy). Table 1 below presents the specific thresholds for the planimetric and altimetric accuracies.

Class	Planimetry		Altimetry	
	Tol. (mm)	SD (mm)	Tol.	SD
A	0.5	0.3	1/2 Eq.	1/3 Eq.
B	0.8	0.5	3/5 Eq.	2/5 Eq.
C	1.0	0.6	3/4 Eq.	1/2 Eq.

Tol.: Tolerance; SD: Standard Deviation; Eq.: Altimetry difference between contour lines (Contour Interval)

Table 1. Limits of planimetric and altimetric tolerances indicated by the PEC

It is emphasized that, despite the PEC not being indicated for assessment of digital products and inputs used in the cartographic production process, this standard was adopted in the studies here performed for being the only existent official norm in Brazil for assessment of cartographic products. The use of the PEC serves mainly, to have a national reference for comparison of the planialtimetric accuracy of cartographic products.

It is stressed also that the assessments refer only to the geometry related accuracy. Most of the assessed products - if not all of them - present information extraction capacity for lower scales than those scales for which its geometry is compatible.

All assessments should be treated as case studies, in view that they refer to the assessment of a unique scene/acquisition for each product, with the possibility of the existence of different results for other study areas. The assessment of planimetric accuracy was always carried out using the generated orthoimage, while the altimetric accuracy evaluation was made directly on the DEM and/or DTM obtained from the sensor images.

The results for each group are presented below, describing the types of evaluated sensors, their common features, besides the results themselves found in the case studies.

4. Scales greater than 1:25,000

4.1. IKONOS 2: Planialtimetric evaluation of the orthoimage and DEM

Ikonos 2 satellite was launched in 1999, being the first one with high-definition available on the market, with a spatial resolution of 1 m for the panchromatic band and

4 m for the multispectral bands (blue, green, red and near infrared), and radiometric resolution of 11 bits. Another novelty offered was the acquisition of the stereo pair in the same orbit.

The Ikonos product used in this assessment comprises four scenes (two stereoscopic pairs), in the GeoOrtho Kit mode (with RPCs - Rational Polynomial Coefficients), acquired on April 11, 2009. The assessment only used the panchromatic band.

Study area is located near the border between São Sebastião and Caraguatatuba, both municipalities of the northern coastline of the state of São Paulo. This area is inserted in the Serra do Mar (Sea Mountain Range), which in this area presents height differences of up to 800 m, with vegetable cover of the tropical forest type.

Scenes referring to the panchromatic band were modeled in Orthoengine 10.2, of PCI Geomatics, making use of RPCs and seven GCPs (Ground Control Points). The orthoimage was generated with the nominal resolution of the raw scene (1 m) and the DEM was generated with resampling, taking the pixel to 2 m. For assessment, 32 points were used, all surveyed with GNSS, through relative static positioning, with single frequency (L1) trackers.

Regarding planimetric accuracy, the results indicated the orthoimage as **class C for 1:2,000 scale**, having in view that 96.9% of the points used in the assessment presented errors of up to 2 m and Standard Deviation (SD) of 0.5 m, demonstrating a very low internal error.

Regarding altimetric accuracy, the generated DEM from the panchromatic band could be fitted in **class C for 1:10,000 scale**, because 91.2% of the points used presented errors of up to 3.75 m and SD of 2.0 m.

4.2. GEOEYE 1: Planialtimetric assessment of orthoimage and DEM

Images of the Geoeye satellite are among those with highest spatial resolution available at present. The acquisition of stereoscopic pairs is also made along the same orbit. The sensor presents a panchromatic band of 0.5 m spatial resolution and multispectral bands (blue, green, red and near infrared) of 2.0 m spatial resolution, and radiometric resolution of 11 bits. The stereoscopic pairs used in this assessment where acquired with RPCs.

Study area comprises the Ilha Grande Island, located in the municipality of Angra dos Reis, in the south of the state of Rio de Janeiro. The prevailing relief is mountainous, with a vegetable cover of the tropical forest type, presenting height differences of up to 1,000 m.

The stereoscopic Geoeye pair, acquired on March 24, 2011, was modeled by Orthoengine 10.2 of PCI Geomatics, using RPCs, 6 GCP and seven Tie Points. For assessment 24 points were used, all of which had their coordinates determined by relative static positioning, using 2-frequency (L1L2) GNSS trackers.

The orthoimage, generated with nominal resolution of 0,5 m, presented planimetric accuracy compatible with the specification, according to PEC, for **class B for 1:2,000 scale**, since 90% of the points used in the assessment presented errors of up to 1 m and SD of 0.3

m. The DEM was generated with resampling, taking the pixel to 2 m. The altimetric accuracy was consistent with **Class C for 1:10,000 scale**, since 90% of points used in the evaluation showed errors of up to 3.2 meters and SD of 2.3 meters.

4.3. LIDAR: Planialtimetric assessment of orthoimage, DTM & DEM

The LIDAR (LIght Detection And Ranging) used in this assessment comprised not only DTM and DEM generated by LIDAR proper, but also orthophotos generated from photos acquired during the survey with a non-metric camera. The point cloud was not assessed, since the ESPAÇO laboratory only received the models generated and the orthophotos, which presented a 25 cm (10") resolution. The LIDAR sensor used was Optech ALTM 2050 (Airborne Laser Terrain Mapper), configured for a point density equivalent to one point at each 80 cm. The airplane used, besides outfitted with a double frequency GNSS tracker, also used an IMU (Inertial Measurement Unit), responsible for recording the airplane altitude variations. The great interest in LIDAR, besides the good accuracy expected and speed in data processing, is the possibility of penetration into the vegetation, due to its high point density.

Thirty-six (36) evaluation points were used, with 29 identified in the orthophotos. All points had their coordinates determined by relative static positioning, with single frequency GNSS trackers.

Study area is located near the border between São Sebastião and Caraguatatuba, municipalities of the northern coastline of the state of São Paulo. This area is inserted in the Serra do Mar Range, which in this area presents height differences of up to 800 m, with vegetable cover of the tropical forest type.

Orthoimages showed excellent planimetric accuracy, presenting a CE90 (Circular Error at 90%) of 0.86 m and SD of 0.3m. With these values it is possible to say that it presented an accuracy compatible with **class A for 1:2,000 scale** because, besides the SD being below the limit for this scale, 93.1% of the 29 points used in the assessment were below the limit for this scale.

The **DEM** presented good altimetric accuracy, with a LE90 (Linear Error at 90%) of 1.33 meters and SD of 0.73 m, presenting itself compatible with the indication by PEC for **class C for I:5,000 scale**, assuming that the contour lines present 2 m of contour interval. A total of 33 (91.7%) of the 36 points used in the assessment presented errors below specification (1.5 m).

The **DTM** presented an even better performance, presenting a LE 90 of 0.58 m and SD of 0.4 m, which would make it compatible with the indication for a **class B for I:2,000 scale,** since 91.6% of the points used in the assessment presented errors up to the specified value (0.6 m) for this class.

The results obtained in the assessment of the LIDAR data were very good, but it was not expected that the altimetric accuracy of the DEM would be lower than the DTM accuracy. It is possible that there was a better treatment of the point cloud during data processing regarding the last return (responsible for DTM generation) in detriment of the data of the first return (that lead to the DEM generation).

5. 1:25,000 scale

5.1. Cartosat-1: Planialtimetric assessment of orthoimage and DEM

The Cartosat-1 satellite - or IRS P5 - is aimed for cartographic applications, especially reference applications, for having only a panchromatic band, but with 2 different acquisition angles in the same passage. Its radiometric resolution is of 10 bits.

The stereoscopic pair analyzed covers the surroundings of São Sebastião, the same area mentioned before.

The stereoscopic pair, acquired on May 23, 2009, was processed with Orthoengine 10.2, by PCI Geomatics, using RPCs, with 7 GCPs and 8 Tie Points. The orthoimage was generated with nominal resolution equal to the rated resolution (2.5 m) and the DEM was resampled for 10 m. For assessment 30 points were used. All points were surveyed with GNSS equipment, with relative static positioning, with single frequency trackers.

The results were quite satisfactory, with the orthoimage reaching **a class B for 1:10,000 scale**, with 93.3% of the points presenting errors of up to 8 m and SD of 2.7 m. Altimetric accuracy was compatible with the specification for a **class A for 1:25,000 scale**, since 94.1% of the assessment points presented errors of up to 5 m and SD of 2.8 m.

5.2. ALOS/PRISM: Planialtimetric assessment of orthoimage and DEM

PRISM sensor onboard ALOS[1] satellite acquired data only into panchromatic band, but with telescopes in three different incidence angles (Backward, Nadir and Forward), with a spatial resolution of 2.5 m. Operating in this mode, called Triplet, it presented optimal conditions for DEM generation, considering that different view combinations could be used, according to relief variation or any other needs.

PRISM data, likewise those from AVNIR2 were distributed by IBGE (*Brazilian Institute for Geography and Statistics*) to noncommercial users for much lower prices than those practiced by the market for compatible products. It explains the interest in such data and the need for assessments.

The results presented here referred to the assessment report of PRISM data in [16] and the work presented by some IBGE employees in [17], who assessed the accuracy of the geometry from a PRISM Triplet acquired with a 1B2R processing level covering the surroundings of the municipality of Itaguaí, in Rio de Janeiro.

The data was processed by Orthoengine 10.1, of PCI, without RPCs, but using 7 GCPs and 11 Tie Points. The orthoimage was generated with a Nadir view of 2.5 m and the DEM was

---

[1] ALOS satellite experienced trouble in April 2011 and was disconnected in May of the same year.

generated using the backward and forward views, resampled for 10 m. For assessment ends, 88 points were used, whose coordinates were determined by relative static positioning, making use of double frequency GNSS trackers.

The authors initially made a planimetric accuracy assessment of an orthoimage generated with 7 GCPs and SRTMDEM. The results showed that, even using a SRTM DEM, with 90 m pixels, the orthoimage would be compatible with an indication for **class A for 1:25,000 scale**, since the 40 points used in this initial assessment presented errors of up to 9.6 m, with SD of 2.1 m. Subsequently, the assessment was complemented using a DEM generated from the Triplet PRISM data. The results were also satisfactory and similar to those obtained with Cartosat1. The orthoimage presented planimetric accuracy compatible with **class B for 1:10,000 scale**, despite the authors' mentioning only attendance to scale 1 to 25,000 (which was the target of the assessment), since all points presented errors up to 7.9 m and SD of 1.9 m. The altimetric accuracy of the **DEM** was compatible with **class A for 1:25,000 scale** since 92% of the assessment points presented errors of up to 5 m and SD of 2.3 m.

6. Scales 1:50,000 or less

6.1. SRTM DEM: Altimetric assessment

The DEMs generated from the data of this mission by NGA (National Geospatial Intelligence Agency), NASA (National Aeronautics and Space Administration), have been widely used in Brazil - as well as in other countries - for being free, of good quality and with almost global coverage. Since its first divulging, in version 1, the immense potential for its use was perceived, mainly in locations without altimetric data of better quality and coverage in digital media. In function of the different versions available, four of them were chosen to be evaluated:

* Version 1: Original DEM with pixel resampled to 90 m, with negative values and many areas without data, distributed by NASA. Additional information may be obtained on the site http://www2.jpl.nasa.gov/srtm/;
* Version 2: Includes additions made to version 1, with delimitation of a coastline, elimination of negative values and reduction of areas without data, also distributed by NASA. Additional information may be obtained on the site http://www2.jpl.nasa.gov/srtm/;
* Version 4: Derived from version 1, maintains a 90 m pixel, as well as some negative values, but completes the areas without data. This version is divulged by CGIAR (Consortium of International Agricultural Research Centers). Additional information may be obtained at: http://srtm.csi.cgiar.org/SELECTION/inputCoord.asp;
* Topodata: This version was generated and distributed by INPE in [18], with reprocessing of version 1 data for the Brazilian territory, resampling the pixel for 30 m by kriging. Additional information can be obtained at: http://www.dsr.inpe.br/topodata/index.php;

The assessment was made in the same way for the four versions, with small variation in the number of assessment points (between 90 and 92 points, depending on the DEM version) -

all surveyed with GNSS, through relative static positioning, with single frequency trackers - in function of the variation of the pixel size and the coastline. The study area corresponds to part of the municipality of São Sebastião on the north coast of the state of São Paulo, in southeast Brazil.

Table 2 presents a summary of the assessment results from the SRTM DEM in the four versions. In it, the number of assessment points used in each product are shown, as well as the percentages of points with errors up to the limit for a **class C for scale 1:50,000** (15 m) and **class A for scale 1:100,000** (25 m) as well as the Standard Deviation (SD) of each set of points used in the assessment for each DEM SRTM version.

DEM SRTM	Assessment Points	Class C 1:50,000	Class A 1:100,000	SD (m)
Version 1	90	91.10%	100%	6.82
Version 2	91	90.50%	100%	6.85
Version 4	91	90.10%	100%	6.84
TOPODATA	92	83.30%	93.40%	10.38

Table 2. Summary of assessments of the 4 versions of DEM SRTM

From the results presented in table 2 one perceives that versions 1, 2 and 4 have very similar performances and present altimetric accuracy very near to the limits indicated for **class C of scale 1:50,000**. Only Topodata is indicated only for **class A of scale 1:100,000**. As the main objective of Topodata is to obtain a better detailing of the morphometric information derived from SRTM DEM, a small loss of accuracy is tolerated in favor to gain better detailing of the morphometric representation. Additionally, the goal and SD observed should be emphasized, mainly in versions 1, 2 and 4, showing that the DEM SRTM presents homogeneity and consistency in results.

6.2. ASTER/VNIR: Planialtimetric assessment of the orthoimage and DEM

ASTER images (Advanced Spaceborn Emission and Reflection Radiometer), onboard Earth satellite, are very interesting for being of low-cost and providing good spectral resolution, besides allowing the generation of DEMs from the stereoscopic pair of the VNIR (Visible and Near Infrared) sensor. This sensor provides a spatial resolution of 15 m, counting with the green, red and near infrared bands (the latter being acquired with two different incidence angles in a same satellite passage). Due to these features, those images can be useful both for thematic as for reference cartography.

Study area again was the vicinities of the municipality of São Sebastião. An assessment was carried out based on the results made from the images and DEMs generated according to two different approaches:

- With Orthoengine 10.2 by PCI Geomatics based on 2 scenes (same date) acquired September 19, 2008, with processing level L1A, making use of 12 control points and 9 Tie Points per scene.

- With ENVI / ASTER MDT: using the same scenes (same date), with the same processing level and control points; and using the same scene, however with L1B processing level and without control points.

The orthoimages and DEMs generated for assessment had the same resolution as raw data: 15 m, despite having carried out tests of the DEM with pixels resampled up to 60 m, without perceiving significant difference.

There was no disagreement between the results obtained for planimetry, placing the VNIR orthoimages in **class A for 1:50,000 scale**, with 92.7% of the 41 assessment points presenting errors of up to 40 m, with a standard deviation of 11.6 m.

In the case of altimetry, however, results were different depending on the approach used in DEM generation. Using Orthoengine, the worst results were obtained, since DEM meets the specifications for **class A for 1:100,00 scale**, with 94.2% of the assessment points presenting errors of up to 25 m, with a standard deviation of 10.5 m. When using the **ENVI /ASTERDTM,** with the same scenes, processing level and same 12 control points, the result changes to a **class C for 1:50,000 scale**, with 92.7% of the assessment points presenting errors of up to 15 m, with standard deviation of 8.6 m. Using the scenes with processing level L1B with **ENVI /ASTERDTM,** without any ground control point, the specification for a **class C for 1:50,000 scale** is reached, with 96.3% of the 41 assessment points presenting errors of up to 15 m, with a standard deviation of 6.1 m.

These results ensued a questioning to the PCI representative in Brazil, with forwarding of all the data used, being answered by PCI that there was no failure or problem. It is therefore understood that the ASTERDTM modeling - which is a specialized ASTER module - is more appropriate.

6.3. ALOS/AVNIR-2: Planimetric assessment of orthoimage

The ALOS/AVNIR-2 data, as well as those of the PRISM sensor of the same satellite, are of great interest for Brazil, since its data was distributed by IBGE (*Instituto Brasileiro de Geografia e Estatística*) for noncommercial users at a much lower price than those practiced by the market for products with the same features. It remained to know its accuracy in order to be able to fully explore its real capacity. The AVNIR-2 sensor acquired data in 4 spectral bands (blue, green, red and near infrared) with a spatial resolution of 10 m. Since the satellite contained a very advanced set of orbital and altitude control systems, its use would be very appropriate both for reference as for thematic cartography.

The results presented here refer to the work published by a group of IBGE employees in [19] who assessed the orthoimage geometry of this sensor.

Orthoimage was generated in Orthoengine 10.2, by PCI, with a resolution of 10 m, based on a scene acquired with processing level 1B2R, making use of seven ground control points and DEM SRTM version 4.0. The control points and the 34 assessment points were obtained from orthoimages in scale 1:25,000 generated by IBGE from photos in the scale 1:35,000. This study area covered the cities of Uberlândia, Indianópolis and Araguari, state of Minas Gerais, in Southeast Brazil. The topography of this area is less mountainous with height differences not exceeding 400 m.

The results indicate that the orthoimage presents a planimetric accuracy equivalent to the **class A of scale 1:50,000**, since 90% of the assessment points presented errors of up to 21.7 m (already including the external error of the orthophotos - of up to 12.5 m), with standard deviation of 3 m. The authors of the referred work emphasize that the assessment only focused on geometry, like the remaining assessments here presented.

6.4. Landsat 7 ETM+: Planimetric assessment of the orthoimage

This assessment was the first research on the geometric accuracy of remote sensors carried out in 2004, by Laboratory ESPAÇO, with the specific purpose to obtain the planimetric accuracy of the Landsat images when orthorectified using the SRTM DEM (a novelty at that time). Despite being a quite well-known sensor, we consider its inclusion in this chapter appropriate for historical reasons and for its resemblance with the TM images from Landsat 5. Presently (2012) Brazil is concluding the updating of reference maps in the scale 1:250,000 using TM sensor images as main input.

Orthoimage was produced from the band 3 (red) of a scene (Path/Row 218/076) of sensor ETM+ (Enhanced Thematic Mapper Plus), Landsat 7 satellite, acquired in August 1999, with processing level 1G. Orthorectification was made with Orthoengine 9.0, by PCI Geomatics, using 10 ground control points (GCPs) - whose coordinates were determined by a navigation GPS (C/A - Coarse Acquisition Code) - and SRTM DEM version 1 (90 m pixel). The choice for the navigation GPS was due to the ETM plus image characteristics - 30 m pixel - and the lack of availability of single or double frequency trackers at that time.

In the planimetric accuracy assessment of orthoimage from Landsat 7 ETM+ only 4 assessment points were used, with coordinates determined by relative static positioning, through single frequency GNSS receivers. It was chosen to work with a small number of points for the assessment, but to determine their coordinates with greater accuracy.

The study area corresponds to the part of the municipality of Angra dos Reis, in Rio de Janeiro, Southeast Brazil. Relief is mountainous in most of the scene, with height differences over 1,000 m, for being part of the Serra do Mar Range. The prevailing vegetable cover consists of tropical forests, with presence of grasses/pastures in some areas.

It was observed that the 4 assessment points presented positioning errors of up to 70 m with an SD of 13 m. The results obtained suggest that the orthoimage presents a planimetric accuracy compatible with the specification for a **class B for 1:100,00 scale**.

6.5. SPOT 4: Planimetric assessment of the orthoimage

Likewise as in the assessment of the Landsat Orthoimage, this was one of our first planimetric accuracy assessments. The SPOT 4 images were renowned as having a very good geometry and it was habitual to use the panchromatic band of that sensor (HRVIR) in scales 1:25.000 and 1:50.000.

The orthorectification of the panchromatic band (in fact it covers the red band – 0.61 to 0.68μm), with a 10 m spatial resolution, was carried out with Orthoengine 9.0, by PCI Geomatics, using 7 GCPs and SRTM DEM.

The assessment used 7 points, whose coordinates were determined by relative static positioning (likewise as for the GCPs), with single frequency GNSS trackers.

The study area was also the same used in the assessment of the Landsat Orthoimage and corresponds to part of the municipality of Angra dos Reis, Rio de Janeiro, Southeast Brazil.

The results **suggest** that the Orthoimage presents an accuracy compatible with the **Class A for 1:50,000 scale** or **Class C for 1:25.000 scale**, since all points presented errors up to 20.5 m and SD of 7.4 metros. Likewise as in the Landsat assessment, the use of a small amount of points in the assessment does not allow to surely warrant the accuracy, but it is an indicator of its quality in terms of geometry.

Finalizing this part of the chapter, we present a summary table (table 3) with results from the accuracy assessments in terms of geometry for each of the sensor products analyzed:

Sensor	Product	Scale	Classe/PEC
LIDAR	Orthophoto	1:2.000	Class A
	MDT	1:2.000	Class B
	MDS	1:5.000	Class C
Geoeye-1	Orthoimage	1:2.000	Class B
	DEM	1:10.000	Class C
IKONOS-2	Orthoimage	1:2.000	Class C
	DEM	1:10.000	Class C
CARTOSAT-1	Orthoimage	1:10.000	Class B
	DEM	1:25.000	Class A
ALOS / PRISM	Orthoimage	1:25.000	Class A
	DEM	1:25.000	Class A
ALOS / AVNIR	Orthoimage	1:50.000	Class A
ASTER / VNIR	Orthoimage	1:50.000	Class B
	DEM	1:50.000	Class C
SRTM (DEM)	Version 1	1:50.000	Class C
	Version 2	1:50.000	Class C
	Version 4	1:50.000	Class C
	Topodata	1:100.000	Class A
SPOT 4 / HRVIR: Pan	Orthoimage	1:50.000	Class A
Landsat 7 / ETM+ / Multiespectral Band	Orthoimage	1:100.000	Class B

Table 3. Summary results of the assessments

7. Issues (still) remaining: Some limitations persisting and potentialities observed

Besides being somewhat limited, the evaluations using PEC made sense for analog mapping, whose construction processes were more vulnerable to graphic issues, such as the tracing of the elements properly. Considering altimetry, it becomes even less valid for the digital data, because reference is made to the contour lines interval, without mentioning specific scales. In several applications, nowadays, it is possible to have a DTM or DEM and, based on them, automatically extract the contour lines, in case required. But what matters is that the primary product tends to be the digital model and not the contour lines, which are dispensable for most of the applications, except in the case when it is desired to print the base with the altimetry represented in a way to allow quantitative estimates. According to this reasoning, it is necessary to assess the accuracy of the DTM or DEM, because they normally are inputs from derived products, although the official Brazilian norm does not make any reference to this type of product. Obviously one should maintain the assessment of the contour lines in the cases where they represent the altimetry.

Even considering the PEC as reference, it can only be applied for assessment of the geometry of reference cartography products. There is no official norm for assessment/evaluation of thematic cartography products in Brazil. Therefore, thematic mappings present accuracies – both in terms of geometry as contents – which are either not assessed or which remain to be decided by the specifications of their producers. The importance of the establishment of norms is emphasized once more, or, at least the establishment of national standards that facilitate data interchange with levels of accuracy that can be known and compared. Brazil is presently making an endeavor to divulge its National Spatial Data Infrastructure (INDE), expanding the utilization of geospatial data in an appropriate manner. Such an endeavor also focuses on the adoption and dissemination for the use/generation of metadata. In this context, the knowledge of the accuracy of the products, whether referential or thematic, is mandatory.

It is clear that both for the reference cartography as for the thematic cartography the assessment should not be restricted to the accuracy of the geometry, but should consider also the reliability of what is being represented – i.e., its actual contents. In this context, some items to be considered are: what is represented, the amount of elements presented, the taxonomic level reached, the level of detail, etc. Obviously, for reference cartography some elements may have greater relevance while in the thematic cartography others may deserve more attention. It is necessary to disseminate among the users and producers of cartographic data the importance to know the quality, both of the geometry as the contents, the detailing of the mapping, always assessed considering the scale adopted. Despite seeming obvious, it is quite common to find the distribution of products presented as compatible with a certain scale, but in the assessment made, one finds that the concern was merely with the accuracy for geolocation, i.e., with the geometry. Among the non-specialist users the problem is even greater, because they rely more easily on the information provided and end up acquiring products which do not correspond to their needs and, when trying to extract an amount of information proportional to the desired scale, they perceive that it is impossible. In some of these cases the user may reach the conclusion that the maximum scale that can be reached for the product in terms of geometry can be very different from the maximum scale reached in terms of information contents available in that product. However, in other cases it is also possible that the user does not even perceive such difference and will end up making inappropriate use of that data.

The authors have no intention to criticize the performance of assessments regarding only accuracy. Actually, what we have been doing mostly, in a systematic way, is this type of assessment. The interest lies in calling attention for the need to complement the assessments with considerations regarding accuracy of the represented contents. As the existence of a critical mass regarding the need for assessments for accuracy of geometry is already perceived, it is desired to create a demand for investigations regarding the contents of what is represented.

As an example of such duality in the assessment of accuracy for the case of orbital images, we can mention the ALOS/PRISM in [16] planialtimetric data assessment report and in [19] on planimetric assessment of AVNIR2 images. Both texts present the maximum scales in terms of planialtimetric and planimetric accuracy, respectively, mentioning that the products may not allow the extraction of elements in the same scale. In order to know this last information it would be necessary to carry out an assessment of the images interpretability.

It is also important, to minimize the intensity of a statement – which is sometimes mentioned – that for thematic cartography, geometry may be neglected, while for reference cartography geometry is all that matters. Due to this, many thematic maps present serious geometry problems, which makes them incompatible for the foreseen scale, likewise as in reference cartography one may find maps – often adjacent to each other – that present different information densities, or even, a lack of standardization of what is being represented.

It is necessary to think about the problem, to suggest solutions and create a form of assessment that is really viable for the use by several producers of cartographic data – whether referential or thematic, whether originated from official mapping agencies or private companies, research groups, etc. who handle these types of data. What matters is to invest in the proposition of standards for assessment of such data. These standards should be appropriate for the new techniques for data acquisition and handling, considering the digital context and dealing with accuracy in terms of geometry and contents.

When we consider the use of remote sensing data – whether originated from aerial or orbital levels – a set of cares is necessary, starting with the appropriate sensor choice, the time of the year for data acquisition, but treatment of the data and extraction of the desired information.

People often say that the only optimal solution is the one obtained through photogrammetric survey. And for many applications it really is the best. But a survey carried out in the scale and dates desired is not always available. And for other applications the optimum solution may be a survey carried out by radar. Other times we find people who do not even think of using photographs, because beforehand they state that the solution is through the use of orbital data. But it may happen that in this type of survey you may not get the acquired image with the geometry and/or level of detailing most appropriate for your application. In other words, the choice of the sensor should be made by an evaluation comparing costs and benefits, focused on the application. Each option normally presents benefits and drawbacks which should be evaluated case-by-case.

Since the appearance and dissemination of high resolution (spatial) sensors, they became the source of the most desired images for several applications that work with medium to large scales. However, not always the spatial resolution is what matters. For several thematic applications, such as mapping of vegetation or land coverage and land use, the spectral resolution is a factor equally or more important than the spatial resolution. On the other hand, one problem which is often overlooked in the planning is the time that it may take to acquire an image (or pair of images) for a certain location. Even with sensors presenting a small revisiting period, for meteorological issues or for great data demand, months may pass until the data is effectively acquired, even if paying a priority rate. For several applications this may be a limiting factor.

There are other basic issues that are sometimes neglected: when choosing to pay an expressive amount for high spatial resolution images, it is assumed that the application is for scales that need greater detailing which, at their turn, demand also greater accuracy in terms of geometry. However it is not rare to find users who acquire such data and do not make the appropriate geometric modeling/correction – the orthorectification, with RPCs supplied by the satellite owner, GCPs (ground control points) with compatible accuracy and appropriate DEM. Even images with good internal geometry will present relatively high external errors if they are not well modeled, especially in areas with great attitude differences. And this is not acceptable in the immense majority of applications that effectively require data with great detailing.

Users of images that allow a more regional coverage – sensors normally included in the category of medium spatial resolution – should be feeling more difficulty to update data since Landsat 5 presented problems in November 2011 in [20], since there is practically no alternative that offers the same possibilities (especially considering cost and data availability). Environmental studies have a great demand for data of this category and need to search for alternatives, as the use of sensors of greater spatial resolution, normally losing spectral resolution – with highlight for the lack of availability of the SWIR band – in most of those sensors.

While the producers of thematic data are experiencing a certain period of low availability of these sensors of medium spatial resolution and highest spectral resolution, the opposite has been occurring for producers of reference cartographic data. These sensors of medium to high spatial resolution (between 2.5 and 5 meters), high (between 1 and 2.5 metros) and very high (below 1 m) can rely on a great offer of inputs. New sensors with higher spatial resolutions are becoming available almost every year. In Brazil, as mentioned before, photogrammetric flights for systematic cartography are again being contracted in greater quantity, making more data aimed at greater scales available (mainly 1:25,000 or above), in addition to a growing offer of LIDAR and RADAR data. Regarding this latter sensor - the project "Amazon Radiography" is presently ongoing, with the use of airborne interferometer RADAR, in the X and P bands, for production of a cartographic reference base in scale 1:50,000 and 1:100,000, among other products in [21].

One feature that may soon become more common for orbital sensors is the planning of Earth observation missions that count with a constellation of satellites as for instance, Rapideye. Obviously, besides technical issues, the constellations tend to favor the reduction of interval revisits, increasing the chance of success in the acquisition of data in the area desired.

Another important remark is the appearance of new spectral bands of high spatial resolution, such as the coastal, yellow and red edge, which offer new investigation possibilities for the enhancement of thematic maps.

With increasingly better orbital and attitude control systems, image suppliers have been offering products with geolocation which are increasingly accurate, at least in areas with flat topography. In some cases those images may even dispense the need for geometric correction for users who work in scales which do not demand great accuracy. Orthorectified images can even be acquired for some products, without the need for the user to supply ground control points.

The possibilities seem endless, with new offers of data and applications arising constantly. However, a critical eye should be kept on each input or product, in order to know their real potentialities and limitations.

8. Conclusion

With the systematic offer of new and re-paged sensors, it is increasingly difficult to close a discussion on the subject. It is a fact that Remote Sensing is becoming indispensable for many cartographic applications, for all its already mentioned facilities. With all this, in the present phase, it is extremely important that the investments for assessment of so many products offered are made, in order to direct them to a conscious use, with awareness of the problems involved, thus avoiding loss of time and resources. The existent shift from analog to digital paradigm is also emphasized, which becomes an additional requirement in the reassessment of methods and processes, not only of products.

Still regarding the new sensors, one may state that the present scarcity of sensors of medium resolution has become a serious problem in mapping and monitoring of natural phenomena and events, mainly in countries with great territorial extensions such as Brazil. It is fundamental and urgent that investments be made in the sector. On the other hand, we emphasize our present concern with the maintenance and the investments in remote sensors that seek to meet the requirements of global studies (low resolution and great coverage), as in the case of climate changes, and local studies (high resolution), as for instance sensitivity and risk maps in support of natural disasters.

It is interesting that we are quickly reaching the point of discussing ethical and legal issues, such as the new limitations for the progress of orbital imaging at still greater detail levels, which reflect problems such as invasion of privacy and national security.

In the case of Brazil, as in any other country, there is a choice of sensors which for several reasons become interesting, and which naturally prevail over the remaining, responding for the great majority of applications. In this investigation, we seek to focus those cases, trying to contribute for greater clarification regarding the potential and limitations of each option, limiting ourselves in this first approach to geometric issues. The 10 sensors presented here constitute the first phase of our investigation line, which will have continuity in the assessment of new sensors, such as Rapideye, Terrasar-X and Worldview 2. Another effort that is being developed, but which is still in an initial phase, is the diligent assessment of the potential of those products for extraction of the cartographic elements, a term which we call interpretability, whose scale of attendance tends to be less than the one defined by geometry.

The organization of the assessments carried out in three classes of scales was considered important for a better orientation of applications by the users. The limits adopted, using as important threshold the 1:25,000 scale, represent mapping challenges in Brazil, whose scarcity at this level of detail is immense. Considering the country's growth, the demand for reliable georeferenced information has increased a lot, which may translate a greater concern for its attendance by specialized institutions of the sector and, on the other hand, unfortunately, the explosion of individualized, and standalone solutions, in most cases, without the necessary control and knowledge. It is sought therefore, to contribute for the awareness of the community in general regarding those issues.

Author details

Carla Bernadete Madureira Cruz and Rafael Silva de Barros
Departamento de Geografia, Instituto de Geociências, Universidade Federal do Rio de Janeiro – UFRJ, Rua Athos da Silveira Ramos, CCMN, Cidade Universitária, CEP 21941-590 – Rio de Janeiro, Brasil

Acknowledgement

Our thanks to the Research Center Leopoldo Americo Miguez de Mello (CENPES), Petrobras, which supported this research by providing the resources necessary for its preparation.

9. References

Journal

[8] TOUTIN, T. & CHENG, P. (2002) Comparision of Automated Digital Elevation Model Extraction Results Using Along-Track ASTER and Across-Track SPOT Stereo Images. SPIE Journal, Optical Engeneering, 41 (9), p. 2102-2106.

[9] TOUTIN, T. (2004) Comparison of Stereo-Extracted DTM from Different High-Resolution Sensors: SPOT-5, EROS-A, IKONOS-II, and QuickBird. IEEE Transactions on Geoscience and Remote Sensing. 42(10):2121-2129.

Online journal

[1] INPE. Notícias. (2010) Site: http://www.inpe.br/noticias/noticia.php?Cod_Noticia=2148. Último acesso 29/03/2012.

[2] GOOGLE. Google History. (2012) Site: http://www.google.com/about/company/history. html#2005. Último acesso 29/03/2012.

[14] BRASIL, Decreto 89.817 de 20 de junho de 1984. (1984) Estabelece as Instruções Reguladoras das Normas Técnicas da Cartografia nacional. Diário Oficial da República Federativa do Brasil, Brasília, n120, 22 de junho de 1984. Site: http://www.planalto.gov.br/ccivil_03/decreto/1980-1989/D89817.htm. Último acesso: 10 de abril de 2012.

[16] IBGE. Avaliação Planialtimétrica de Dados ALOS/PRISM Estudo de Caso: Itaguaí - RJ. (2009) Disponível em http://www.ibge.gov.br/alos/RelatoriodeAvaliacaoAlos.pdf. Último acesso: setembro 2009.

[20] NASA. (2012) Landsat 5. Site: http://landsat.gsfc.nasa.gov/about/landsat5.html. Último acesso: 10 de abril de 2012.

Theses

[3] FERNANDES, M. C. (2004) Desenvolvimento de Rotina de Obtenção de Observações em Superfície Real: uma aplicação em Análises Geoecológicas. Tese de Doutorado. IGEO / UFRJ.

[4] VICENS, R. S. (2003) Abordagem geoecológica aplicada `as bacias fluviais de tabuleiros costeiros no Norte de Espirito Santo: uma contribuição para avaliação e gestão de recursos hídricos. [Rio de Janeiro], 252p.(IGEO/ UFRJ), D.Sc., Geografia, Tese – Universidade Federal do Rio de Janeiro, IGEO.

[5] GOMES, R. A. T. (2002) Modelagem de previsão de movimentos de massa a partir da combinação de modelos de escorregamentos e corridas de massa. [Rio de Janeiro], 102p.(IGEO/ UFRJ), D.Sc., Geografia, Tese – Universidade Federal do Rio de Janeiro, IGEO.

[6] GUIMARÃES, R. F. G. (2000) Utilização de um modelo de previsão de áreas susceptíveis à ocorrência de escorregamentos rasos com controle topográfico: adequação e calibração em duas bacias de drenagem. [Rio de Janeiro], (IGEO/ UFRJ), D.Sc., Geologia, Tese – Universidade Federal do Rio de Janeiro, IGEO.

[7] CORREIA, J. D. (2008) Mapeamento de feições deposicionais quaternárias por imagens orbitais de alta resolução espacial – Médio Vale do Paraíba do Sul. Tese Doutorado, IGEO/UFRJ.

[11] SANTOS, P. R. A. (2005) Avaliação da Precisão Vertical dos Modelos SRTM em Diferentes Escalas: Um Estudo de Caso na Amazônia. Dissertação de Mestrado. IME. Rio de Janeiro.

[12] BARROS, R. S. (2006) Avaliação da Altimetria de Modelos Digitais de Elevação Obtidos a Partir de Sensores Orbitais. [Rio de Janeiro], 2006 XIX, 172p.(IGEO/ UFRJ), D.Sc., Geografia, Tese – Universidade Federal do Rio de Janeiro, IGEO.

[13] CRUZ, C. B. M. (2000) As bases operacionais para a modelagem e implementação de um banco de dados geográfico - um exemplo aplicado à bacia de Campos, RJ. [Rio de Janeiro], IGEO/ UFRJ), D.Sc., Geografia, Tese – Universidade Federal do Rio de Janeiro, IGEO.

Annals

[10] GONÇALVES, G. A.; da SILVA, C. R.; MITISHITA, E. A. (2005) Comparação dos Dados do SRTM com as RNs da Rede Geodésica Altimétrica do IBGE para Região Sul do Brasil. *In*: IV Colóquio Brasileiro de Ciências Geodésicas. Curitiba – PR.

[15] SANTOS, S. D. R.; HUINCA, S. C. M.; MELO, L. F. S.; SILVA, M. T. Q. S.; DELAZAR, L. S. (2010) Considerações sobre d Utilização do PEC (Padrão De Exatidão Cartográfica) Nos Dias Atuais. In: III Simpósio Brasileiro de Ciências Geodésicas e Tecnologias da Geoinformação. Recife - PE, 27-30 de Julho de 2010. p. 1–5.

[17] BARROS, R. S.; COELHO, A. L.; OLIVEIRA, L. F.; MELO, M. F.; CORREIA, J. D. (2009) Avaliação Geométrica de Imagens ALOS/PRISM níveis 1B2G e 1B2R ortorretificada – estudo de caso: Itaguaí-RJ. In: XIV Simpósio Brasileiro de Sensoriamento Remoto. Florianópolis.

[18] VALERIANO, M. M. (2005) Modelo digital de variáveis morfométricas com dados SRTM para o território nacional: o projeto TOPODATA. In: XII Simpósio Brasileiro de Sensoriamento Remoto, 2005, Goiânia, GO. Anais do XII Simpósio Brasileiro de Sensoriamento Remoto. p. 1-8.

[19] BARROS, R. S.; COELHO, A. L.; MELO, M. F.; CORREIA, J. D.; OLIVEIRA, L. F. (2010) Avaliação Planimétrica de Dados ALOS/AVNIR-2. Estudo de caso: Uberlândia – MG. In Congresso Brasileiro de Cartografia. Sergipe.

[21] CORREIA, A. H. (2011) Metodologias e Resultados Preliminares do Projeto Radiografia da Amazônia. In: XV Simpósio Brasileiro de Sensoriamento Remoto. Curitiba. Brasil.

Chapter 9

Contribution of SAR Radar Images for the Cartography: Case of Mangrove and Post Eruptive Regions

Janvier Fotsing, Emmanuel Tonye, Bernard Essimbi Zobo, Narcisse Talla Tankam and Jean-Paul Rudant

Additional information is available at the end of the chapter

1. Introduction

With the advent of new satellite sensors of type synthetic aperture radar (SAR) (ERS-1 and ERS-2, JERS-1, and RADARSAT), a large number of satellite images are currently available. However, radar remote sensing has a major drawback which is the difficulty to extract the information it contains. Since 1992, several works are conducted with the radar images on Cameroon to study the ecosystem of the coastal zone (Baltzer et al., 1996; Rudant et al., 1997) and the Mount Cameroon region (Akono et al., 2005, 2006; Talla, 2008).

Texture analysis is a robust approach of processing satellite radar images. It is a set of mathematical techniques to quantify the different gray levels present in an image in terms of intensity or roughness and distribution. Several methods of texture analysis exist and can be classified into two broad categories: structural methods and statistical methods (Haralick et al., 1979). Structural methods are used for describing the texture by defining primitives and "rules" of arrangement between them. Statistical methods are used for study the relations between a pixel and its neighborhood. They are defined according to different orders: 1, 2, 3, ...,n. The second order is the most classic, based on co-occurrence matrices. Obtaining these matrices is very time-consuming calculation, which has prompted researchers to not usually go beyond the second order in the evaluation of textural parameters and provides high order information to lower levels (Li, 1994). The matrices of order greater than two are called matrices of frequency. In texture analysis, the interest of researchers is moving increasingly towards optimization methods of evaluation time statistical parameters. Indeed, (Unser, 1995) replaced the co-occurrence matrix by the sum and histograms difference that define the main axes of the probabilities of second order stationary processes.

(Marceau et al., 1990) propose an approach for their textural and spectral classification of different themes and adopt a reduced level of quantification (16, 32 instead of 256). (Peckinpaugh, 1991) for his part describes an efficient approach for calculating texture measures based on co-occurrence matrix, thereby saving valuable time. (Kourgli et al., 1999) present a new algorithm to calculate the statistical parameters of texture through various histograms. Furthermore, Akono et al. (2003) have proposed a new approach in the evaluation of textural parameters of order 3. The present study is a generalization of the work of Akono et al. (2003). It proposes a generic tree method in the evaluation of textural parameters of order n≥ 2 near a window image which is to explore as if it were a tree, while memorizing the visited nodes.

Several studies have been conducted in the field of classification by texture analysis. (Ulaby et al., 1986) used texture parameters from the method of co-occurrence to identify four classes of land cover in radar images. (Lucieer et al., 2005) propose a segmentation method based on texture parameters varied for multi object recognition on an image. The authors of this study include an operator called "Local Binary Model," modeling the texture, in a hierarchical segmentation to identify regions with homogeneous texture in an image. In (Linders, 2000), three methods (the method of fuzzy logic, regression analysis and principal component analysis) are used to select significant texture parameters for discrimination of different forest canopies. Recognition of forest cover is then performed by the method of neural networks. Puissant et al. (2005) (Puissant et al., 2005) examine the utility of the textural approach to improve the classification accuracy in an urban context. Texture analysis is compared to multispectral classification. In this study, textural parameters of Haralick (Haralick et al., 1973) of the third order are used. More, (Jukka & Aristide, 1998) have used first-order textural statistics in classifying land used in urban areas by means of Landsat TM and ERS-1. (Franklin & Peddle, 1989) used a mixture of spectral data, topographic (elevation, slope, aspect, curvature, relief) and statistical co-occurrence of the second order for the classification of SPOT images and radar in the boreal. Their work showed that the co-occurrence matrices of second order contain important textural information that improves the discrimination of classes with internal heterogeneity and structural forms. Homogeneous classes of soil are characterized adequately by spectral information alone, but the classes containing mixtures of vegetation types or structural information were characterized more accurately by using a mixture of texture and spectral data. Methods in the literature usually consist create neo-channels of the original image by calculating various parameters of the image texture. The neo-channels created are then combined with each other and the original image, for the production of a classified image. In another study, an analysis of different estimators for the characterization of classes of texture on SAR image is performed (Oliver, 1993). In this study, estimators under the maximum likelihood method are evaluated. Note that this method requires knowledge of the shape of the probability density data to be processed. In (Oliver, 1993), it is also considered a probabilistic description of the texture classes according to the law K and to the Weibull model, which are often applied to characterize classes in SAR images. Other extraction methods of textural parameters have been proposed in the literature (Randen &

Husoy, 1999; Reed & Hans Du Buf, 1993). For most of these methods, a single texture parameter is applied to the discrimination of classes. In this chapter, which applies to images from a radar sensor, we introduce the notions of vector texture, patterns and valleys of the histogram and textural signature for the characterization of classes of land, and show that textural parameters of order higher than 2 are more effective for discrimination of these classes.

In the following, we will present the arborescent method of textural parameters evaluation, followed by the presentation of notion of mode and valley of histogram in SAR image analysis. The criteria of choosing textures parameters are also presented. Once the characterization of the various training zones is done and the classification algorithm is presented. Finally, we present some experimental results.

2. Problem context

The usefulness of image classification is not to be demonstrated today. A good classification requires a better identification of information classes on the image. This identification requires the selection of good feature parameters.

3. Methodology

Our methodology is divided into two parts: the first part concerns the improvement of the computational time required for the evaluation of textural parameters. The second part deals with an approach of SAR images classification.

3.1. Formulation of high order of statistical textural parameters

Basically, statistical textural parameters are function of the occurrence frequency matrix (OFM), which is used to define the occurrence frequency of n-ordered gray levels in the image.

3.1.1. The occurrence frequency matrix (OFM)

In an image with $Ng+1$ levels of quantification, the OFM of order $n > 1$ is a $(Ng+1)^n$ size matrix. In this matrix, each element $P_{i_1 i_2 \ldots i_n}$ expresses the occurrence frequency of the n-ordered pixels $(i_0, i_1, \ldots, i_{n-1})$ following the connection rule $R_n(d_1, d_2, \ldots d_{n-1}, \theta_1, \theta_2, \ldots \theta_{n-1})$. This connection rule defines the spatial constraint that must be verified by the various pixels of the n-ordered pixels $(i_0, i_1, \ldots, i_{n-1})$ used in the occurrence frequency matrix evaluation. This rule means that the pixel i_{k+1} $(0 < k < n)$ is separated to the pixel i_k by d_{k-1} pixels in the θ_k direction. For the sake of simplicity, $R_n(d_1, d_2, \ldots d_{n-1}, \theta_1, \theta_2, \ldots \theta_{n-1})$ will be noted by R_n in the following.

3.1.2. The textural parameters

A parameter of texture $Para_n$ in any order n is a real function defined in general manner by the equation given after:

$$Para_n = F \times \{R_n\} \to \Re \tag{1}$$

where F is an image or window of image on which one is evaluated texture parameter and R_n is a rule of connection associated.

Let's consider an image window F of size NLxNC, where NL is the number of lines and NC is the number of columns. The classical expression of textural parameters is given by the following expression:

$$Para_n = \sum_{i_0=0}^{N_g} \sum_{i_1=0}^{N_g} \cdots \sum_{i_{n-1}}^{N_g} \left(\phi(i_0, i_1, \cdots, i_{n-1}) \times P_{i_0 i_1 \cdots i_{n-1}} \right) \tag{2}$$

where $\left(P_{i_0 i_1 \cdots i_{n-1}} \right)$ is the OFM and ϕ is a real function defined in N^n.

The synthesis of the generalisation of texture parameters is conciliated on the Table 1 given below.

Parameter		
Order 2	Classical formulation of order n	Arborescent formulation in order n
1- Contrast		
$\dfrac{1}{N}\sum_{n=0}^{L-1} n^2 \sum_{i=0}^{L-1}\sum_{j=0}^{L-1} \cdot \left[(i-j)^2 P_{ij}\right]$	$\dfrac{1}{N}\sum_{i_0=0}^{L-1}\sum_{i_1=0}^{L-1}\cdots\sum_{i_{n-1}=0}^{L-1}\left[n^2\sum_{k=0}^{n-2}\sum_{l=k+1}^{n-1}(i_k-i_l)^2 P_{i_0 i_1 \cdots i_{n-1}}\right]$	$\dfrac{1}{N}\sum_{(p,q)\in D}\left[n^2\sum_{u=0}^{n-2}\sum_{v=l+1}^{n-1}(i_u-i_v)^2\right]$
2- Correlation		
$\dfrac{1}{N}\sum_{i=0}^{L-1}\sum_{j=0}^{L-1}\cdot\dfrac{(i-\mu_x)(j-\mu_y)}{\sigma_x\sigma_y}P_{ij}$	$\dfrac{1}{N}\sum_{i_0=0}^{L-1}\sum_{i_1=0}^{L-1}\cdots\sum_{i_{n-1}=0}^{L-1}\left[\dfrac{\prod_{k=0}^{n-1}(i_k-\mu_{i_k})}{\prod_{k=0}^{n-1}\sigma_{i_k}}P_{i_0 i_1 \cdots i_{n-1}}\right]$	$\dfrac{1}{N}\sum_{(p,q)\in D}\left[\dfrac{\prod_{u=0}^{n-1}(i_u-\mu_{i_u})}{\prod_{u=0}^{n-1}\sigma_{i_u}}\right]$
3- Covariance		
$\dfrac{1}{N}\sum_{i=0}^{L-1}\sum_{j=0}^{L-1}\cdot(i-\mu_x)(j-\mu_y)P_{ij}$	$\dfrac{1}{N}\sum_{i_0=0}^{L-1}\sum_{i_1=0}^{L-1}\cdots\sum_{i_{n-1}=0}^{L-1}\left[\prod_{k=0}^{n-1}(i_k-\mu_{i_k})P_{i_0 i_1 \cdots i_{n-1}}\right]$	$\dfrac{1}{N}\sum_{(p,q)\in D}\left[\prod_{u=0}^{n-1}(i_u-\mu_{i_u})\right]$
4- Inverse Difference		
$\dfrac{1}{N}\sum_{i=0}^{L-1}\sum_{j=0}^{L-1}\cdot\dfrac{P_{ij}}{1+\lvert i-j\rvert}$	$\dfrac{1}{N}\sum_{i_0=0}^{L-1}\sum_{i_1=0}^{L-1}\cdots\sum_{i_{n-1}=0}^{L-1}\left(\dfrac{P_{i_0 i_1 \cdots i_{n-1}}}{1+\sum_{k=0}^{n-2}\sum_{l=k+1}^{n-1}\lvert i_k-i_l\rvert}\right)$	$\dfrac{1}{N}\sum_{(p,q)\in D}\left(\dfrac{1}{1+\sum_{u=0}^{n-2}\sum_{v=u+1}^{n-1}\lvert i_u-i_v\rvert}\right)$
5- Dissymmetry		
$\dfrac{1}{N}\sum_{i=0}^{L-1}\sum_{j=0}^{L-1}\cdot\lvert i-j\rvert P_{ij}$	$\dfrac{1}{N}\sum_{i_0=0}^{L-1}\sum_{i_1=0}^{L-1}\cdots\sum_{i_{n-1}=0}^{L-1}\left(\sum_{k=0}^{n-1}\sum_{l=k+1}^{n}\lvert i_k-i_l\rvert P_{i_0 i_1 \cdots i_{n-1}}\right)$	$\dfrac{1}{N}\sum_{(p,q)\in D}\left(\sum_{u=0}^{n-1}\sum_{v=k+1}^{n}\lvert i_u-i_v\rvert\right)$

6- Standard Deviation

$$\frac{1}{N}\sqrt{\sum_{i=0}^{L-1}\sum_{j=0}^{L-1}(i-\mu_x)^2 P_{ij}} \qquad \frac{1}{N}\sqrt{\sum_{i_0=0}^{L-1}\sum_{i_1=0}^{L-1}\cdots\sum_{i_{n-1}=0}^{L-1}\left[(i_0-\mu_x)^2 P_{i_0 i_1\cdots i_{n-1}}\right]} \qquad \frac{1}{N}\sqrt{\sum_{(p,q)\in D}(i_0-\mu_x)^2}$$

7- Cluster Shade

$$\frac{1}{N}\sum_{i=0}^{L-1}\sum_{j=0}^{L-1}(i+j-2\mu_x)^3 P_{ij} \qquad \frac{1}{N}\sum_{i_0=0}^{L-1}\sum_{i_1=0}^{L-1}\cdots\sum_{i_{n-1}=0}^{L-1}\left(\sum_{k=0}^{n-1}i_k - n\mu_x\right)^3 P_{i_0 i_1\cdots i_{n-1}} \qquad \frac{1}{N}\left(\sum_{u=0}^{n-1}i_u - n\mu_x\right)^3$$

8- Importance of Great Numbers

$$\frac{1}{N}\sum_{i=0}^{L-1}\sum_{j=0}^{L-1}\left(i^2+j^2\right)P_{ij} \qquad \frac{1}{N}\sum_{i_0=0}^{L-1}\sum_{i_1=0}^{L-1}\cdots\sum_{i_{n-1}=0}^{L-1}\left[\left(\sum_{k=0}^{n-1}(i_k)^2\right)P_{i_0 i_1\cdots i_{n-1}}\right] \qquad \frac{1}{N}\sum_{(p,q)\in D}\left(\sum_{u=0}^{n-1}(i_u)^2\right)$$

9- Importance of Small Numbers

$$\frac{1}{N}\sum_{i=0}^{L-1}\sum_{j=0}^{L-1}\frac{P_{ij}}{\left(i^2+j^2\right)+\varepsilon} \qquad \frac{1}{N}\sum_{i_0=0}^{L-1}\sum_{i_1=0}^{L-1}\cdots\sum_{i_{n-1}=0}^{L-1}\frac{P_{i_0 i_1\cdots i_{n-1}}}{\sum_{k=0}^{n-1}(i_k)^2+\varepsilon} \qquad \frac{1}{N}\sum_{(p,q)\in D}\frac{1}{\sum_{u=0}^{n-1}(i_u)^2+\varepsilon}$$

10- Inverse Differential Moment

$$\frac{1}{N}\sqrt{\sum_{i=0}^{L-1}\sum_{j=i+1}^{L-1}\frac{P_{ij}}{1+(i-j)^2}} \qquad \frac{1}{N}\sqrt{\sum_{i_0=0}^{L-1}\sum_{i_1=0}^{L-1}\cdots\sum_{i_{n-1}=0}^{L-1}\frac{P_{i_0 i_1\cdots i_{n-1}}}{1+\sum_{k=0}^{n-2}\sum_{l=k+1}^{n-1}(i_k-i_l)^2}} \qquad \frac{1}{N}\sum_{(p,q)\in}\frac{1}{1+\sum_{u=0}^{n-2}\sum_{v=u+1}^{n-1}(i_u-i_v)^2}$$

11- Mean

$$\frac{1}{N}\sum_{i=0}^{L-1}\sum_{j=0}^{L-1}iP_{ij} \qquad \frac{1}{N}\sum_{i_0=0}^{L-1}\sum_{i_1=0}^{L-1}\cdots\sum_{i_{n-1}=0}^{L-1}\left(i_0 P_{i_0 i_1\cdots i_{n-1}}\right) \qquad \frac{1}{N}\sum_{(p,q)\in D}i_0$$

12- Variance

$$\frac{1}{N}\sum_{i=0}^{L-1}\sum_{j=0}^{L-1}(i-\mu_x)^2 P_{ij} \qquad \frac{1}{N}\sum_{i_0=0}^{L-1}\sum_{i_1=0}^{L-1}\cdots\sum_{i_{n-1}=0}^{L-1}\left[(i_0-\mu_x)^2 P_{i_0 i_1\cdots i_{n-1}}\right] \qquad \frac{1}{N}\sum_{(p,q)\in D}(i_0-\mu_x)^2$$

Table 1. Classical and new formulations of textural parameters of order $n \geq 2$.

3.2. Aborescent method of textural parameters evaluation

This method consists in reducing the number of operations necessary for the calculation of the co-occurrence matrix. From each pixel of the image, all others pixels involved in the computation of the co-occurrence matrix are reached and operations are directly made on these pixels. This process allows avoiding the evaluation and the stocking of the co-occurrence matrix in the main memory of the computer. In the tree approach, the route of the image pixels is not made any more line after line and pixel after pixel, but rather by following a generic tree. From a pixel and according to its position, one reaches directly, by following a rule of connexion, the neighbouring pixels which are involved in the evaluation of the parameter of texture. After that, the necessary operations are made on these pixels.

3.2.1. Generic tree

The generic tree for the computation of the textural parameters (figure 1) has for seed a pixel positioned in some coordinates (a, b) of the image, with the condition $a \in [0, NL]$ and $b \in [0, NC]$, where NL and NC designate the number of lines and the number of columns of the image, respectively. The different conditions specify the borders of variation of a and b.

From the pixel positioned in coordinates (a, b) of the image window one can reach the pixel positioned in $f_0(a, b, d_1)$ (respectively in $f_1(a, b, d_1)$) by respecting the rule of connexion $R_1(d_1, \theta_1)$ (respectively $R_1(d_1, \theta_1 + 180°)$), if the condition C_0 (respectively the condition C_1) is verified. From the pixel positioned in $f_0(a, b, d_1)$ (respectively in $f_1(a, b, d_1)$) one can reach the pixel positioned in $f_{00}(f_0(a, b, d_1), d_2, \theta_2)$ (respectively in $f_{01}(f_0(a, b, d_1), d_2, \theta_2 + 180°)$) by respecting the rule of connexion $R_2(d_2, \theta_2)$ (respectively $R_2(d_2, \theta_2 + 180°)$), if the condition C_2 (respectively the condition C_3) is verified. Let us call back that a rule of connexion $R_3(d_2, \theta_2)$ means that one considers an inter-pixel distance of d_2 and an angle of θ_2 (and $\theta_2 + 180°$) degrees for the evaluation of the textural parameters. At the order three, one obtains the seven pixels i, j, k, l, m, n, o, presented on the equation 3:

$$
\begin{cases}
i = (a, b) \\
j = f_0(i, d_1, \theta_1) \\
k = f_{00}(j, d_2, \theta_2) \\
l = f_{01}(j, d_2, \theta_2 + 180°) \\
m = f_1(i, d_1, \theta_1 + 180°) \\
n = f_{10}(m, d_2, \theta_2) \\
o = f_{11}(n, d_2, \theta_2 + 180°)
\end{cases}
\tag{3}
$$

In the equation 3:

i is the current pixel, identified by its coordinates (a, b) in the image;

$f_0(i, d_1, \theta_1)$ is the neighbouring pixel of i respecting the rule of connexion $R_0(d_1, \theta_1)$;

$f_{00}(j, d_2, \theta_2)$ is the neighbouring pixel of j respecting the rule of connexion $R_2(d_2, \theta_2)$;

$f_{01}(j, d_2, \theta_2 + 180°)$ is the neighbouring pixel of j respecting the rule of connexion $R_3(d_2, \theta_2 + 180°)$;

$f_1(i, d_1, \theta_1 + 180°)$ is the neighbouring pixel of i respecting the rule of connexion $R_1(d_1, \theta_1 + 180°)$;

$f_{10}(m, d_2, \theta_2)$ is the neighbouring pixel of m respecting the rule of connexion $R_2(d_2, \theta_2)$;

$f_{11}(n, d_2, \theta_2 + 180°)$ is the neighbouring pixel of m respecting the rule of connexion $R_2(d_2, \theta_2 + 180°)$.

We then generalize by saying that, starting from the pixel $f_n(a, b, d_1, d_2, \cdots, d_n, \theta_n)$, one can reach the pixel $f_{n0}(f_n(a, b, d_1, d_2, \cdots, d_n, \theta_n), d_{n+1}, \theta_{n+1})($ respectively the pixel $f_{n1}(f_n(a, b, d_1, d_2, \cdots, d_n, \theta_n), d_{n+1}, \theta_{n+1} + 180°))$ by respecting the rule of connexion $R_{n0}(d_n, \theta_n)$(respectively the rule of connexion $R_{n1}(d_n, \theta_n + 180°)$), while respecting also the conditions $C_{2^n-1} \cdots C_{2^{n+1}-2}$ for each value of n ($n = 1,2,3, \cdots$). The conditions $C_{2^n-1} \cdots C_{2^{n+1}-2}$ concern the constraints that have to respect $a, b, d_1, d_2, \cdots, d_n$ and the size of the image window. These constraints are major in the processing of the pixels placed in border of the image or in border of the image window. The order of the textural parameter to estimate is the depth of the generic tree. This order is equal here to $n + 1$. On the figure 2, we present the generic for the calculation of textural parameters of order two and on figure 3, we

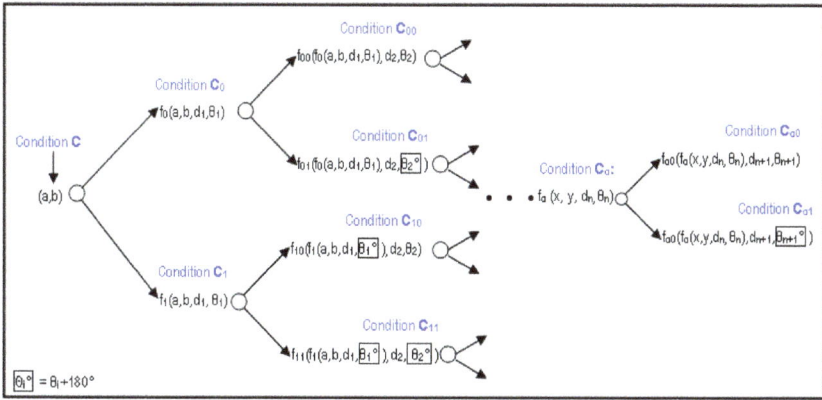

Figure 1. Generic tree for the evaluation of textural parameters of order n>1

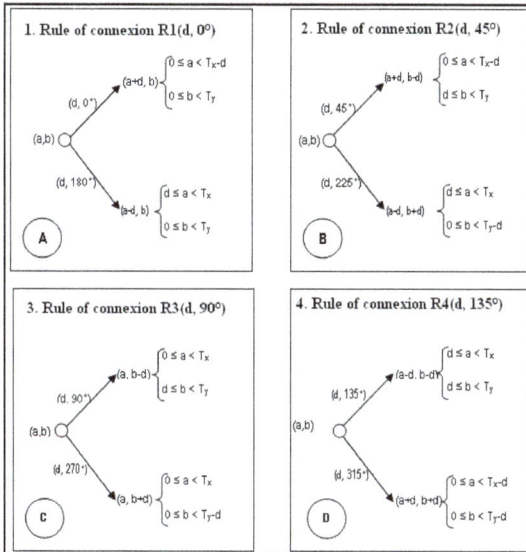

Figure 2. Exploration of the generic tree for the calculation of textural parameters of order two.

present the generic tree for the calculation of textural parameters of order three, for various rules of connexion. The conditions $C_{2^n-1} \cdots C_{2^{n+1}-2}$ are detailed on each of these figures.

- (a, b) is a position of pixel in the image window;
- d is a distance between pixels;
- T_x is a width of window image;
- T_y is a height of window image.

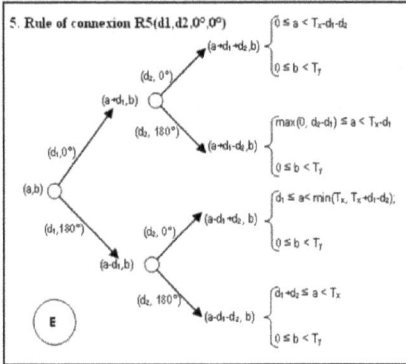

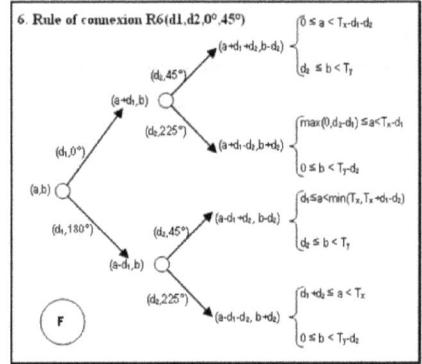

Figure 3. Exploration of the generic tree for the calculation of textural parameters of order three. E : rule of connexion R1 (d1, d2, 0°, 0°) ; F : rule of connexion R2 (d1, d2, 0°, 45°).

3.2.2. Formulation of the textural parameters of order n>1 from the generic tree

On table 1, the expressions of some textural parameters are presented. They are expressed by the approach of generic tree at the order n >1. The corresponding classical formulations are also presented there.

3.2.3. Evaluation of textural parameters from the generic tree

We interest, for example, in the evaluation of the textural parameter «asymmetry» in the order two, in some pixels of an image. The classical formulation of this parameter expresses itself by the equation 4:

$$Asym2 = \sum_{i=0}^{N_g}\sum_{j=0}^{N_g}|i - j| \times p(i,j) \tag{4}$$

In this equation, N_g is the maximal level of gray in the image window and $p(i,j)$ is the occurrence frequency of the pair of levels of gray (i,j) in the image, respecting the rule of connexion $R(d,\theta)$. This parameter can still express itself by the equation 5 :

$$Asym2 = \sum_{i=0}^{N_g}\left(\sum_{j=0}^{N_g}\underbrace{|i - j| + |i - j| + \cdots + |i - j|}_{p(i,j)\ times}\right) \tag{5}$$

The approach using generic tree in this study contains the following stages:

a. create a vector $ASS[N_g + 1]$ (that is a vector of size $N_g + 1$) in a dimension such as the relation below (equation 6) is verified:

$$Asym2 = \sum_{i=0}^{N_g} ASS[i] \tag{6}$$

That means that the vector $ASS[\ \ ]$ is calculated by the following expression (equation 7):

$$ASS[i] = \sum_{j=0}^{N_g} \left(\underbrace{|i - j| + |i - j| + \cdots + |i - j|}_{p(i,j)\ times} \right); \tag{7}$$

b. initialize the vector $ASS[\ \ ]$ with 0 (the value 0 is affected in all the entries of the vector);
c. for the entry i of vector $ASS[\ \ ]$, one adds the value $|i - j|$, provided that pixel j exist and respect the rule of connexion $R(d, \theta)$.

i being pixel in position (a, b) on the generic tree (figure 1), if the condition C_1 (respectively the condition C_2) is verified j will be the pixel in position $f_0(a, b, d_1, \theta_1)$ (respectively in position $f_1(a, b, d_1, \theta_1 + 180°)$). In other words, the entry i of vector $ASS[\ \ ]$ will be evaluated in the following way (equation 6):

$$ASS[i] = |i - NG(a, b)| + |i - NG(f_0(a, b, d_1, \theta_1))| + |i - NG(f_1(a, b, d_1, \theta_1 + 180°))| \tag{8}$$

$NG(a, b)$, $NG(f_0(a, b, d_1, \theta_1))$ and $NG(f_1(a, b, d_1, \theta_1 + 180°))$ being levels of gray of the pixels positioned in (a, b), $f_0(a, b, d_1, \theta_1)$ and $f_1(a, b, d_1, \theta_1 + 180°)$ on the generic tree, respectively;

d. repeat the process of stage 3 for all the entries j of the vector $ASS[\ \ ]$, with $j = 0, 1, \cdots, N_g$.

These entries j represent the levels of gray of the pixels being in the considered image window. It is about the image window centered on the current pixel, which is the pixel for which one wants to determine the value of the textural parameter;

e. make the summation of all the elements of the vector $ASS[\ \ ]$. The result of this summation represents the value of the textural parameter for the current pixel (equation 9):

$$Asym2 = \sum_{j=0}^{N_g} ASS[j]. \tag{9}$$

The results produced by this new approach and those obtained by the classical approach are exactly the same, but the last method requires the evaluation of the co-occurrence matrix, which is very expensive in time of calculation and memory space of the computer. This is due to the fact that the new formulation needs only the evaluation of a one-dimension vector the size of which being equal to a side of the co-occurrence matrix. Furthermore, the evaluation of the co-occurrence matrix makes intervene a lot of operations of multiplication linked in multiple reminders, and these operations are expensive in time machine. It is important to notice that the complexity of calculation increases with the order of the textural parameter.

3.2.4. Example

Let us consider for example the image window below, centered on a pixel having the level of gray 2. Let us estimate the textural parameter «asymmetry» on this window, with the rule of connexion $R(2,45°)$.

0	1	2	4	3
4	0	0	2	3
4	4	2	0	1
4	3	2	1	2
4	2	4	4	4

1. Computation by the generic tree approach
    *   The maximal level of gray of this image window is 4. A vector $ASS[\ ]$ of size 5 is then created and initialized to the value zero.
    *   The five following values are calculated: $ASS[0]$, $ASS[1]$, $ASS[2]$, $ASS[3]$ and $ASS[4]$ according to the equation (8), by respecting the rule of connexion $R(2,45°)$ and the neighbouring conditions $C_{2^n-1} \cdots C_{2^{n+1}-2}$, with $n = 1$ (order 2). The following values are then obtained :

$$\begin{cases} ASS[0] = |0 - 4| + |0 - 2| = 6 \\ ASS[1] = |1 - 4| = 3 \\ ASS[2] = |2 - 4| + |2 - 3| + |2 - 4| + |2 - 3| + |2 - 3| + |2 - 0| = 9 \\ ASS[3] = |3 - 2| + |3 - 2| + |3 - 2| = 3 \\ ASS[4] = |4 - 4| + |4 - 2| + |4 - 4| + |4 - 0| + |4 - 2| + |4 - 1| = 11 \end{cases}$$

    *   The textural parameter $Asym2$ is equal to the result of the summation of the vector $ASS[\ ]$ elements, according to the equation 9. One obtains:

$Asym2 = ASS[0] + ASS[1] + ASS[2] + ASS[3] + ASS[4] = 32$

2. Computation with classical approach

Let us make the same calculation by the classical approach of the co-occurrence matrix of levels of gray. In that case the asymmetry is calculated in the considered window by the classical formula: $Asym2 = \sum_{i=0}^{N_g} \sum_{j=0}^{N_g} |i - j| \times P_{ij}$, P_{ij} being the number of time when the pair of levels of gray (i,j) appears in the window, by respecting the rule of connexion $(2,45°)$. The following result is obtained:

$$\begin{aligned} Asym2 = & (|0 - 1| \times P_{01}) + (|0 - 2| \times P_{02}) + (|0 - 3| \times P_{03}) + (|0 - 4| \times P_{04}) \\ & +(|1 - 0| \times P_{10}) + (|1 - 2| \times P_{12}) + (|1 - 3| \times P_{13}) + (|1 - 4| \times P_{14}) \\ & +(|2 - 0| \times P_{20}) + (|2 - 1| \times P_{21}) + (|2 - 3| \times P_{23}) + (|2 - 4| \times P_{24}) \\ & +(|3 - 0| \times P_{30}) + (|3 - 1| \times P_{31}) + (|3 - 2| \times P_{32}) + (|3 - 4| \times P_{34}) \\ & +(|4 - 0| \times P_{40}) + (|4 - 1| \times P_{41}) + (|4 - 2| \times P_{42}) + (|4 - 3| \times P_{43}) \\ = & [(1 \times 0) + (2 \times 1) + (3 \times 0) + (4 \times 1)] \end{aligned}$$

$$+[(1 \times 0) + (1 \times 0) + (2 \times 0) + (3 \times 1)]$$
$$+[(2 \times 1) + (1 \times 0) + (1 \times 0) + (2 \times 2)]$$
$$+[(3 \times 0) + (2 \times 0) + (1 \times 3) + (1 \times 0)]$$
$$+[(4 \times 1) + (3 \times 1) + (2 \times 2) + (1 \times 0)]$$

$Asym2 = 6 + 3 + 9 + 3 + 11 = 32$

One sees that the result is the same that of the generic tree approach. However, the classical approach is much more complex in the term of calculations.

4. Classification

The methodological approach adopted for the classification consists of four phases: an initial pre-processing of SAR data, a second selection phase of the textural parameters relevant for classification, a third phase mode and valleys selection of the texture image and finally the automatic classification of cluster centers identified on the histogram of image.

4.1. Pre-processing of SAR images

The first phase of treatment consisted of speckle reduction and geometric correction. SAR images are indeed affected by multiplicative noise should be discontinued, or at least reduced to a better interpretation of them (Girard, 2004).The noise is spatially decorrelated from the rest of the image and this is translated by a signal of higher frequency than the image (Pratt, 1991). The Lee filter (Lee, 1981) is applied to reduce speckle (keep low) and improve the readability of structures on the images. This filter preserves the high frequencies (discontinuities) in ERS SAR images (Caloz et al., 2001). Then, the geometric correction of SAR images is performed using topographic map 1: 200000 available on the study site. The radar image is rectified to external data made stackable.

4.2. Analysis and choice of the textural parameters

The notion of the order of statistical texture parameter is very important in matter of analysis of SAR image texture. In fact, the order 2 which is the most classical does not always allow one to identify all the components of a SAR image texture. Above the order 5, Li Wang (1994) proved that the quality of discrimination of the image is deteriorated. Various orders beginning from 2 to 5 will thus enable one to better discriminate a SAR image (Talla, 2003). In fact, experimentally we notice that each order highlight complementary structures to others orders.

Absolutely, there is no universal criterion concerning the selection of texture parameters for image classification. Most often, a succession of texture parameters tests is done and the selection is made empirically. This selection is based on several criterions, notably the capacity of the texture parameter to enhance image discontinuities (Talla et al., 2006a, 2006b), its aptitude to enhance image darkened regions and its aptitude to darken image lit regions. The choice of a parameter is jointly linked to the image itself and the usage.

Experimentally, one parameter has been selected. The mean parameter in order 3 after several tests on our studies site has been selected. The method of choice of the index and order of textural parameters is largely presented in (Fotsing et al., 2008).

4.3. Principle of detection modes and valleys of the histogram

4.3.1. Histogram modeling

The histogram of an image is a graphic representation having abscissa values of gray levels, and the ordinate the number of pixels associated with each gray level value. The mode is a local maximum and valley a local minimum of the histogram. The maximum and minimum (no zero) of a histogram indicate a group of pixels and is used to detect cluster centers. Kourgly et al. (Kourgly et al., 2003) exploit observed nesting on the experimental variogram textures for segmentation urban image.

How to extract the classes contained in a SAR image? A good method for extracting classes is that will arrive at a correct interpretation. To achieve this goal, we used thresholding techniques. We go with the principle that, the thresholding has aim to segment an image in to several classes using only histogram. This assumes that the information associated with the image alone allows the segmentation, which is to say that a class is characterized by its gray level distribution. At each peak of the histogram has an associated class.

There are numerous methods of thresholding a histogram (Diday et al., 1982; Otsu, 1979). Most of these methods are applied correctly if the histogram actually contains separate peaks. Moreover, these methods have often been developed to treat the particular case of segmentation in two classes (that is to say moving to a binary image) and generality face multi-class case is rarely warranty. In this work, we assume that each class corresponds to a different range of gray level. The histogram is then m-modal. The position of minima and maxima of the histogram H can set the thresholds $(m - 1)$ to separate the m classes.

In mathematical terms, the thresholds S_i are obtained by equations given below:

$$H(S_i) = Min[H(k)] \ with \ k \in \ ]m_i, m_{i+1}[\tag{10}$$

$$H'(S_i) = Max[H(k)] \ with \ k \in \ ]m_i, m_{i+1}[\tag{11}$$

Equations (10) and (11) indicate the thresholds for valleys and modes of the histogram, respectively. Similarly, in these expressions m_i and m_{i+1} are the mean values (modes or valleys) of the light intensity in the classes C_i and C_{i+1}. The range $]m_i, m_{i+1}[$ is obtained on the basis of average values of the valleys, these when the threshold is calculated by the equation (10) and by (11) otherwise.

The histogram gives comprehensive information on the distribution of gray levels in the image. If we note x the value of gray level, another way to represent the histogram can be to search for a mathematical expression $y = f(x)$ with y the number of pixels whose gray level x. The form of the function f determines the signature of the analyzed image. Based on the form m-modal histograms in general, we approached a curve of type polynomial regression. These polynomials have formulated by the equation 12:

$$y = a_n x^n + a_{n-1} x^{n-1} + \cdots + a_1 x + a_0 \qquad (12)$$

where n is the degree of polynomial.

The continuous function f (the continuous line passing through the vertices of each peak of the histogram) adjusted using the least squares method, allows to set the degree of the number of classes centers.

4.3.2. Determination of degree n

The choice of the number of cluster centers in a radar image is difficult. In his thesis, (Lorette, 1999) uses the entropy criterion for determining the number n of cluster centers for the analysis of urban areas on satellite images.

As part of this project, we used a map of ancient plant formation of the study site. For this, we proposed to re-issue the said card with latest radar image from the perspective of studying the direction of evolution of different forest communities. In general, we decided to use the same number of class with maps of land of the study sites in our possession.

4.3.3. Principle of threshold detection

Detection modes and valleys take place on the transformed histogram, based on the concept of change in concavity of the curve of a function. Indeed, a curve changes concavity through a local minimum (valley) or by a local maximum (mode). The points where the curve changes concavity points are very sensitive and rich in information. The detection of these on the histogram of the transformed SAR image is used to inform the analyzer on the existence of thematic classes. Thematic classes are groups of pixels with similar characteristics (or almost) with respect to their brightness values in the different thematic data. The analyst has the role to determine the usefulness of different thematic classes.

4.3.2.1. Search valleys "inter-modal thresholds"

As part of this work, identification of thresholds is conducted by analyzing the histogram of gray levels and looking for local minima.

Our detection of these thresholds used Fisher's method which uses the criterion of minimizing the sum of the inertia of each class. The calculation algorithm is the dynamic type and evaluating an optimal sequence of partitions according to the scheme described in (Cocquerez et al., 2001).

4.3.2.2. Search modes

We used to detect patterns by Bhattacharya method (Bhattacharya, 1967) that models the histogram by a weighted sum of Gaussian and identifies each mode by its mean and variance. The algorithm of the method of deployment is widely presented and discussed (Cocquerez et al., 2001).

4.4. Algorithm classification

The final image is classified by the method of detection modes and valleys of the histogram. The classification algorithm implemented is summarized by the following steps:

1. construct the histogram of the image (Table $H[\ ]$);
2. detect n local extrema of Table $H[\ ]$, where n is the number of classes desired. The abscissas of the n local extrema represent the nuclei or centers of different classes of formation;
3. group pixels of the image according to the criterion of minimum distance to the various cluster centers. Each pixel is placed in the class whose center is closest;
4. assign one color to the pixels belonging to the same class and display the resulting image.

The details of the third point of this classification algorithm can be found in (Akono et al., 2003).

5. Application

5.1. Data used

The mangrove is a type of vegetation that grows in water or in mud. We tested the proposed method on a SAR image of E-SAR program (Figure 4), registered on C-band (wavelength 5.66 cm) and VV polarization with a resolution of 6m acquired on the mangrove coastal region of Cameroon.

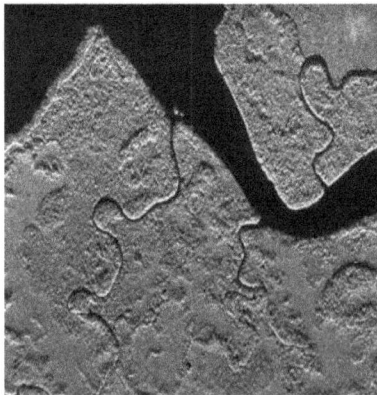

Figure 4. Experimental image SAR image of the mangrove region

The image obtained from Mount Cameroun region is also used. The studied site is situated in the south west of Cameroon. The Mount Cameroon is a volcano in activity. The image used (Figure 5) is a SAR image, acquired by ERS-1 satellite in C band (λ = 5.66cm) with VV polarization and SLC (Single Look Complex) format. Its spatial resolution is about 25m and the side of a pixel is 12.5m. This image of 8000 columns and 8269 lines has been acquired on the 7th November 1998.

Due to the lack of a *priori* knowledge on these test sites, various litho-structural maps, formation plant maps including topographic maps and lithographs at 1: 200 000 of scale allowed us to identify the various themes.

Figure 5. Original image of the Mount Cameroun region

5.2. Histogram and thresholds detection

The histogram of original image of the mangrove region (Figure 4) is illustrated on figure 6. The previous histogram is approximated by the regression curve f which is shown on Figure 7.

The previous approach, which has the advantage of being simple and fast, is well suited to images having forms regularly distributed.

The texture image is obtained from the parameter "Mean" in order 3. The calculation was performed on a window of size 7x7 around each pixel. The choice of this window size is justified by the fact that we get results like the original image with best representation of thematic classes. In addition, we found that the larger of window size was wide, there were more smoothing of the resulting image with absence of fine structure in the image.

The histogram of the texture image is shown in figure 8 and corresponding signature is shown on the figure 9. The irregularities observed on the histogram of texture images

(Figure 8) obtained from the texture index of Haralick introduced on the transformed histogram (Figure 9) many irregularities that can make the fault detection of local minima. This is the main reason that pushed us in this work to adopt a parallel approach to detect patterns of valleys.

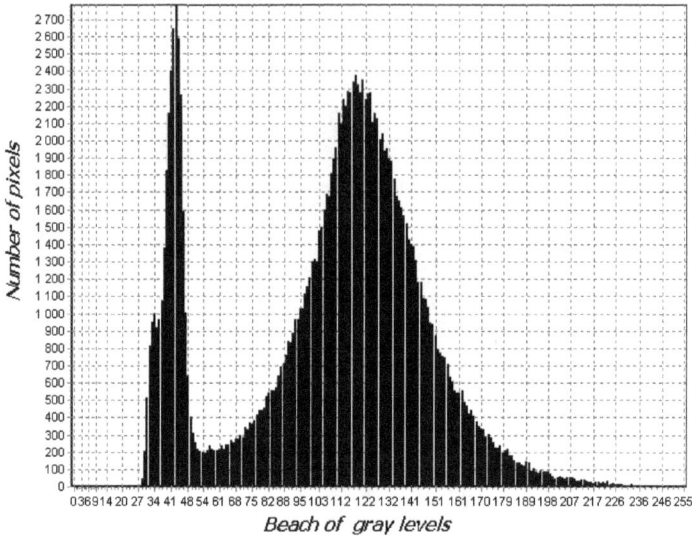

Figure 6. Filtered image histogram illustrating the presence of two classes of intensity in the image of the mangrove region.

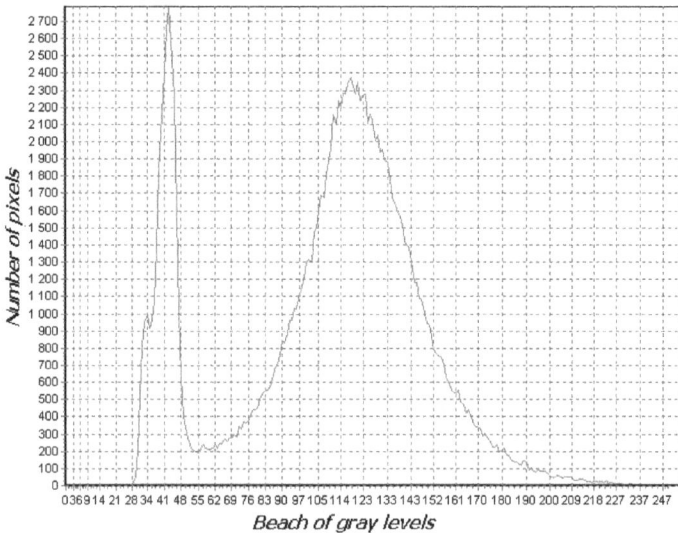

Figure 7. Corresponding signature of the filtered image illustrating the appearance of a polynomial function approximation by the least squares sense.

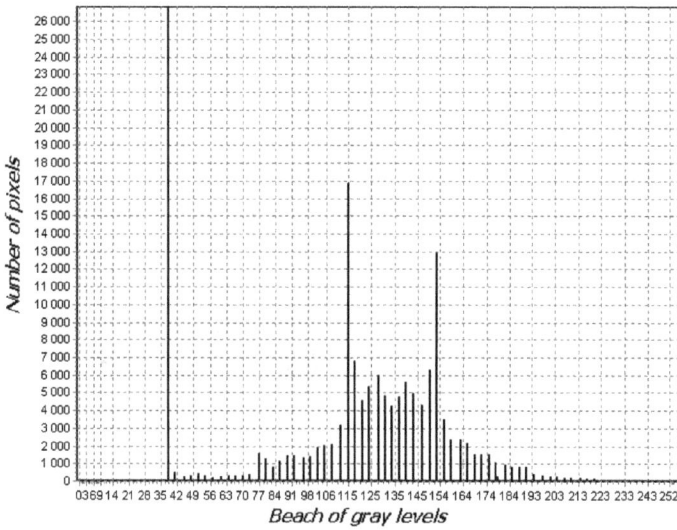

Figure 8. Texture image histogram obtained by index parameter of Haralick "Mean" on the mangrove region

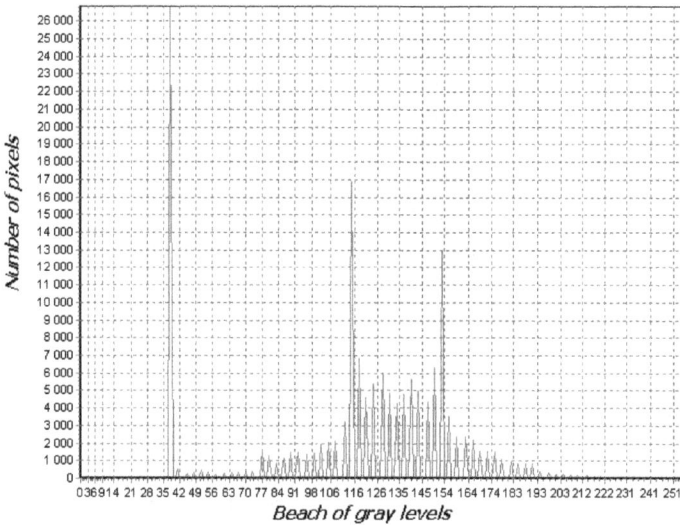

Figure 9. Corresponding signature of the texture image histogram

Table 2 shows the modes and valleys that were used in classification thresholds. It was obtained on the basis of exploitation of the histogram and the signing of the texture image which can be seen representations in figures 8 and 9.

For reference, the histogram of the original image and the texture image obtained on the basis of the radar image taken on the Mount Cameroon region are shown on figures 10 and

11. Signatures (Figures 12 and 13) corresponding to each of the histograms are also immediately followed. It is interesting to note that the texture image was obtained from the parameter "Mean" with a window of size 5x5. These parameters were got after several experimental tests.

	Thresholds / Color codes						
Modes	38	50	76	89	115	153	172
Color codes							
Valleys	45	57	83	96	121	137	179
Color code s							

Table 2. Detection thresholds for classification and color coding of thematic classes: case of mangrove region

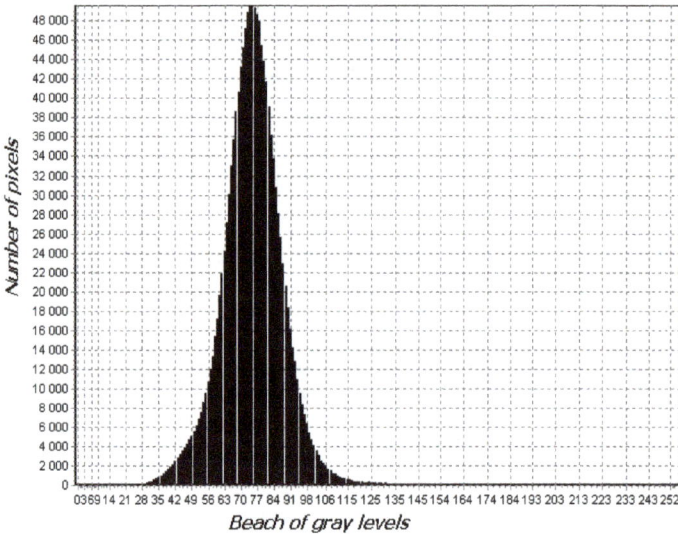

Figure 10. Filtered image histogram illustrating the presence of one main class of intensity in the image of Mount Cameroon region.

It can be seen visually that the histogram of the image filtered even not enough to make a good partition of the base image (Figures 6 and 10). This explains the poor performance of these filters in scenes that contain fine structures such as lineaments, which are generally not well preserved by these filters. Similarly, the histogram of the filtered image does not favor the detection of local extrema accurately. To remedy this shortcoming, a proposal method for modifying the histogram is implemented. It consists of a transformation of the representation of the histogram of the SAR image. For this, a histogram of the envelope curve passing through the ends of each peak is plotted (Figures 7 and 11) using the method of least squares regression. To keep up the properties of both representations, we plan to keep the same scale in both cases representation. The transformed histogram (Figures 9 and 13) favors the detection of local extrema of peaks from the accentuation of the regression line well represented.

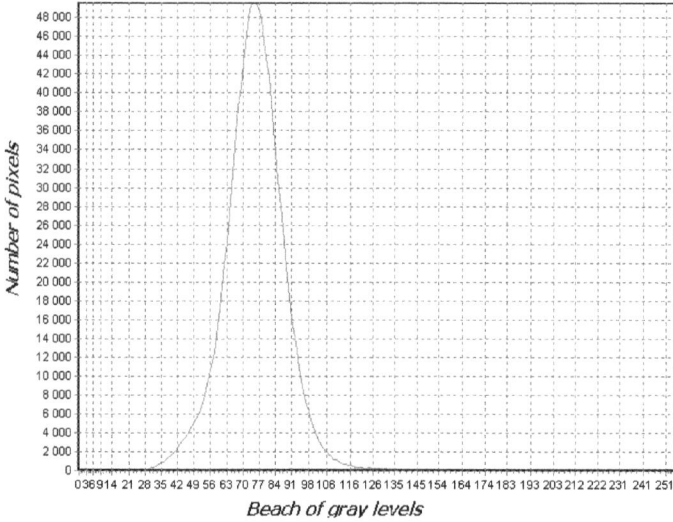

Figure 11. Signature of the filtered image illustrating the appearance of a polynomial function approximation by the least squares sense.

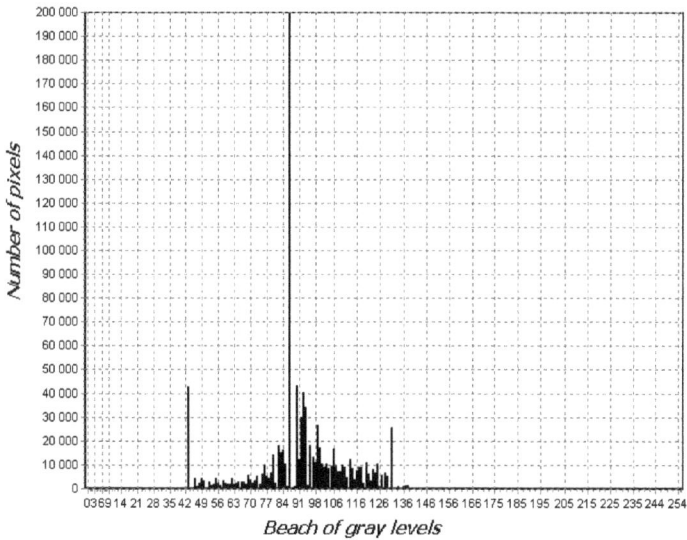

Figure 12. Texture image histogram obtained by index parameter of Haralick "Mean" on the Mount Cameroon region

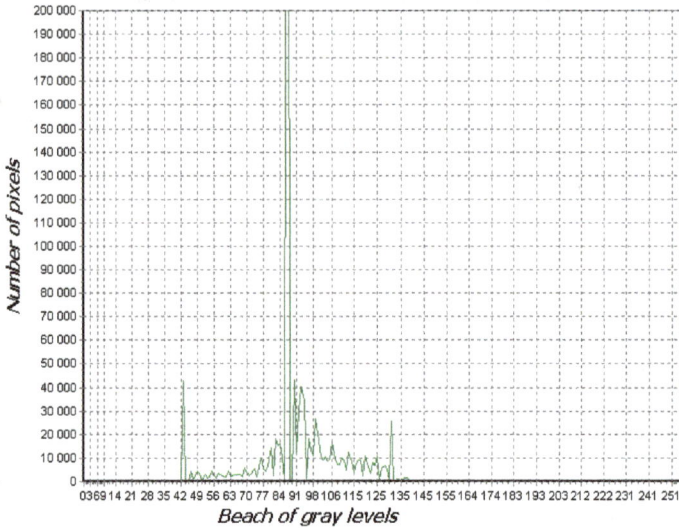

Figure 13. Corresponding signature of the texture image histogram

The image texture enhances the visual interpretation. Indeed, it contains information on the spatial distribution of color variations. This is observed in the texture image by the presence of dark shades, clear, smooth and gray. The texture image obtained does not allow partitioning of the image into separate classes because its histogram (Figure 7 or 11) does not present specific modes and valleys. It is again transformed using the histogram. This second representation offers the advantage of facilitating the visualization of local maxima and minima. Each representation has them peculiarities and shortcomings; however, the combined use of two methods of representation facilitates the detection thresholds for classification (Tables 2 and 3).

Table 3 shows the modes and valleys that were used in classification thresholds. It was obtained following the same methodological approach followed with the image data of the mangrove area.

	Thresholds / Color codes			
Modes	43	87	99	131
Color codes				
Valleys	55	90	97	103
Color codes				

Table 3. Detection thresholds for classification and color coding of thematic classes: case of Mount Cameroon region

5.3. Results of thematic maps

5.3.1. Case of mangrove region

The final classified image presents the results of unsupervised classification obtained with different thresholds identified above (Table 2). This result is obtained on the basis of 14 cluster centers. Thus, the classified image traces an occupancy map of the study site highlighting 14 thematic classes whose characterization using data from field missions, old maps and the lithographic charts, aerial photographs allowed us to establish different classes of information. The exploitation of these data led to the realization of the space map shown in figure 14.

Detecting the number of class by this approach remains a major challenge. However, after several experiments, we found that we could not indefinitely increase the number of classes. Indeed, it proved that beyond a certain value (14 in this case) noted almost no more change the final result. Moreover, for low numbers of classes, there was more of a smoothing of the final result with a merging of small units in large.

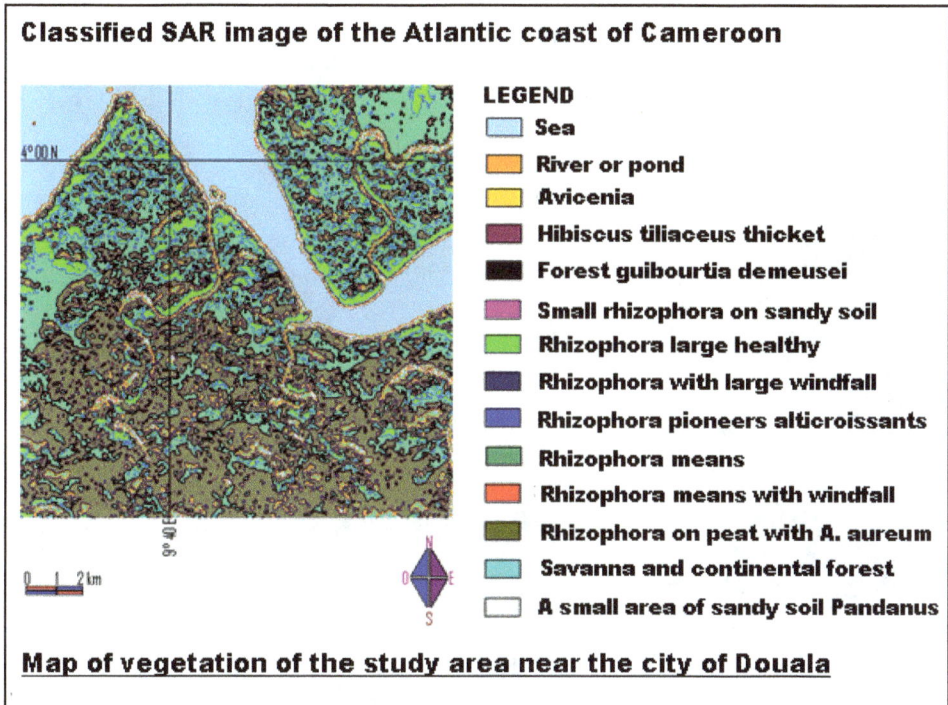

Classified SAR image of the Atlantic coast of Cameroon

LEGEND
- Sea
- River or pond
- Avicenia
- Hibiscus tiliaceus thicket
- Forest guibourtia demeusei
- Small rhizophora on sandy soil
- Rhizophora large healthy
- Rhizophora with large windfall
- Rhizophora pioneers alticroissants
- Rhizophora means
- Rhizophora means with windfall
- Rhizophora on peat with A. aureum
- Savanna and continental forest
- A small area of sandy soil Pandanus

Map of vegetation of the study area near the city of Douala

Figure 14. Satellite map from the mangrove region of the estuary of Douala, Cameroon

The classified image obtained from the 14 threshold and observing the color code (Table 2) is shown in Figure 14. This classification method belongs to the family of unsupervised classification. Indeed, the spectral classes are first formed. These classes are based solely on digital information data. As a result, the classification algorithm presents below is used to determine statistical natural groups of data. We obtain quite detailed thematic classes. The 14 classes provide a map of land of the study site with good delineation of the different classes (Figure 14).

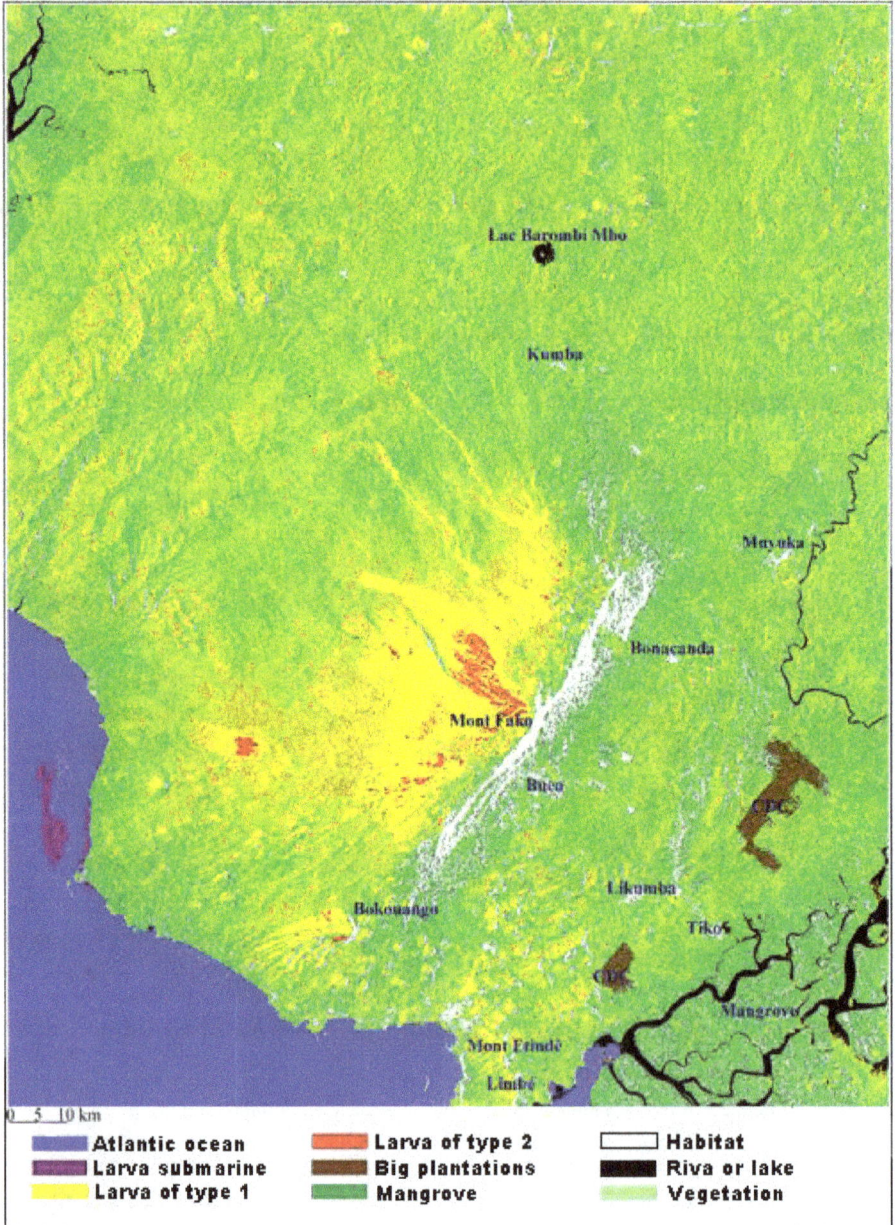

Figure 15. Satellite map from the Mount Cameroon region

Thematic classes are categories of interest that the analyst tries to identify in the images, as different types of crops, forests or species of trees, different types of rocks or geological features, etc.

The result obtained in this study presents a double advantage. The first advantage is at the level of accuracy in the identification of certain classes of information like the class sea. The second one is in the identification of certain thematic classes within certain classes of information and therefore a precise characterization can help in highlighting other information, as for example, different canopies in terms of vegetation class information.

5.3.2. Case of mount Cameroon region

In Figures 10 and 11, we have uni-modal aspect of the histogram of the image after filtering. This makes difficult any exploitation for segmentation. To overcome this drawback, it then uses to texture images.

Figures 12 and 13 show them with now m-modal shape observed on the histogram of the image texture. After several experiments, we retained eight cluster centers summarized in table 3 with the color codes used for each class center.

For the Mount Cameroon region, the same approach as that used previously on the mangrove is applied and the result is presented on Figure 15. The use of maps available and research work on this site (Akono et al., 2005, 2006) are used to characterize the eight thematic classes. The use of this information provides the classified image of the Mount Cameroon region including the specification of each class of information is summarized on the legend of Figure 15. As can be seen at the legend, we have nine more thematic classes instead of eight. This is because when classifying all pixels are not classified. All unclassified pixels were grouped in class vegetation. Furthermore, a broad thematic class (e.g. forest) may contain multiple spectral classes with spectral variations. Using the example of the forest, the spectral sub-classes can be caused by variations in age, species, tree density or simply the effects of shadowing or variations in illumination. The analyst's job is to determine the usefulness of different thematic classes and their correspondence to the thematic classes useful.

6. Conclusion

The purpose of this study was the production of space maps with the synthetic aperture radar (SAR) images. To achieve this, we proceed by adopting approaches optimized of texture analysis of images, using the statistical parameters of Haralick generalized at the order n. The approach is based on the concept of generic tree. It has the advantage of being less time consuming calculation from the conventional approach which frequently uses co-occurrence matrices for texture analysis and especially the processed images are generally very large sizes. In classification, the approach relies on the concept of detecting "modes" and "valleys" of histograms in a SAR image using classification of type unsupervised. For each of the SAR images, the histogram of the convolution image obtained at base of texture parameter is represented and approximated by a regression line called "signature" using least square method. The operation of the histogram and the signature of the texture image can facilitate the detection of classification thresholds. The main interest of the proposed approach is that we have results that are approaching the best of the reality field; it also does not require a serial multi-date SAR data for the realization of satellite image maps. The method was successfully tested on two satellite images from two different sensors: one from the ESAR program obtained at the resolution 6m and one other from the ERS-1 sensor of resolution 25 m.

A limitation of the classification approach lies at the empirical detection of local extrema. A perspective would then be to automate the detection of the number of classes and local extrema.

Author details

Janvier Fotsing
Corresponding Author
University of Buea, Faculty of Science/Department of Physics, Cameroon

Emmanuel Tonye
University of Yaounde I, National Advanced School of Engineering,
Department of Electrical and Telecommunications Engineering, Cameroon

Bernard Essimbi Zobo
University of Yaounde I, Faculty of Science, Department of Physics, Cameroon

Narcisse Talla Tankam
University of Dschang, Fotso Victor Institute of Technology, Department of Computer Sciences,
Cameroon

Jean-Paul Rudant
University of Marne-La-Vallée, Institut Francilien des Géosciences, France

Acknowledgement

This work was supported by the LETS laboratory of the National Advanced School of Engineering of the University of Yaoundé I. We are also grateful unto the European Spatial Agency (ESA) for the grant of SAR ERS-1 image used in this study.

7. References

Akono, A.; Tonyé, E.; Ndi Nyoungui, A. (2003). Nouvelle méthodologie d'évaluation des paramètres de texture d'ordre trois. *Internationnal Journal of Remote Sensing*, vol.24, n°9, pp. 1957-1967.

Akono, A.; Talla Tankam, N.; Tonyé, E.; Dzepa, C. (2005). Nouvel algorithme d'évaluation des paramètres de texture d'ordre n sur la classification de l'occupation des sols de la région volcanique du Mont Cameroun. *Télédétection*, vol. 5, n° 1-2-3, pp. 227-244, Avril 2005.

Akono, A.; Talla Tankam, N.; Tonyé E.; Ndi Nyoungui, A.; Dipanda, A. (2006). High Order Textural Classification of two SAR ERS images on Mount Cameroon. *Geocarto International*, vol. 21, n° 3, pp.1-16.

Baltzer, F.; Rudant, J.-P.; Tonyé, E. (1996). Applications de la télédétection micro-ondes en bande C à la cartographie des mangroves de la région de Douala. *Proceedings of the second ERS applications Workshop*, ESA SP 383, pp. 455-461.

Bhattacharya, C., G. (1967). A simple method of resolution of a distribution into gaussian components. *Biometrics*, vol. 23, pp.115-135.

Caloz, R.; Collet, C. (2001). *Précis de télédétection, traitements numériques d'images de télédétection* (Presses de l'Université du Québec / AUF, Québec). Vol.3, 386 pages.

Cocquerez, J. -P.; Philipp, S.; Maître, H. (2001). *Analyse d'images: filtrage et segmentation* (Masson). Paris, Milan, Barcelonne.

Diday, E.; Lemaire, J.; Pouget, J.; Testu, F. (1982). *Elements d'analyse de données* (Dunod, Paris).

Fotsing, J.; Tonyé, E.; Talla Tankam, N.; Kanaa, T., F., N.; Rudant, J.-P. (2008). Classification non supervisée d'image RSO à l'aide d'extremums locaux d'histogramme: application à la cartographie de la mangrove littorale camerounaise. *Revue Française de la Photogrammétrie et de Télédétection*, vol. 1, n°189, pp. 28-39.

Franklin, S., E; Peddle, R., D. (1989). Spectral texture for improved class discrimination in complex terrain. *International Journal of Remote Sensing*, vol. 10 n°8, pp. 1437-1443.

Girard, M-C. (2004). *Traitement des données de télédétection* (Dunod). 530 pages.

Haralick, R., M.; Shanmugam, K.; Dinstein, I. (1973). Textural features for image classification. *IEEE Transactions on Systems, Man and Cybernetics*, vol. 3, n° 6, pp. 610 - 621.

Haralick R., M. (1979). Statistical and structural approaches to texture. *Proceedings of the IEEE*, vol. 67, n°5, pp. 786-804.

Jukka, H.; Aristide, V. (1998). Land cover/land use classification of urban areas: a remote sensing approach. *Internationnal Journal of Pattern Recognition and Artificial Intelligence*. Vol.12, n°4, pp. 475-489.

Kourgli, A.; Belhadj-Aissa, A. (1999). Nouvel algorithme de calcul des paramètres de texture appliqué à la classification d'images satellitaires. *Acte des 8 èmes Journées Scientifiques du Réseau Télédétection l'AUF*. Pp. 109-118, Lausanne, Suisse, 22-25 novembre 1999,. Ed. AUPELF-UREF, Analyse critique et perspectives. ISBN 2-920021-92-3.

Kourgly, A.; Belhadj-Aissa, A. (2003). Segmentation texturale des images urbaines par le biais de l'analyse variographique. *Revue Télédétection*, vol. 3, n° 2-3-4, pp. 337–348.

Lee, J.,S. (1981). Speckle analysis and smoothing of synthetic aperture radar images. *Computer Graphics and Image Processing*, vol. 17, pp. 24-32.

Li, W. (1994). Vector choice in the texture spectrum approach. *International Journal of Remote Sensing*, vol. 15, n°18, pp. 3823-3829.

Linders, J. (2000). Comparison of three different methods to select feature for discriminating forest cover types using SAR imagery. *International Journal of Remote Sensing*, vol. 21, n° 10, pp. 2089 - 2099.

Lorette, A. (1999). Analyse de texture par méthodes markoviennes et par morphologie mathématique: application à l'analyse des zones urbaines sur des images satellitales.

Thèse présentée à l'Université de Nice- Sophia Antipolis pour obtenir le titre de Docteur en Sciences Spécialité Sciences de l'Ingénieur. 162 pages.

Lucieer, A.; Stein, A.; Fisher, P. (2005). Multivariate texture-based segmentation of remotely sensed imagery for extraction of objects and their uncertainty. *International Journal of Remote Sensing*, vol. 26, n° 14, p. 2917 - 2936.

Marceau, D.,J.; Howarth, P., J.; Dubois, J-M., M.; Gratton, D., J. (1990). Evaluation of the gray-level co-occurrence matrix method for land-cover classification using SPOT imagery. *IEEE Transactions on Geoscience and Remote Sensing*, vol. 28, n° 4, pp. 513-518.

Oliver, C., J. (1993). Optimum texture estimators for SAR clutter. *Journal of Physics D: Applied Physics*, vol. 26, pp. 1824 - 1835.

Otsu, N. (1979). A threshold selection method from gray-level histograms. *IEEE Trans. On SMC*, vol.11, pp. 191-204.

Peckinpaugh, S., H. (1991). An improved method for computing gray-level co-occurrence matrix based texture measures. *Graphical models and Image Processing*, vol.53, pp. 574-580.

Pratt, W. (1991). *Digital image processing* (2nd edition, Willey-interscience). 698 pages.

Puissant, A.; Hirsch, J.; Weber, C. (2005). The utility of texture analysis to improve per-pixel classification for high to very high spatial resolution imagery. *International Journal of Remote Sensing*, vol. 26, n° 4, p. 733 - 745.

Randen, T.; Husoy, J., H. (1999). Filtering for texture classification: a comparative study. *IEEE Transactions PAMI*, vol. 21, n° 4, pp. 291-310.

Reed, T., R.; Hans Du Buf, J., M. (1993). A review of recent texture segmentation and feature extraction techniques. *CVGIP: Image Understanding*, vol. 57, n° 3, pp. 359-372.

Rudant, J-.P.; Baltzer, F.; Tupin, F.; Tonyé, E. (1997). Distinction entre formations végétales littorales et continentales dans leur rapport avec la géomorphologie: intérêt des images ERS-1. *Symposium ERS-1, Publications ESA*, pp. 1069-1073, Florence, 14-21 mars 1997.

Talla Tankam, N. (2003). Nouvelle méthodologie d'évaluation des paramètres de texture d'ordre n>1: Application aux images radar du Mont Cameroun. *Mémoire de DEA*, Université de Yaoundé I (Cameroun), 60 pages.

Talla Tankam, N.; Dipanda, A.; Tonyé, E.; Akono, A. (2006a). Classification d'images satellitaires radars RSO par valeurs propres de texture. Application à la mangrove littorale Camerounaise. *Actes du Colloque Africain sur la Recherche en Informatique (CARI'06)*, Cotonou (Bénin), 6-9 novembre 2006.

Talla Tankam, N.; Kouamé Koffi, F.; Dipanda, A.; Akono, A.; Bernier, M.; Tonyé, E.; Affian, K. (2006b). Caractérisation des discontinuités-images par l'approche de vecteur de texture: application à des images RSO d'ERS-2. *Journal des Sciences pour Ingénieurs (JSPI)*, n° 7, ISSN 0851-4453.

Talla Tankam, N. (2008). Une nouvelle approche d'analyse automatique de texture d'images : application à l'étude de la dynamique d'occupation spatiale sur le Mont Cameroun et ses environs. *Cotutelle de thèse de Doctorat/PhD en Informatique*, Université de Yaoundé I, Cameroun et l'Université de Bourgogne, France, 194 pages.

Ulaby F., T.; Kouyate, F.; Brisco, B.; Lee Williams T., H. (1986). Textural information in SAR images. *IEEE trans. Geosci. Remote Sens.* GE-24, pp. 235-245.

Unser, M. (1995). Texture classification and segmentation using wavelet frames. *IEEE Trans. On Image Processing*, vol. 4, n°11, pp. 1549-1560.

Chapter 10

GIS-Based Models as Tools for Environmental Issues: Applications in the South of Portugal

Jorge M. G. P. Isidoro, Helena M. N. P. V. Fernandez,
Fernando M. G. Martins and João L. M. P. de Lima

Additional information is available at the end of the chapter

1. Introduction

Geographical Information Systems (GIS) simulate a given geographic space in a computational environment, allowing to store, map and analyse large amounts of georeferenced data (*e.g.*, Umbelino *et al.*, 2009). GIS were converted into a powerful tool in regional natural resources assessment, as it permits a speedy integration and representation of several biophysical attributes (*e.g.*, Bastian, 2000; Bocco *et al.*, 2001) from diverse origins such as, *e.g.*, topographic, cartographic, photogrametric, GPS and remote sensing.

The integration of Digital Terrain Models (DTM) in GIS leads to the emergence of methodologies to represent and simulate the real-word, complementing the use of thematic environmental information (*e.g.*, Felicísimo, 1999). A DTM is a numerical representation of a variable obtained from a discrete set of points, with well-known cartographic coordinates, which distribution allows calculating, by interpolation, that variable for any arbitrary point (Fernandez, 2004). If the mesh of points is altitude-related, the DTM is designated as a Digital Elevation Model (DEM). From a DEM, it becomes easy to attain topographic-derivate models (*e.g.*, slopes, orientations, curvature and visibility, shadowed and exposed areas).

Many authors have used DEM processing techniques to automatically extract geographic features (*e.g.*, Herrington & Pellegrini, 2000; MacMillan *et al.*, 2000; Burrough *et al.*, 2001; Jordán *et al.*, 2007b; Zavala *et al.*, 2005a, 2007), hydrologic structures (*e.g.*, Flanagan *et al.*, 2000; Maidment, 2000), erosive processes (*e.g.*, Zavala *et al.*, 2005b), vegetation habitats (*e.g.*, Anaya-Romero, 2004; Anaya-Romero *et al.*, 2005; Jordán *et al.*, 2007a; Pino *et al.*, 2010), among other uses.

In this chapter three GIS-based models were used, relying on different methodologies and techniques, to illustrate visualization and quantification of the geomorphic processes. This valuable input for decision-makers is attained through the versatility of GIS. Geo-form mapping in a coastal lagoon catchment, rainfall erosivity on a mountain ridge and urban flood delimitation show the potential usefulness of DEM/DTM based cartographic models for helping to solve environmental issues. All case studies presented are from the south of mainland Portugal.

The first case study is used to illustrate the main geo-form classes of the catchment including the Ria Formosa. The Ria Formosa is a shallow coastal lagoon covering an area of about 16,000 ha in the south of Portugal. It is protected by EU and national Laws, and is classified as a Wetland of International Importance under the RAMSAR convention (PORTUGAL Ramsar Site 212). This study aims to establish a method based on the Hammond hierarchical criteria and geographical information related to soft-slopes, local topography and terrain profiles, to locate and classify the geo-forms present in the Ria Formosa catchment.

The second case study focuses on the use of DEM/DTM based on climate models to obtain and analyze isohyetal maps, and to identify how rainfall distribution influences water erosion. Rainfall distribution, which is highly variable in space and time, is difficult to study, due to the lack of good quality data (*e.g.*, insufficient or poorly-distributed gauges in the study areas; non-homogeneous rainfall data series; dubious readings from non-automated gauges; lack of radar coverage). This issue may be partially addressed by geographic models (*e.g.*, DEM/DTM) and climate data related to rainfall. Parameters such as curvature, slope and orientation of hillslopes, which influence local climate, can also be obtained from these geographic models. Isohyetal maps, through multilinear regression analysis, can then be created using the DEM/DTM and their derivative models, the hillshading potential and some specific data (*e.g.* distance to coastline). From these elements and using the Modified Fournier Index (MFI), rainfall erosivity can finally be quantified. This technique was used to assess the soil erosion risks in Serra de Grândola, which is a north-south oriented mountain ridge with an altitude of 383 m, located in southwest mainland Portugal.

The third case study demonstrates the use of cartographic information to produce flood delimitation maps. The city of Tavira (10,600 inhabitants) in the south of Portugal embraces the outfall of the Séqua/Gilão River into the Atlantic Ocean. A GIS-based hydrologic model of the 221 km² Séqua/Gilão river catchment was first created to obtain soil use and type regional parameters. Afterwards, and to identify the maximum water heights for those return periods, a hydraulic model of the rivers' last 9.5 km was produced. These maximum water heights were compared with the observed values (flood level marks, photographs and video records) of the 3rd December 1989 flood and used to validate the model. Mean sea level changes due to climate change were also considered. With this procedure, it was finally possible to produce flood delimitation maps for the Tavira urban area. This type of modelling may provide a useful tool for urban planners and city authorities.

2. Location and classification of geo-forms in the Ria Formosa estuary

2.1. Introduction

The study of the risk of soil degradation is the starting point for development and sustainable land management. The global warming and the land use changes expected in the XXI Century predict loss of quality and reduction of soil's productivity (Cerdan *et al.*, 2010). In the Mediterranean area the natural and semi-natural vegetation is sclerophyllous, which is well suited for the local climatic conditions, however, extreme weather and human activity can cause imbalances in the ecosystem (Kosmas *et al.*, 2000). The southern Portugal is a region where the balance between the natural environment and human activity is very sensitive to erosion and desertification (Gonçalves *et al.*, 2010). Thus, it is necessary to employ control, prevention and correction measures to preserve the soil and prevent the emergence and intensification of desertification processes, which can become irreversible, as has happened in other Mediterranean areas (Kosmas *et al.*, 2000).

This study emphasizes geomorphologic processes, because they describe the natural space, the dynamics of occupations and the anthropogenic changes. According to Hammond (1954, 1964), geomorphologic studies of the earth's surface can be carried out over large areas (small scale) and based on the analysis of the land features, topographic maps or directly through field measurements. The variables considered should be quantitative, been used for a hierarchical classification.

Later, with the appearance of the GIS, Dikau (1989) and Dikau *et al.* (1991), the Hammond procedures were rectified and automated in some regions in the United States. The DEM becomes the most important tool in the process of identifying landforms. Using the "moving-window" process and through algorithms based on local operators (spatial filters) it is possible to create models (such as slope, local relief and relative position). This allows a more detailed study, with a large number of variables in a large scale. The landforms hierarchy is made in terms of size, order and geometric complexity.

Many researchers have used this methodology, although with some modifications as explained by Martínez-Zavala *et al.* (2004) and Jordan *et al.* (2005), in studies conducted in Spain and Mexico in a detailed scale. Drăgut and Blaschke (2006), and Gerçek *et al.* (2011) classified the landforms in regions of Transylvania (Romania), Germany and Turkey, respectively based in decision rules of fuzzy logic. Another technique widely used was the non-supervised classification based on levels of information. For example, Sallun *et al.* (2007) in the catchment of Alto Rio Paraná in Brazil extracted information by applying the Principal Component Analysis (PCA) in multispectral satellite images. Oliveira and Santos (2009), applied spatial filters of high pass and low pass in DEM in Feira de Santana (state of Bahia, Brazil). Teng *et al.* (2009) extracted the landforms in Shaanxi Province (China) with hillslope units from DEM.

In the present study we intend to carry out the mapping of landforms in the catchment of the Ria Formosa located in the south of Portugal (Algarve), using the methodology followed by Jordán *et al.* (2005), modified in order to take in consideration the specific features of this region.

This study contributes to a better characterization of the region allowing the preparation of regional plans to control the processes of soil degradation, with an indication of possible uses and restrictions.

2.2. Study area

The Ria Formosa catchment is limited by the WGS84 coordinates 37º 15' N to 36º 57' N and 7º 28' W to 8º 4' W. It has an area of 864 km² and a perimeter of 166 km, including a shallow coastal lagoon with an area of about 16,000 ha. It is protected by EU and Portuguese Laws, and is classified as a Wetland of International Importance under the RAMSAR convention (PORTUGAL Ramsar Site 212). It covers the municipal areas of Tavira, Faro, Olhão, São Brás, Loulé, Vila Real de Santo António and Castro Marim. The topography of the region is regular and continuous without abrupt changes in altitude. The average slope is 11% and the elevation varies between 0 and 530 meters above sea level. Mean annual rainfall of the catchment ranges between 400 and 800 mm. The mean annual temperature is 17 ° C.

2.3. Methodology

The DEM has been used as a basic source of information on the catchment of the Ria Formosa and was obtained from a geostatistical study with a resolution of 10×10 m², which was based on cartography at the scale 1:25,000 from the Geographical Institute of the Portuguese Army (IGeoE, 2004). From this model, other information about other terrain features were obtained. The analysis and mapping of the data has been performed with the IDRISI Taiga software (Eastman, 2009). With this software, several terrain variables such as slope, curvature, relative position, and local relief were modelled for each point relatively to the DEM. Finally, the automatic classification of landforms was carried out, as established by Jordan *et al.* (2005).

2.3.1. Slope and curvature

Maps of slopes and curvatures are commonly used to describe the hydrologic drainage structure of a region or a catchment. Soil properties and the characteristics of the hillslopes are factors that combined determinate a higher or lower resistance to soil erosion, especially due to rainfall. The inclination, the length and the shape of a slope are associated to the velocity of runoff and to the water infiltration into the soil.

The slope at point (M, P), in the azimuth direction A, is given by calculating the inner product between the gradient of the surface H and the unit vector $\bar{w}_A$ with the components ($\alpha = SinA$; $\beta = CosA$):

$$\nabla H \mid \bar{w}_A = \frac{\partial H}{\partial X}\alpha + \frac{\partial H}{\partial Y}\beta = \|\nabla H\| = \sqrt{\left(\frac{\partial H}{\partial X}\right)^2 + \left(\frac{\partial H}{\partial Y}\right)^2} \tag{1}$$

When the ground is represented by a matrix H (m, n), X and Y are coordinate axes, and the gradient of the surface H represents the topography of a square that includes point (M,P) defined by a bi-linear polynomial expression:

$$H = aMP + bM + cP + d \tag{2}$$

The coefficients a, b, c and d are determined at the expense of the coordinates of the square-defined four vertices.

The determining of slope of each pixel in line l and column k is based on elevation values of neighbouring pixels and the spatial resolution of the model, E (distance between pixels). The slope is calculated using Eq. (3).

$$\delta = \sqrt{\left(\frac{H(l,k+1) - H(l,k-1)}{2E}\right)^2 + \left(\frac{H(l-1,k) - H(l+1,k)}{2E}\right)^2} \tag{3}$$

From the analysis a smooth slope map was produced, using the "Moving Window" technique with a size of 4,900 m² (*i.e.*, 7×7 matrix). For each window was determined a percentage of soft slope (considered below 4%), and that value was assigned to the central pixel of the window. After this process, the map was reclassified (Table 1).

Code	Percentage of soft slope
1	More than 80%
2	50% at 80%
3	20% at 50%
4	Less than 20%

Table 1. Reclassification of the slopes by percentage of soft slopes.

The curvature is defined according to the rate of change of the slope, which is determined by the partial derivatives of second degree of surface H:

$$C_A = \nabla^2 H = \frac{\partial^2 H}{\partial X^2} + \frac{\partial^2 H}{\partial Y^2} \tag{4}$$

As we are working on a discrete space, it is possible to approximate the Laplacian in two dimensions, using the finite difference method, being the size equal to the unity (one cell):

$$\nabla^2 H = H(l+1,k) + H(l-1,k) + H(l,k+1) + H(l,k-1) - 4H(l,k) \tag{5}$$

Therefore, the same expression in the matrix form, can be written as follows:

$$\begin{bmatrix} 0 & 1 & 0 \\ 1 & -4 & 1 \\ 0 & 1 & 0 \end{bmatrix} \tag{6}$$

This matrix 3×3 is called the Laplacian filter, which was used in a spatial convolution process on the MDT, wherein each central cell of the window was assigned a curvature value. The negative values indicate concavity (sedimentation basins, valleys, etc.), while positive values indicate convexity (massifs, domes, peaks, upper parts of slopes, etc.). The values equal or very close to zero correspond to flat surfaces.

After extraction of the curvature, it was necessary to make a spatial convolution in order to filter the unhelpful and inconsistent information and highlight the most important formations. We used a Gaussian filter with 10×10 m² cells within a 7×7 matrix.

2.3.2. Local relief and relative position

The local relief can be expressed as the vertical difference between the highest point and the lowest points, of a surface, within a certain horizontal distance or in a determined area of analysis (Figure 1; Left). The relative position sets up the flat shapes of the terrain, in uplands and lowlands, separating the plateaus from the plains with hills or mountains. In this study we considered that all concave and convex areas were, respectively, lowlands and highlands (Figure 1; Right).

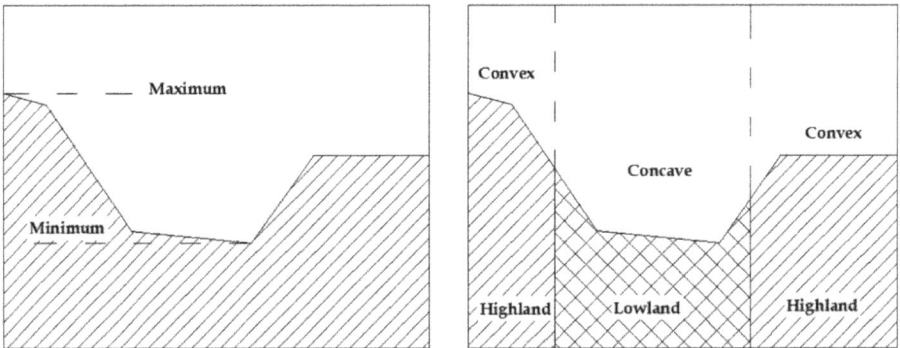

Figure 1. Left: Local relief; Right: Relative position.

The determination of the local relief was calculated directly from the MDE, through two spatial convolution processes of a (7×7) matrix. The maximum and minimum elevations were determined and replaced in the respective central cells. At the end the results were subtracted. After this process, the map was reclassified as follows (Table 2).

Code	Class	Classification
1	0 - 15 m	Very smooth
2	15 - 30 m	Smooth
3	30 - 90 m	Localized
4	90 - 150 m	Moderate
5	150 - 220 m	Rough

Table 2. Classes of the local relief.

To calculate the relative position, the curvature and the smooth slopes (<4%) were integrated. The classification was used to define the classes presented in Table 3.

Code	Class
a	> 75% of smooth slopes on concaves hillsides
b	50 – 75 % of smooth slopes on concaves hillsides
c	50 – 75 % of smooth slopes on convexes hillsides
d	> 75% of smooth slopes on convexes hillsides

Table 3. Classes of the relative position.

2.3.3. Landforms

The mapping of landforms was constructed by crossing three levels of information: smooth slope, local relief and relative position. The five main forms considered were: plains, plateaus, plains with hills, and open hills. In turn, these classes were divided into twenty sub-classes.

2.4. Results and discussion

The main forms and the respective subclasses, for the region of Ria Formosa are shown in Figure 2 and Table 4.

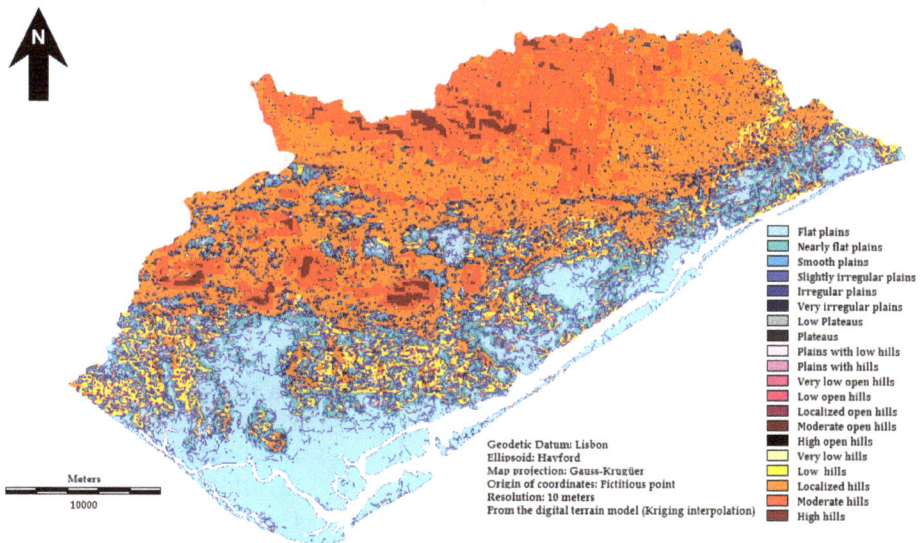

Figure 2. Map of landforms in the Ria Formosa catchment.

As can be seen, the *localized hills* and *moderate hills* (Table 4) cover most of the catchment area (43%), mainly on areas such as Mountain of Caldeirão, and the mountain formations of carbonated rocks (Monte Figo, Malhão, Guilhim and Nexe).

In particular, in the Mountain of Caldeirão, there is a very dense drainage network due to the formation of Flysch from Baixo Alentejo region, consisting of turbidites (greywackes, silts and shales) which, because of its layered structure, hamper the infiltration of surface runoff. On the other hand the vegetation is typically mediterranean, composed by quercíneas and sclerophyllous over Eutric Leptosol. Therefore on more rugged slopes, there is little capacity for water storage, making it difficult to sustain the vegetation. Furthermore, poor agricultural practices have been destroying the natural vegetation, causing in the rainy season a deterioration of these soils and turning them into skeletal. So the predictable risk of erosion ranges from moderate to high.

The flat plains are the second class of the highest representation (25.3%), distributed on the littoral, next to the Ria Formosa and along the leeward coast. These flat surfaces, with less dense dendritic drainage systems, are composed mainly of alluvium, sand dunes and pebbles (slightly cohesive soils and sediments). This system flows in parallel form to the arms of the estuary and may be subject to flooding. These areas are fluvisol, luvisols and cambisols and are fertile for agriculture. They have dense vegetation, with orchards and complex cultural systems providing a low or moderate risk of soil erosion.

Class	Subclass	Code	Area (km²)
Plains	Flat plains	11a 11b 11c 11d	142.93
	Nearly flat plains	12a 12b 12c 12d	65.92
	Smooth plains	13a 13b 13c 13d	16.04
	Slightly irregular plains	21a 21b 21c 21d	27.51
	Irregular plains	22a 22b 22c 22d	52.62
	Very irregular plains	23a 23b 23c 23d	23.97
Plateaus	Low Plateaus	14c 14d 24c 24d	0.52
	Plateaus	15c 15d 25c 25d	0.02
Plains with hills	Plains with low hills	14a 14b 24a 24b	0.63
	Plains with hills	15a 15b 25a 25b	0.01
Open hills	Very low open hills	31a 31b 31c 31d	3.84
	Low open hills	32a 32b 32c 32d	32.33
	Localized open hills	33a 33b 33c 33d	37.38
	Moderate open hills	34a 34b 34c 34d	3.48
	High open hills	35a 35b 35c 35d	0.13
Hills	Very low hills	41a 41b 41c 41d	2.22
	Low hills	42a 42b 42c 42d	46.44
	Localized hills	43a 43b 43c 43d	226.67
	Moderate hills	44a 44b 44c 44d	128.85
	High hills	45a 45b 45c 45d	14.45

Table 4. Classification of 20 classes of landforms and their cover area (Ria Formosa).

2.5. Conclusions

Using an automatic hierarchical method for classification the Ria Formosa drainage basin has been subdivided in twenty landforms. The area included in each class is characterized by a number of geomorphologic characteristics, which distinguishes them from neighbouring areas. The localized and moderate hills and flat plains are the classes with the highest representation. Allied to information on soil type and vegetation cover, the former, appears to have a moderate to high erosion risk and the latter might have a low to moderate risk. However, in future research, it is intended to create a more accurate map of erosion risks, by matching the satellite images, climatic data, mapping of land use, geological and pedological features in an appropriate scale. Moreover, methodology used in this study for landform mapping, can also be validated by elaborating a descriptive mapping of a sample area, based in photo interpretation and field observations. The aerial photography, with 60% overlap, allows the creation of stereoscopic pairs which facilitate the characterization of the terrain. Fieldwork will also be useful for add and/or confirm the information obtained by stereo restitution. Comparison of the pixels in each unit of land allows validating the model.

3. Mapping of rainfall erosion in Serra de Grândola

3.1. Introduction

Erosion is a global scale threat to sustainability and productive capacity of the soil (*e.g.*, Yang *et al.*, 2003; Feng *et al.*, 2010). It is estimated that about 10 million hectares of farmland are lost annually in the world due to soil erosion (Yang *et al.*, 2003; Pimentel, 2006).

Climate change may have a great influence in soil erosion (Pruski and Nearing, 2002). Changes in the erosive power of rainfall can be hazardous in terms of soil erosion (Favis-Mortlock and Savabi, 1996; Williams *et al.*, 1996; Favis-Mortlock and Guerra, 1999; Pruski and Nearing, 2002).

Erosion, the most common type of soil degradation, should be considered as the main symptom of desertification. Since the first half of the XX Century numerous studies have been carried out and gave a strong contribution to the knowledge on the mechanical processes leading to erosion and how these processes interact in the environment. However, studies on how social, economic, political and institutional factors are affected by erosion, have been developed only during the last decades.

According to the digital Soil Map of the World (FAO, 1989) and a climate database Eswaran *et al.* (2001) the vulnerability to desertification of the Mediterranean area countries, it is considered that more than 600,000 km^2 of the Mediterranean basin are at risk of desertification. Project DesertWatch, presented at the 10th Conference of the Parties to the United Nations Convention to Combat Desertification, states that the 33% of the Portuguese territory is at risk of desertification, being the Alentejo the most affected area.

The main objective of this work is the development of a GIS to determine the risk of erosion in Serra de Grândola (Alentejo, Portugal).

3.2. Study area

The study area is delimited by the UTM coordinates: Zone 29S, M_{min}=512,930.44 m, P_{min}=4,205,893.13 m, $M_{máx}$=540,965.44 m, $P_{máx}$=4,230,328.13 m. Its area, with 675 km^2, includes the Grândola, Sines and Santiago do Cacém municipalities, and the mountain Serra de Grândola (maximum altitude: 383 m). Serra de Grândola extends up to the West coast, in a regular and continuous form, without major abrupt changes in topography. Annual rainfall ranges from 600 to 1,200 mm. The Atlantic-influenced climate is moderate, with average annual temperatures of 17^0 C. Lithologically there are three important groups (1:50,000 Portuguese Geological Map; DGM, 1984): (i) in the highlands, the Flysch formation of the lower Alentejo, (ii) in the highlands surrounding areas, sandstones and gravel of the littoral of Lower Alentejo and Vale do Sado and, (iii) in the coastal zone, the beach and sand dunes. Pedologically there is a predominance of Eutric Lithosols (highlands) and Podzols (coastal zone).

3.3. Methodology

Based on the intersection of the soil's erosive status with the rainfall aggressivity, the latter classified according to the Modified Fournier Index (MFI), the risk of erosion at the Serra de Grândola was assessed (Figure 3) using IDRISI Taiga software (Eastman, 2009).

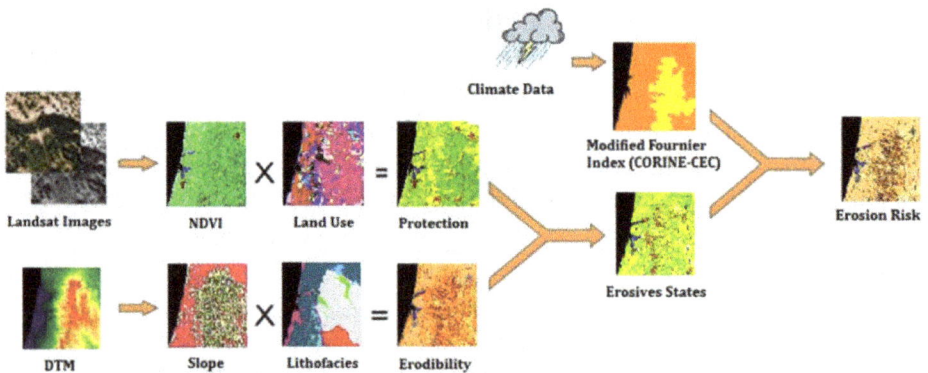

Figure 3. Schematic of the methodology used in this application.

3.3.1. Mapping of the erosive status

The vegetation coverage map was obtained by applying vegetation indexes developed in order to simplify the number of parameters present in the multi-spectral measurements. These indexes, generated from remote sensing data, constitute an important way to include anthropic activity in the ecosystems. Although there are many vegetation index available in the literature, because vegetation has a high reflectance in Near Infra-Red (NIR) and a low reflectance in R (*e.g.*, Lillesand *et al.*, 2004), this study used the Normalized Difference Vegetation Index (NDVI), a technique introduced by Rouse *et al.* (1974) which enables to know the density and the state (greenness) of the vegetation cover.

The land use information was obtained from the CORINE Land Cover map (CORINE-CLC, 2006) at a 1:100,000 scale. The map, based on satellite images SPOT-4, SPOT-5 and IRS-P6

LISS III, represents 44 classes of land use with a 150 m of positional accuracy resolution, with a minimum mapping unit of 25 ha.

The soil protection map was obtained by crossing the land use map with the vegetation coverage map. According to the classification proposed by Zavala (2001) the soil protection was classified into the following classes: 1-Very High, 2- High, 3- Moderate, 4- Low , 5- Very Low 6- Unprotected.

A DEM was obtained by ordinary kriging of 32,000 points, which were retrieved from a set of 526,770 points obtained from discretization of 10 m equidistant contour lines (military 1:25,000 map).

The hillslopes map was obtained from the DEM (see Eq.(3)). The slopes were classified into ranges: 0-3%, 3-16%, 16-21%, 21-31% and over 31%.

The lithofacies map was created from a 1:50,000 scale Portuguese Geological Map (DGM, 1984). Five lithofacies classes were defined based on the PAP/RAC, (1997) classification.

The erodibility map was obtained by crossing the slope and lithofaceis maps. Accordingly to Zavala (2001) five levels of erodibility were set: 1- Very Low, 2- Low, 3- Moderate, 4- High and 5- Very High.

The erosive status map was obtained by crossing erodibility and soil protection maps. According to Zavala (2001) five levels of erosive status were set: 1- Very Low, 2- Low, 3- Moderate, 4- High and 5- Very High. Soil erosive states are expressed in terms of protection and erodibility (Table 5).

Protection	Erodibility				
Classes	1	2	3	4	5
1	1	1	1	2	2
2	1	1	2	3	4
3	1	2	3	4	4
4	2	3	3	5	5
5	2	3	4	5	5
Unprotect	3	4	5	5	5

Table 5. Classification of soil erosive status.

3.3.2. Modified Fournier Index model

The Modified Fournier Index (MFI), an improved version of the Fournier Index (FI); Fournier (1960), is used to estimate the rainfall aggressivity for all months as in this area it occurs along the year, since the FI is only used in regions characterized by dry seasons. The MFI is calculated by Eq. (7) (e.g., Arnouldus, 1978).

$$MFI = \sum_{i=1}^{n} \frac{P_i}{P_t}. \tag{7}$$

where: P_i is the monthly rainfall at month i (mm) and P_t is the annual rainfall (mm).

Monthly rainfall data observed from 01.01.1911 to 31.12.2010 on 30 weather stations were used as an input to the *MFI* model.

Rainfall distribution is highly variable in space and time. In this study the precipitation data are insufficient and poorly distributed. Thus, for creating the MFI model other variables which are correlated with precipitation were used such as hillshading, aspect, distance to coastline, latitude and elevation. The software Statistica 6.0 (Statsoft, 2001) was used to establish a multilinear regression with t critical=1.711 for a significance of 95% (α = 0.05). Aspect model was created based on DEM, by Eq. (8).

$$Aspect = \text{atan2}\left(\frac{dz}{dy}, -\frac{dz}{dx}\right) \tag{8}$$

where: $\frac{dz}{dy}$ is the variation of height in latitude and $\frac{dz}{dx}$ is the variation of height in longitude.

Models of hillshading created refer to solstice and equinox. These models represent the hillshading for a given solar declination and azimuth. These variables were calculated using the equations described by Díez-Herrero *et al.* (2006). Hillshading models were calculated with Eq. (9).

$$HS = 255\left[\left(Sin(\gamma)\,Cos(D) + Cos(\gamma)\,Sin(D)\,Cos(\phi - A)\right)\right] \tag{9}$$

where: *HS* is the hillshading, γ (°) is the elevation of the sun above the horizon, D (°) is the declination of the sun, ϕ (°) is the solar azimuth and A (°) is de aspect.

While mapping of the MFI various models were created and the respective outputs analysed such as Cook's distance, consistency, independence and normality. The best model found was that in which the independent variables were latitude, elevation, and distance to coastline, spring hillshading and aspect:

$$R = 1.1708\,E + 0.0009\,L + 708.8582\,SHS - 0.1225\,A - 0.0032\,DC - 3{,}432.4703 \tag{10}$$

where: R (mm) is rainfall, E (°) is elevation, L (°) is latitude, SHS is spring hillshading, A (°) is aspect and DC (m) is a distance to coastline.

Classes	Range	Classification
1	<60	Very low
2	60-90	Low
3	90-120	Moderate
4	120-160	High
5	>160	Very high

Table 6. *MFI* values according CORINE – CEC (1992).

3.3.3. Erosion risk mapping

Finally, the erosion risk map (Figure 4) was obtained by relating the erosive status and MFI classified according to CORINE-CEC (1992). According to Zavala (2001), five levels of erosion risk were set: 1- Very Low, 2- Low, 3- Moderate, 4- High and 5- Very High. Erosion risk was expressed in terms of erosive status and MFI (Corine-CEC, 1992) (Table 7).

Erosive status	MFI (Corine-CEC)				
	1	2	3	4	5
1	1	1	1	2	3
2	1	2	2	3	4
3	1	2	3	4	5
4	2	3	4	4	5
5	3	4	5	5	5

Table 7. Determination of the levels of erosion risk as a function of the erosive status and the rainfall aggressivity.

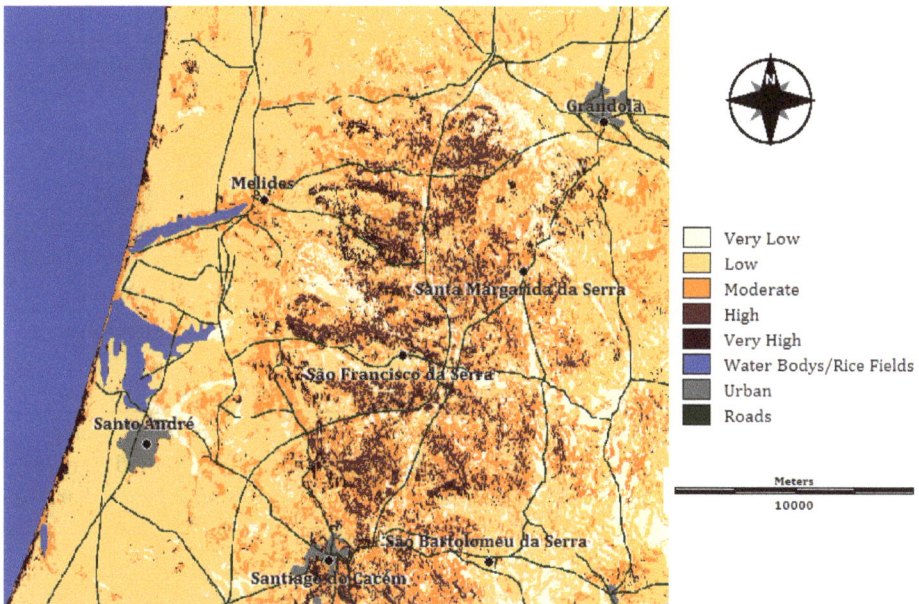

Figure 4. Digital model of the annual risk of erosion of the Serra de Grândola.

3.4. Results and discussion

3.4.1. Erosive status map

Each erosive status class was characterized according to the slope, soil and vegetation cover (Table 8).

Table 8 shows that most representative classes have low risk (28%) and moderate risk (44%) of soil erosion, usually on the highlands or at the coastal areas. The coastline consists mainly of cliffs with non-cohesive materials (sand and gravel). In the highlands the material is more resistant (Flysch group from lower Alentejo) but the amount of rainfall is higher and the steep slopes favour soil erosion processes. Soils with material from class D (soil or poorly resistant or deeply altered rocks) and E (soils or sediments that are poorly cohesive or detritus materials) are the most representative, with slopes ranging between 0-3% (slow runoff) and 3-16% (moderate runoff).

3.4.2. Modified Fournier Index

All study areas, classified according to CORINE - CEC (1992), present values of *MFI* of low risk (71%) and moderate risk (29%) of erosion, which means a small aggressiveness of rainfall.

Erosive State	Occupation	Slope	Soil	Land Cover
Very Low	10%	0-3%	Schists, phyllites, silltites, quartzites and Flysch group	Hardwood forest
			Dunes	Permanently irrigated land
Low	28%	0-3%	Sand, sandstone and gravel	Coniferous forest
		3-16%	Flysch group	Hardwood forest
Moderate	44 %	3-16%	Sand, sandstone, gravel and dunes	Coniferous forest
			Flysch group	Hardwood forest
High	16%	3-16%	Limestone, dolomite, sand, sandstone gravel and dunes	Cultures /systems fragmented complex / non-irrigated arable land
		16-31%	Flysch group	Hardwood forest/agro forestry
Very High	3%	21-31%	Limestone, dolomite, sand, sandstone gravel and dunes	Non-irrigated arable land/ forest or shrub vegetation transition/beaches
			Flysch group	Agro-forestry

Table 8. Characterization of the erosive status.

3.5. Conclusions

Environmental biophysics requires knowledge of the resources and processes affecting ecological systems conservation, as well as planning and land management. In completion of the erosion cartography it was possible to develop topographic models (slope, orientation of slopes and distance to the coastline), climate (rainfall, hillshading, Modified Fournier index) and lithofaces. These models represent quantitatively the environmental variables that affect the process of erosion.

Moderate erosive status is the most frequent class in the study area. Highlands, where the soil material have moderate resistance (flysch formation) and the precipitation is higher, and the coastline, essentially composed of cliffs with low cohesive material (sand gravel), are the most sensitive areas to erosion. However, using this erosive model it is difficult to justify the risk of water erosion at the coastline. In coastal areas infiltration prevails (sand dunes) and the wind action is the most important factor in the erosion process. Therefore, in future studies, the model should be capable of including wind erosion.

The largest amount of precipitation is falling in December and January and the lowest in July and August. Through the mapping of the rainfall erosivity, it was found that the aggressiveness of rain in the coastline and in the highlands is higher in these months.

4. Flood delimitation mapping of the Tavira urban area

4.1. Introduction and objective

This case study illustrates the use of GIS as a tool to establish hydrologic regional parameters for urban flood mapping purposes. Cartographic elements, hydrological and hydraulic models, and boundary conditions used to establish the maximum flood levels of a 10- and a 100-years return period flood and a 100-years climate change scenario are described. Cartographic information was completed by *in-situ* measurements and observations.

Hydrologic regional parameters are extensively used in flood simulation, since drainage basins are characterized by natural variability in land-surface features (*e.g.*, Wooldridge and Kalma, 2001). Prasad (1997) refers that the improved accuracy of GIS-based hydrologic simulation comes from the capability that these models have to integrate hydrologic regional parameters; updating or modifying GIS data to study the impact of changes in a drainage basin (*e.g.*, land use) becomes a relatively easy task.

This application is focused on the simulation of fluvial-originated urban flooding. The area selected for this study is the town of Tavira. This town is situated in the southernmost region of Portugal – Algarve. The Séqua/Gilão River crosses throughout the Tavira urban area until it flows into the Ria Formosa coastal lagoon. As the Séqua/Gilão River is intrinsically connected with the urban fabric, an overtopping of the margins always has negative consequences to people and assets. An example of a severe flood event was the 3rd December 1989 flood which caused extensive damage in the city.

4.2. Study area

Tavira (12,000 inhabitants) is one of the southernmost towns of mainland Portugal. The town origins come back since about 2,000 BC and from these years until the half of the last century Tavira has had in agriculture and fishing its major economical activities. Like many areas in the Algarve region, in these last decades, tourism has gained a sound importance in the local economy and lifestyle. The Séqua/Gilão River, which crosses the entire town, is crossed by 6 bridges.

This study characterizes the whole river basin of the Rio Séqua/Gilão with length of the river valley of 9.5 km from the outlet. The drainage basin of the Séqua/Gilão River considered for this study is located immediately upstream of the northern limit of Tavira's urban area (top left of Figure 5).

4.3. Materials and methods

This section presents the hydrologic and hydrodynamic models used to determine the peak flows and the delimitation of the flooded areas. Particular attention was given to obtaining the hydrologic parameters related to soil use and type (CN values; Brunner, 2006). The flood delimitation methodology and the verification of the modelled flood – by comparison with observed data – are also presented.

Figure 5. Urban area of Tavira and the Séqua Gilão River, flowing from left to right.

4.3.1. Hydrologic model

The Séqua/Gilão drainage basin has an area of 227 km² immediately upstream of Tavira. This drainage basin can be divided into 4 sub-basins, namely: Alportel, Asseca, Fornalha and Séqua (Figure 6). Characteristics of the sub-basins are shown in Table 9.

Characteristics of the drainage sub-basins and rivers	Sub-basins			
	Alportel	Asseca	Fornalha	Séqua
Sub-basin area (km²)	93.12	61.35	40.06	32.31
Main river length (km)	49.07	19.77	15.06	14.25
Equivalent slope (m/km)	6.5	9.4	17.6	5.9
Time of Concentration (h)	9.2	4.0	2.5	3.7
Lag-time (h)	5.5	2.4	1.5	2.2

Table 9. Major hydrological and physical characteristics of the drainage sub-basins and rivers of the Séqua/Gilão River.

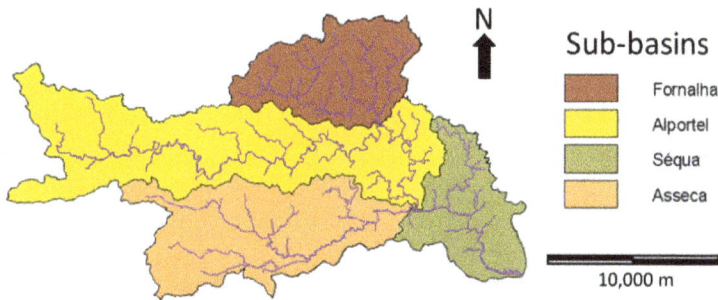

Figure 6. Drainage basin of the Séqua/Gilão River immediately upstream of Tavira. The 4 sub-basins are identified.

According to the Portuguese Soils Map (SROA, 1970) most of the soils of the Séqua/Gilão drainage basin may be classified as "Ex - Lithosols of xeric regime climate, schists or greywackes" (Cardoso, 1965). Soils properties were defined using field tests data (Koop *et al.*, 1989). The soils hydrologic groups were defined (*e.g.*, Lencastre and Franco, 1992) as B, C and D (US Soil Conservation Service Curve Number method). Group D was by far the most common in the drainage basin. Soil use was acquired via the Corine Land Cover chart and largely consists of forests and farmlands. According to the Curve Number method, information on the soil types and uses were combined to obtain the CN II values, to which soil types and uses determine the relationship between rainfall and effective rainfall for soil moisture conditions between the wilting point and field capacity. However, in flood-related studies, the soil moisture conditions should correspond to soils with high water content which are able to reach or surpass field capacity, a situation that leads to the origin of larger floods. Adjusted values (CN III) used in this study are shown in Figure 7.

The design hyetographs were obtained by using Intensity-Duration-Frequency curves referred in the Portuguese Law (DR 23/95 of 23rd August) and the "alternating block method" (Chow, 1988). Kirpich formula (*e.g.*, Guo, 2006; de Lima, 2010) was used do estimate the Time of Concentration of each of the sub-basins; the Lag-times were approximated as 60% of the latter (Table 9).

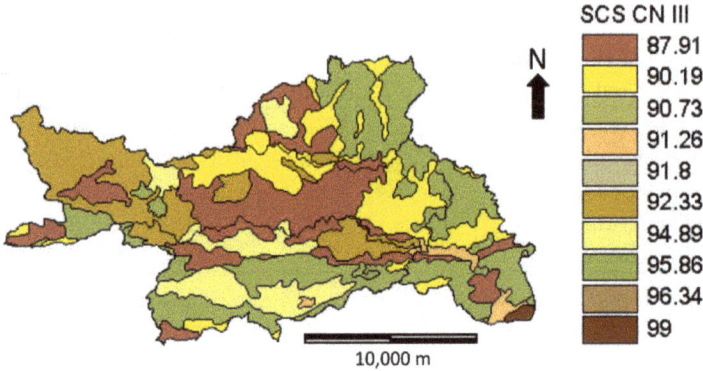

Figure 7. Soil Conservation Service CN III values for the drainage basin of the Séqua/Gilão River.

The information above served as an input to the HEC–HMS model (Brunner, 2006). Data for the drainage basin watershed were introduced into the model, differentiating the four sub-basins and their topological connections, the reaches and the final section of the drainage basin (sink). Resulting Séqua/Gilão River basin model is shown schematically (Figure 8).

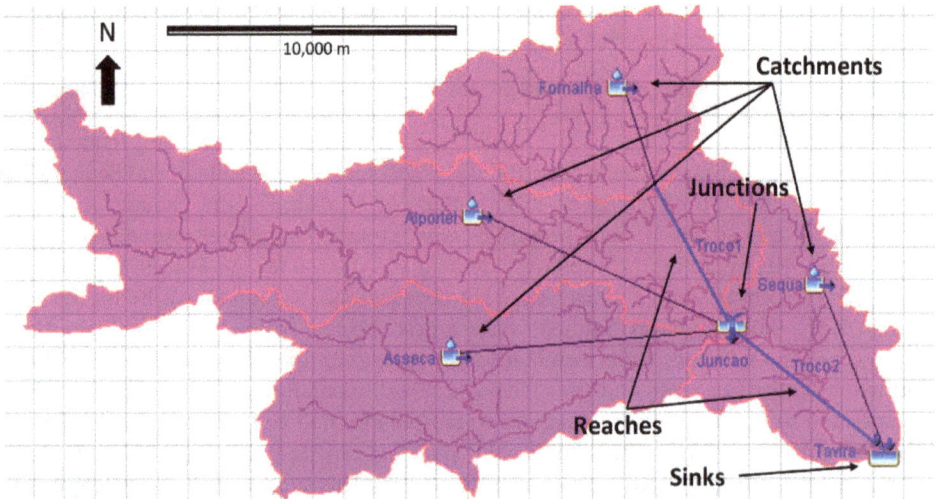

Figure 8. Topological scheme of the Séqua/Gilão River drainage basin.

The resulting design hydrographs produced peak flows for the 10- and 100-years return period, respectively, of 520.6 m³/s and 928.0 m³/s.

4.3.2. Hydrodynamic model

The HEC–RAS model (Hydraulic Engineering Center – River Analysis System) was used to define, for the full extension of the Séqua/Gilão River, the maximum flood levels to be expected for 10- and 100-years recurrence periods. In the simulation it was assumed (i) a 1D unsteady flow, (ii) inflow in the upstream boundary condition is defined by the previously obtained design hydrographs, (iii) water level at the downstream boundary condition is defined by expected spring tide levels at Faro bar and an additional climate change scenario with a mean sea rise of 0.91 m was also used (mean sea rise value was imposed by national authorities), (iv) flow resistance is approximated by the Manning-Strickler equation, (v) densely urbanized areas are considered as non-effective flow areas, *i.e.*, where water overflows and returns in the same river section and (iv) head loss in hydraulic structures (*e.g.*, bridges) is due both to the head loss during flow along the hydraulic structure and the localized head loss.

The main flow line was discretized by cross sections (STs) of the river, based in topographic and bathymetric surveys. These STs were set where the river geometry showed important changes and near hydraulic structures. Geometrical information of the latter was retrieved from the structures final drawings and *in situ* measurements made for this purpose. Riverbed material was classified by local observations (Chow, 1959). Maximum flood levels in the Tavira urban area were obtained for the worst case scenario, *i.e.*, when the downstream-moving river flood wave overlaps the upstream-moving tide wave. This case scenario takes place when the river peak flow reaches the upstream boundary cross section 1 hour before the tidal high water occurs at the downstream boundary cross section.

4.3.3. Flood delimitation

The following cartographical data and the hydrodynamic model results were combined to define the 10- and 100-years recurrence period flood-affected areas:

- Portuguese Army topographical maps 1:25,000 and ortophotomaps;
- Topographical map of Tavira with contour lines every 10 m;
- Bathymetric survey of the Séqua/Gilão riverbed and the Tavira bar, Quatro Águas bar and Gilão River.

The flood-affected areas were delimited for the 10- and 100-years return period and the 100-years climate change scenario by retrieving, for each of the cross sections, the maximum flood levels from the hydrodynamic model. After completion of this process, the areas above the maximum flood which were within the flood delimitated area (island areas) were trimmed out.

4.3.4. Comparison of the model results with observed flood levels

In the night of the 3rd of December 1989 a severe flood occurred in Tavira with 120 mm of daily rainfall registered during that day at the São Brás de Alportel meteorological station.

Making use of an amateur video from that night and from flood level inscriptions still visible on some walls, some flood-affected locations were identified (Figure 9). Since it was possible to recognize from the video the maximum flood levels, by means of a set of *in-situ* measurements the depth of water in those locations was estimated (Table 10).

Location (point)	Observed flood depth (m)	Simulated flood depth (m)
A	2.5	2.9
C	4.4	4.3
D	4.4	4.3
H	2.7	3.1
K	4.1	4.1
M	4.4	4.3

Table 10. Comparison between the simulation results and the observed flood depths attained on the 3rd of December 1989 flood event.

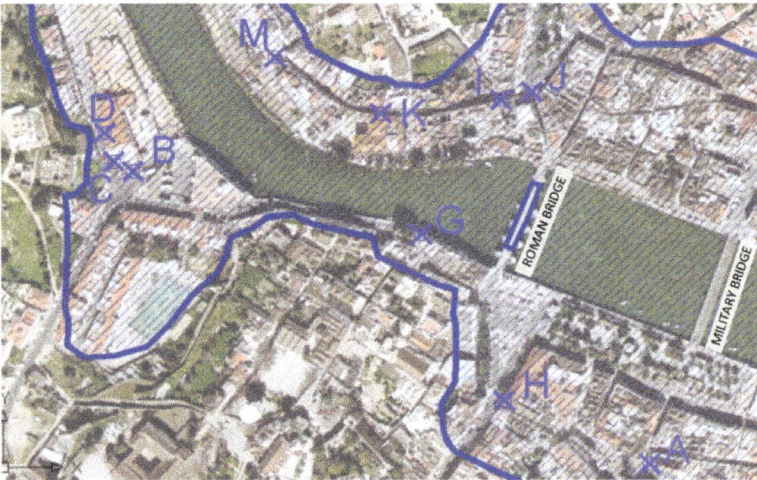

Figure 9. Location of places used to compare the simulation results and the observed flood depths attained on the 3rd of December 1989 flood event.

4.4. Results and discussion

The 10- and 100- year flood area maps of the Tavira urban area show that a significant part of the urban area which is adjacent to the Séqua/Gilão River is within the flood-affected perimeter (Figure 10). The City's centre is severely affected both by the simultaneous occurrence of high spring tide and the 10- or the 100-year rainfall events.

It is clearly visible that after the ST11 section the flood area expands widely. This is because ST11 represents a heritage bridge which causes significant obstruction to the river channel. A hydraulic gradient is formed by this bridge thus allowing water to overflow the river channel.

Figure 10. Flood delimitation for the Tavira urban area. Blue and magenta areas represent, respectively, the 10- and 100-years return period.

4.5. Conclusions

This application shows an example of how GIS-based soil data may be used as an input for flood-modelling purposes. In this example HEC-HMS (hydrological model) and HEC-RAS (hydraulic model) were used to obtain maximum flood levels for 10- and 100-years return periods.

5. General conclusions

This work aimed to illustrate how GIS-based models can be used as tools for environmental studies through three case studies in the south of mainland Portugal. The first dealt with the problem of geo-form classification in the Ria Formosa estuary. The second focused on using DEM/DTM-based climate models to obtain and analyze isohyetal maps and to identify rainfall distribution influence on water erosion at the Serra de Grândola. In the third application, GIS-based models were used to determine hydrological regional parameters for urban flood mapping purposes (this last application focused in the Séqua/Gilão River and the city of Tavira). The applications allowed demonstrating the versatility and usefulness of GIS-based models when used to solve environmental issues.

Author details

Jorge M. G. P. Isidoro

IMAR – Marine and Environmental Research Centre, Department of Civil Engineering, University of Algarve, Campus da Penha, Faro, Portugal

Helena M. N. P. V. Fernandez and Fernando M. G. Martins
Department of Civil Engineering, University of Algarve, Campus da Penha, Faro, Portugal

João L. M. P. de Lima
*IMAR – Marine and Environmental Research Centre, Department of Civil Engineering,
University of Coimbra, Rua Luís Reis Santos, Campus II – University of Coimbra,
Coimbra, Portugal*

Acknowledgement

This research was partly supported by project PTDC/ECM/105446/2008, funded by the Portuguese Foundation for Science and Technology (FCT) and by the Operational Programme "Thematic Factors of Competitiveness" (COMPETE) through the European Regional Development Fund (ERDF). The three first authors wish to acknowledge FCT Ph.D. grants SFRH/PROTEC/49736/2009, SFRH/PROTEC/67603/2010 and SFRH/PROTEC/ 67438/2010, respectively. The second and third authors also wish to express their gratitude to research group MED_Soil of the Department of Crystallography, Mineralogy and Agricultural Chemistry of the Faculty of Chemistry, University of Seville, Spain.

6. References

Anaya-Romero M, Pino R, Jordán A, Zavala LM, Bellinfante NC (2005). Modelización del hábitat potencial de formaciones forestales en la provincia de Huelva. *Edafología*, 12, 65-73 (in Spanish).

Anaya-Romero M, Jordán A, Zavala LM, Bellinfante NC (2004). A comparison of methods to predict the potential area of forest types in southern Spain, In: *Proceedings of the International Symposium on Forest Soils Under Global and Local Changes: From Research to Practice*, Institut Européen de la Forêt Cultivée, Bordeaux, September, 2004, pp. 119-120.

Arnoldus HM (1978). An aproximation of the rainfall factor in the Universal Soil Loss Equation, In: De Boodst M, Gabriels D, eds. *Assessments of erosion*, New York, John Wiley & Sons, Inc., pp. 127-132.

Bastian O (2000). Landscape classification in Saxony (Germany) - a tool for holistic regional planning. *Landscape and Urban Planning*, 50, pp. 145-155.

Bocco G, Mendoza M, Velázquez A (2001). Remote sensing and GIS-based regional geomorphological mapping—a tool for land use planning in developing countries. *Geomorphology*, 39, pp. 211-219.

Brunner GW (2008). *HEC-RAS River Analysis System – Hydraulic Reference Manual v. 4.0*. US Army Corps of Engineers – Hydrologic Engineering Center, Davis, CA, USA.

Burrough PA, Wilson JP, Van Gaans PFM, Hansen AJ (2001). Fuzzy k-means classification of topo-climatic data as an aid to forest mapping in the Greater Yellowstone Area, USA. *Landscape Ecology*, 16, pp. 523-546.

Cardoso J (1965). *Os Solos de Portugal, sua Classificação, Caracterização e Génese*, Direcção Geral dos Serviços Agrícolas, Lisbon (in Portuguese).

Cerdan O, Govers G, Le Bissonais Y, Van Oost K, Poesen J, Saby N, Gobin A, Vacca A, Quinton J, Auerswald K, Klik A, Kwaad F, Raclot D, Ionita I, Rejman J, Rousseva S, Muxart T, Roxo MJ, Dostal T. (2012). Rates and spatial variations of soil erosion in Europe: A study based on erosion plot data. *Geomorphology*, 122(1-2), pp. 167-177.

CORINE-CEC (1992). *CORINE soil erosion risk and important land resources. An assessment to evaluate and map the distribution of land quality and soil erosion risk.* Office for official publications of the European Communities, City of Luxemburg.

CORINE-CLC (2006). *Cartografia CORINE Land Cover 2006 para Portugal Continental.* Instituto Geográfico do Exército, Grupo de Detecção Remota, Lisbon (in Portuguese).

Chow VT (1959). *Open Channel Hydraulics*, New York, McGraw-Hill.

Chow VT, Mays L, Maidment D (1988). *Applied Hydrology*, New York, McGraw-Hill.

DGM (1984). *Carta Geológica à escala 1:50,000.* Direcção-Geral de Minas e Serviços Geológicos. Lisboa.

de Lima JLMP (2010) (ed.). *Hidrologia Urbana: Conceitos Básicos.* ERSAR - Entidade Reguladora de Serviços de Água e Resíduos, Lisbon (in Portuguese).

Díez-Herrero A, Gómez JL, Pérez IG, Azcárate JA, Moral SS, Jiménez JCC (2006). Análisis de la insolación directa potencial como factor de degradación de los conjuntos pictóricos rupestres de Villar del Humo (cuenca). *Proceedings of the IX Reunión Nacional de Geomorfología: Geomorfología y territorio*, Santiago de Compostela, 13-15 September 2006, pp. 903-1008 (in Spanish).

Dikau R (1989). The application of a digital relief model to landform analysis in geomorphology, In: Raper J, ed. *Three dimensional applications in Geographical Information Systems*, Taylor & Francis, London, pp. 51-77.

Dikau R, Brabb EE, Mark R (1991). *Landform classification of New Mexico by computer.* US Geological Survey, Open File Report 91/634.

Drãgut L, Blaschke T (2006). Automated Classification of Landform Elements Using Object-Based Image Analysis. *Geomorphology*, 81(3-4), pp. 330-344.

Eastman JR (2009). IDRISI Taiga, Worcester, Clark University.

Eswaran H, Reich P, Beinroth F (2001). Global desertification tension zones, In: Stott DE, Mohtar RH, Steinhardt GC, eds. 2001. *Sustaining the global farm. Selected papers from the 10th International Soil Conservation Organization Meeting*, 24-29 May 1999, Purdue, pp. 24-28.

FAO (1989). Role of forestry in combating desertification. *FAO Conservation Guide No. 21*, Forest Conservation and Wildlands Branch, Forest Resources Division, Forestry Department, Rome.

Favis-Mortlock DT, Guerra AJT (1999). The implications of general circulation model estimate of rainfall for future erosion: A case study from Brazil. *Catena*, 37, pp. 329-354.

Favis-Mortlock DT, Savabi MR (1996). Shifts in rates and spatial distributions of soil erosion and deposition under climate changes, In: Anderson MG, Brooks SM, eds. *Advances in Hillslope Processes*, Vol. 1, Chichester, Wiley.

Felicísimo AM (1999). La utilización de los MDT en los estudios del medio físico. *150 aniversario de la creación del Instituto Tecnológico Geominero de España*, Universidad de Oviedo, Oviedo (in Spanish).

Feng X, Wang Y, Chen L, Fu B, Bai G (2010). Modelling soil erosion and its response to land-use changes in hilly catchments of the Chinese Loess Plateau. *Geomorphology*, 118, pp. 239-248.

Fernandez HM (2004). *Aplicação da Geostatística na Criação de um Modelo Digital de Terreno* (MSc Thesis). Technical University of Lisbon, Lisbon (in Portuguese).

Flanagan DC, Renschler CS, Cochrane TA (2000). Application of the WEPP model with digital topographic information, In: Parks BO, Clarke KM, Crane MP, eds. *Proceedings of the 4th International Conference on Integrating Geographic Information Systems and Environmental Modeling: Problems, Prospectus, and Needs for Research*, Banff, September, 2000.

Fournier F (1960). *Climat et érosion*. Paris, Presses Universitaires de France (in French).

Gerçerk D, Toprak V, Strobl J (2011). Object-based classification of landforms based on their local geometry and geomorphometric context. *International Journal of Geographical Information Science*, 25 (6), pp. 1011-1023.

Gonçalves MC, Ramos TB, Martins JC, Kosmas C (2010). Use of PESERA and MEDALUS models to assess soil erosion risks and land desertification in Vale do Gaio watershed. *Revista de Ciências Agrárias*, 33(1), pp. 236-246.

Guo JC (2006). *Urban Hydrology and Hydraulic Design*. Water Resources Publication, ISBN-13 978-1-887201-48-3, ISBN-10 1-887201-48-3, Highlands Ranch, CO.

Hammond EH (1954). Small scale continental landform maps. *Annals of the Association of American Geographers*, 44, pp. 34-42.

Hammond EH (1964). Classes of landsurface form in the forty-eight states, USA. *Annals of the Association of American Geographers*, Map Supplement No. 54.

Herrington L, Pellegrini G (2000). An advanced shape of country classifier: extraction of surface features from DEMs., In: Parks BO, Clarke KM, Crane MP, eds. *Proceedings of the 4th International Conference on Integrating Geographic Information Systems and Environmental Modeling: Problems, Prospectus, and Needs for Research*, Banff, September, 2000.

IGeoE (2004). *Carta Militar de Portugal à escala 1:25000*. Instituto Geográfico do Exército, Lisbon (in Portuguese).

Jordán A, Zavala LM, Bellinfante NC, González-Peñaloza FA (2005). Cartografía semicuantitativa del riesgo de erosión en suelos mediterráneos, In: Jiménez-Ballesta, R, ed., *Proceedings of the II Simposio Nacional sobre Control de la Degradación de Suelos*, Madrid, July, 2005, pp. 701-706 (in Spanish).

Jordán A, Zavala LM, Anaya-Romero M, Bellinfante NC (2007a). Propuesta de un Modelo de Distribución de Especies Forestales en el parque Natural Sierra de Aracena y el Andévalo Occidental (Huelva, España), In: Bellinfante NC, Jordán A, eds. *Tendencias Actuales de la Ciencia del Suelo*, Universidad de Sevilla, Seville, pp. 993-1002 (in Spanish).

Jordán A, Zavala LM, Peñaloza FG, Bellinfante NC (2007b). Elaboración de un modelo de geoformas del terreno, In: Bellinfante NC, Jordán A, eds. *Tendencias Actuales de la Ciencia del Suelo*, Universidad de Sevilla. Seville. pp. 792-803 (in Spanish).

Koop E, Sobral M, Soares T, Woerner M (1989). *Os Solos do Algarve e Suas Características*, Direcção Regional de Agricultura do Algarve, Faro (in Portuguese).

Kosmas C, Danalatos NG, Gerontidis S. (2000). The effect of land parameters on vegetation performance and degree of erosion under Mediterranean conditions. *Catena*, 40, pp. 3-17.

Lencastre A, Franco FM (1992). *Lições de Hidrologia*, Monte da Caparica, Universidade Nova de Lisboa (in Portuguese).

Lillesand TM, Kiefer RW, Chipman JW (2004). *Remote Sensing and Image Interpretation* (5th edition), New York, John Wiley and Sons Inc.

MacMillan RA, Pettapiece WW, Nolan SC, Goddard TW (2000). A generic procedure for automatically segmenting landforms into landform elements using DEMs, heuristic rules and fuzzy logic. *Fuzzy Sets and Systems*, 113, 81–109.

Maidment DR (ed.) (2000). *ArcGIS Hydro data model: Draft data model and manuscript, Proceedings of the GIS Hydro 2000*. San Diego, June 2000.

Oliveira AM, Santos RL (2009). Análise comparativa entre fatiamento e a classificação de imagens aplicada ao mapeamento das unidades de vertentes em Feira de Santana-BA. *Proceedings of the XIII Simpósio Brasileiro de Geografia Física Aplicada*, Universidade Federal de Viçosa, Viçosa, Julho, 2009 (in Portuguese).

PAP/RAC (1997). *Guidelines for mapping and measurement of rainfall-induced erosion processes in the Mediterranean coastal areas*, Split, UNEP/FAO.

Pimentel D (2006). Soil erosion: a food and environmental threat. *Environment, Development and Sustainability*, 8, pp. 119-137.

Pino R, Anaya-Romero M, Cubiles de la Vega MD, Pascual Acosta A, Jordán A, Bellinfante NC (2010). Predicting the potential habitat of oaks with data mining models and the R system. *Environmental Modelling & Software*, 25, pp. 826-836

Prasad T (1997). GIS Applications for Flood Simulation and Management, In: Misra B, ed. *Geographic Information System and Economic Development: Conceptual Applications*, New Delhi, Mittal Publications, pp. 49–56.

Pruski FF, Nearing MA (2002). Climate-induced changes in erosion during the 21[st.] century for eight U.S. locations. *Water Resources Research*, 38(12), pp. 1298-1308.

Rouse JW, Hass RH, Schell JA, Deering DW (1974). *Monitoring the vernal advancement and retrogradation (Greenwave effect) of nature vegetation*, NASA/GSFCT Type III Final Report, Greenbelt.

Sallun AEM, Suguio K, Filho WS (2007). Geoprocessing for Alto Rio Paraná Allogroup Cartography (SP, PR e MS). *Revista Brasileira de Cartografia*, 59, pp. 289-299.

SROA (1970). *Carta de Solos de Portugal*. Secretaria de Estado da Agricultura, Lisbon.

StatSoft, Ltd (2001). *STATISTICA (data analysis software system), v.6 users manual*, Statsoft Inc. Tulsa.

Teng Z, Xuezhi C, Ruoyin L, Guoan T (2009). Landform classification based on hillslope units from DEMs. *Proceedings of the 30th Asian Conference on Remote Sensing*, Beijing, October, 2009.

Umbelino G, Carvalho R, Antunes A (2009). Uso da Cartografia Histórica e do SIG para a reconstituição dos caminhos da Estrada Real. *Revista Brasileira de Cartografia*, 61(1), pp. 63-70 (in Portuguese).

Williams J, Nearing MA, Nicks A, Skidmore E, Valentine C, King K, Savabi R (1996). Using soil erosion models for global change studies. *Journal of Soil and Water Conservation*, 51 (5), pp. 381-385.

Wooldridge SA, Kalma JD (2001). Regional-scale hydrological modelling using multiple-parameter landscape zones and a quasi-distributed water balance model. *Hydrology and Earth System Sciences*, 5(1), pp. 59–74.

Yang D, Kannae S, Oki T, Koike T, Musiake K (2003). Global potencial soil erosion with reference to land use and climate changes. *Hidrological Processes*, 17(14), pp. 2913-2928.

Zavala LM, Jordán A, Anaya-Romero M, Bellinfante NC (2004). Cartografía semicuantitativa del riesgo de erosión hídrica. *Edafología*, 12, pp. 79-87 (in Spanish).

Zavala LM, Jordán A, Anaya-Romero M, Gómez IA, Bellinfante NC (2005a). Clasificación automática de elementos geomorfológicos en la cuenca del Río Tepalcatepec (México) a partir de un modelo digital de elevaciones: Clasificación automática de formas del terreno. *Cuaternario y Geomorfologia*, 19, pp. 49-61 (in Spanish).

Zavala LM (2001). Análisis territorial de la comarca del Andévalo Occidental: Una aproximación desde el medio físico (PhD Thesis). University of Seville, Seville (in Spanish).

Zavala LM, Jordán A, Illana P (2007). Aplicación de un sistema de información geográfica al análisis del medio físico en el Parque Natural Los Alcornocales: Aproximación a una cartografía geomorfológica a partir de un modelo digital de elevaciones, *Almoraima*, 35, pp. 245-254 (in Spanish).

Zavala LM, Jordán A, Gómez IA, Anaya-Romero M, Girón V, Segura D (2005b). Estudio del riesgo de erosión potencial en la cuenca alta del Río Hozgarganta. *Almoraima*, 31, pp. 111-118 (in Spanish).

Chapter 11

Cartography of Landscape Dynamics in Central Spain

N. López-Estébanez, F. Allende, P. Fernández-Sañudo,
M.J. Roldán Martín and P. De Las Heras

Additional information is available at the end of the chapter

1. Introduction

Ecological and spatial analysis helps us to characterize the territory and know the spatio-temporal relationship between different components of the landscape. Landscape ecology has developed several methods of assessment and analysis of indicators by using Geographical Information Systems [1-4]. Such methods allow characterization of changes in land structure and land uses, as well as the interpretation of the ecological consequences of these dynamics [5]. They also facilitate analysis of the territory, trying to recognize and compare different spatial configurations, using patches of different shapes, numbers, classes, etc. [6-8].

Several authors have carried out research attempting to integrate the study of territorial dynamics, from an ecological perspective, using Geographic Information Systems [9, 10]. The landscape is influenced by natural and anthropic processes, and the effects of both factors are expressed either at local or regional scale on the territory, showing changes in their structure and composition [11]. Clearly, the landscape appears to us as a complex of many different elements that can reach a great diversity [12]. In Mediterranean areas, the landscape is characterized by a heterogeneous mosaic of land uses and vegetation, where natural subsystems coexist adjacent to other systems at different degrees of perturbation due to human intervention and, therefore, with different degree of ecological maturity, separated by clear boundaries [13, 14]. The intense dynamic of land use changes occurred in these areas over recent decades has caused important changes in the structure of the landscape, as a result of fragmentation processes [14-19]. This influences various ecological processes, including those relating to the matter and energy flows between patches, by altering the composition and distribution of communities, the survival and coexistence of species, and species diversity [20-25].

The region of Madrid, due to its geographical location (centre of the Iberian Peninsula), and physiographic variability, -from the Cordillera Central to the river Tagus depression-, has a variety of lithological, mesoclimatic, edaphic and geomorphological traits, which have resulted in a great diversity of ecosystems, land use types and landscapes, some of them of great natural value. For example, It is noteworthy the contrast between areas of the Guadarrama Range (belonging to the Central Mountain Ranges), where summits and slopes covered by pine and oak forests, dehesas and grasslands of high nature value, are well-preserved, compared to other areas intensely humanized, that have a very deteriorated landscape, such as the metropolitan area of Madrid. In the recent history of this region there is a clear abandonment of traditional agricultural activities that had provide the maintenance of semi-natural systems with a high degree of functionality, which has resulted in a clear instability [26]. Thus, at present, there is a heavily modified landscape, more homogeneous and probably more polluted, that has lost much of the typical positive externalities of the traditional landscape (natural services, basic ecological processes, biodiversity, aesthetic tourist-recreational values, etc).

For this reason, much of the territory of Madrid Region is protected by European Community legislation, as well as national and regional laws, which aim to consolidate the protection and conservation of natural diversity, and at the same time, seek to promote (uphold, improve) sustainable development.

Since 1985, the autonomous government of Madrid has declared seven protected areas, which represent 14% of its surface, so it is the sixth region in Spain in terms of protected territory. In parallel, along decades, there has been a very important socio-economic development in the region, a great population growth and a deep process of change in the use and exploitation of the area's resources. This study was conducted in a Protected Natural Ares stated at 1985.

The aims of this study were, firstly, to analyse the main changes in land use occurred inside the Protected Natural Area (PNA) over a period of 35 years, in order to determine the principal territorial dynamics occurred, and the consequences of these processes of change on the landscape configuration and on the evolutions of territorial structure in this PNA. Secondly, in a smaller geographical context, GIS tools are combined with key ecological parameters such as richness of uses, diversity (by Shannon) [27], evenness, connectivity and fragmentation, to analyze the structure and organization of the territory.

2. Study area

We conducted our study in the PNA (Cuenca Alta del río Manzanares), located quite close to Madrid city (approximately 50 km NW, Figures 1 and 2). This PNA is characterised by a mid-mountain Mediterranean landscape, with altitudes ranging from 660 m to 2,200 m. The summits with slopes of gneiss and granite are covered by oak (*Quercus pyrenaica*) and pine reforestation (*Pinus sylvestris*). On these slopes is located one of the most important granite landscapes of Europe, "La Pedriza", protected since 1930. A rocky piedmont covered by "dehesas" of *Quercus ilex* subsp. *ballota* links to the sedimentary river Tagus basin, where alternate cereal crops with oak forests. The Spanish Committee of UNESCO's MAB

Programme in 1992 designated the area as a Biosphere Reserve, due to the high ecological value of the area as well as for its cultural heritage and agricultural and landscape values.

Demography, recreation and urban and transport infrastructures development are the most important pressures in this PNA [28]. This Park was declared a PNA in 1985 abarcando una extensión de casi 53.000 ha [29]. Biogeographically, this territory belongs to the Mediterranean Iberian-Atlantic region in the Central Mountain Range.

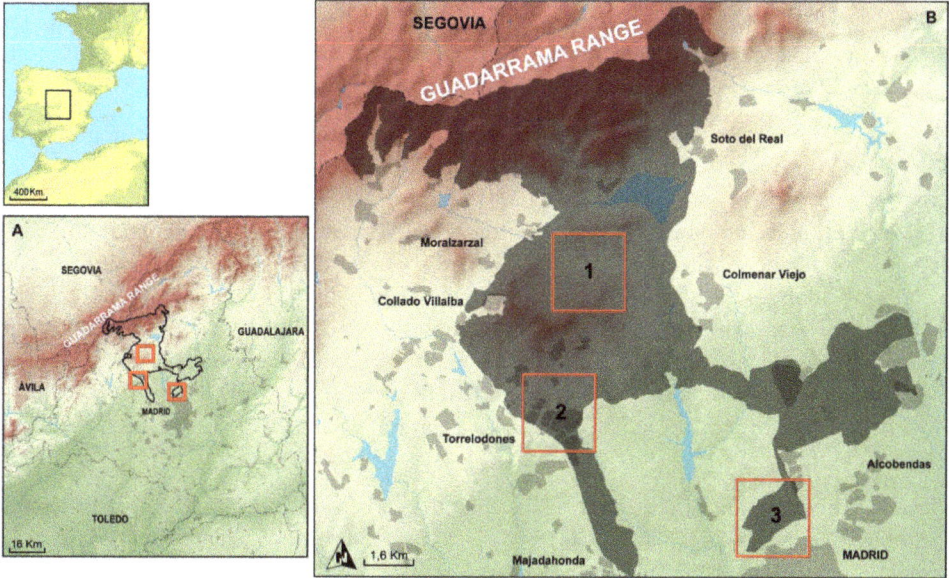

Figure 1. Study area: Location in Madrid region (A) and the location of the three study cases (B): Cerceda (1), Torrelodones (2). Fuencarral (3) into boundary of PNA (shadow).

Figure 2. Guadarrama Range and piedmont with open woodland and urban development in PNA (Cuenca Alta del Río Manzanares)

3. Methodology

We used cartographic techniques that combine remote sensing and GIS, with ecological analysis at different spatial scales. Both methods allowed us to determine the landscape changes and to detect pressures acting upon some areas included in the PNA during the last decades (urban and transport infrastructures development, etc). We reviewed i) previous studies focused on photo-interpretation techniques [30-34]; ii) land uses changes and territorial dynamics in different environmental conditions [35-40] and iii) definition of land uses categories [41-44]. We chose two working scales: the first analyzed the variability of the PNA as a whole (1:50,000), and the second, that use a more detail scale, permitted us to recognize different ecological processes that occurred in the territory (1:12,500).

We used the aerial photography of Spanish Air Force of 1975 and the orthophotography of 2009 (Plan Nacional de Ortofotografía Aérea). ERDAS Imagine 9.1 software was used for processing analogical information (1975 flight, Spanish Air Force). Previously, we had tested ER-MAPPER and ArcGis 9.10 methods to this end. We scanned each of 150 photograms at 600 dpi. Mosaicing was conducted using Mosaic Tool extension. This tool showed the best balanced colour result. Each photo was improved with a Root Mean Square (RMS) and its tolerance was less than 0.5. Distortions of the photos were corrected using a Digital Terrain Model (DTM) at 1:5,000 scale. Re-sampling method applied was the nearest neighbour algorithm employing at least a cubic polynomial fit using 15 Control Ground Points and at least 10% of overlapping areas. The result was a continuous image of the study area with a 5 x 5 m resolution.

Phenological changes mean different colours that depend on season and this colour difference causes errors in the photo-interpretation so, it was necessary to balance colours for the ortophotography of 2009 (2.5 x 2.5 m resolution).. This problem is more evident in agricultural areas. Likewise, existing vegetation and land use maps of Madrid Region were reviewed [45, 46]. All layers were managed in a format compatible with ArcGIS 9.3 (shapefile, coverage or GRID) referred to the WGS84 ellipsoid and UTM coordinates Zone 30 N.

The photo-interpretation considered an accuracy of the reference map unit depending on the type of area, establishing a minimum of 0.5 ha (1:12,500 scale), similar to that used in other studies with this orthophotography scale in forest formations [47] and of 0.61 ha (1:50,000 scale) using cartographic techniques that merge patches (dissolve ie.) Also on the orthophotography were required some data about the phytostructure and percentages of canopy cover. If necessary, we worked with DGPS techniques (GPS Trimble Nomad 6GB) to refine the patches shape.

We employed a touch screen Wacom Cintig 12WX joined to ArcGis editor. Topological errors were processed using ARCEDIT. Finally, we assigned a topology to each layer using ARC/INFO 9.1. Database was designed with three fields: land use, connectivity value and surface area (ha).

The photointerpretation was completed with fieldwork. It took over 2,000 panoramic photographs for checking on field the areas that were more complex during the development of cartography. It was considered of interest to note the state or degree of consolidation of urban or under urban development, as this is one aspect that has changed in the whole Madrid Region from the 70's of last century.

We then crossed the two land-use maps by means of techniques of overlapping and digital layer intersection, thus comparing the information from both years. The result was a new map on which each patch showed the land use observed in both years. Thus, we identified all the types of changes that had taken place in the PNA during the 35-year period. Then it was possible to know if each type of vegetation or land use in each part of the territory has changed or not. Results allowed us to identify a set of changes occurred in the PNA during these 35 years.

Then we calculated the percentage of the area that has changed in the whole PNA and for each type of dynamic [28, 48]. These percentages allowed us to establish categories of dynamism that distinguished zones of more or less changing and map its spatial distribution. Finally, we used categories of dynamism map (1:12,500 scale) to select three locations. We conducted a more detailed study about natural dynamic of territory [49-51], using a set of ecological parameters as indicators of structure and organization of territory.

A more detailed study allowed us to recognize and map 23 types of land uses. These were the result of consider at the same time land uses and vegetation units. Using this criterion we obtained categories as Dehesas *of Quercus ilex* subsp. *ballota* i.e.

In each selected locations, we calculated relative frequencies of each land use type using their abundance, and graphed their relative frequency profile in 1975 and 2009. Each profile has been defined by three parameters: the richness of land uses, $R(u)$, calculated as the number of different categories of land uses in the corresponding year [52]; Shannon's diversity, $H'(u)$, expressed in bits [27]; and Pielou's evenness, $E(u)$ [53].

Evenness is the proportion between observed diversity value and maximum diversity value that would be possible to reach with the registered $R(u)$. High values of $E(u)$ indicate an even land use distribution, that means that there are not a dominating land use in the territory. To the contrary, low values of $E(u)$ are indicating that one or a small set of land uses are more frequent [54].

In order to complete the study of evolution of territorial structure and organization, we analyzed ecological connectivity and fragmentation in each selected location in 1975 and 2009.

We calculated ecological connectivity (c) of territory according to a set of permeability values assigned to each land uses. These permeability values were established taking into account matter and energy flows (genes, seeds, species, etc.) through two adjacent land uses [55-57], and assessing the quality of each land uses according to landscape functionality, forestry value, cultural value, among others.

Finally, we calculated fragmentation of territory (f) according to the variation in patches number occurred between 1975 and 2009 at each selected locations. We focused on boundary effect caused by human land uses. Results were mapped.

4. Results

4.1. Diachronic analysis of land uses in a protected natural area

We obtained two diachronic maps showing different scenarios relating to land uses in this territory for 1975 and for 2009, differentiating 7 types (Table 1, Figure 3).

Land uses	Description
Forest	All formations of deciduous leafy tree species primarily *Quercus pyrenaica*
	All formations of sclerophyllous tree species primarily *Quercus ilex* subsp. *ballota*
	Pinus sylvestris forests
Scrub	High altitude shrubland (*Cytisus oromediterraneus*)
	Scrubland with *Cistus ladanifer, Lavandula stoechas* subsp. *stoechas, Thymus* spp. *Retama sphaerocarpa*
	Scrub with scattered *Quercus ilex* subsp. *ballota* or *Quercus pyrenaica*
Pastures	High-altitude pastures of *Festuca curvifolia*
	Pastures for extensive livestock farming
	Therophytic pastures in abandoned agricultural areas (wasteland)
	Pastures with scrub and scattered trees
Croplands	Dry farming crops
	Irrigated crops
Rocky areas with scrub and trees	Granite landscapes covered by scrubs and trees (*Quercus ilex* subsp. *ballota, Pinus sylvestris,Juniperus oxycedrus* subsp. *oxycedrus*).
Reservoirs	
Urban areas	Areas under urban development
	Consolidated urban areas

Table 1. Classification and description of the seven land uses and vegetation types in the PNA.

In both years, the most abundant type of land use is pasture, which reflects the importance of extensive livestock farming in this area. Nature conservation in Spain should involve protection and conservation of these traditional agrosilvopastoral systems [58, 59]. An essential ecological feature of this kind of system involves a high level of efficiency in energy and nutrient use. As a result, the use of land resources is optimized and the rural activities are adapted to natural production cycles [60, 61]. The next most abundant types of land use by extension are forest, scrub and rocky areas with scrub and trees. These are all characteristics of this PNA, although the latter comprises this territory's unique landscape. We found no noteworthy changes in number of land uses or in the typology within the protected area. We can only highlight the absence in 2009 of areas under urban development as those identified in 1975 have now become consolidated urban areas (Figure 3). The land uses showing the biggest increase in extension are scrub, tree formations and urban areas, whereas croplands show a decrease.

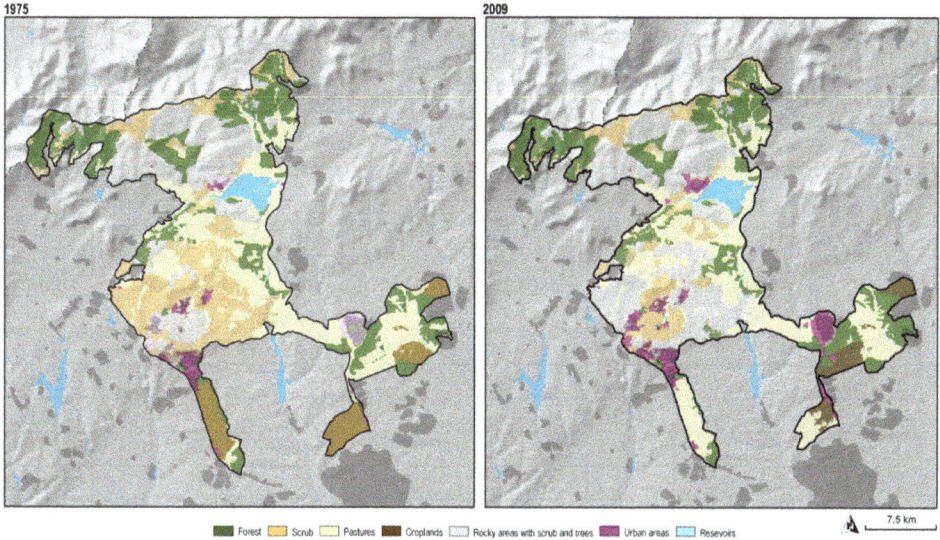

Figure 3. Land uses cartography (PNA)

Land use	1975			2009		
	Ha	% PNA	% Land use	Ha	% PNA	% Land use
Forest	11,326.8	21.4	45.0	12,306.1	23.2	47.1
Scrub	10,427.6	19.7	72.4	11,939.5	22.5	75.6
Pastures	15,185.9	28.7	50.2	15,157.0	28.6	54.4
Croplands	3,587.3	6.8	50.3	1,769.4	3.3	37.5
Rocky areas with scrub and trees	9,329.1	17.6	89.1	8,110.6	15.3	88.1
Reservoirs	1,017.5	1.9	88.7	10,17.5	1.9	88.7
Urban*	1,983.7	3.7	22.6	2,638.8	4.9	21.0

* Includes urban areas and areas under urban development

Table 2. Land uses in 1975 and 2009

4.2. Dynamics and changes in a PeriUrban Park

The comparative analysis of the vegetation and land use maps enabled us to identify and quantify five main change dynamics in the territory: urban development, scrub encroachment, forest encroachment, agricultural abandonment and new crops. Most of these dynamics were consequence of the abandonment of traditional activities or the increase in urbanised areas (Figure 4 and Table 3).

We observed few changes in the study area, indeed, 79.5% of the Park's territory had undergone no change. The total area affected by some type of change from 1975 to 2009 is 10,852.7 ha. Within this changing area, the most significant dynamics were new pastures (29.8%) and forest encroachment (26.6%). Scrub encroachment represents 19.5% and urban development processes 15.7%. New crops represent only 8.8% and occur in the southern sector of the Park (Table 4).

		2009						
		Forest	Scrub	Pastures	Croplands	Rocky areas with scrub and trees	Reservoirs	Urban
1975	Forest	NCh	SE*	AA				URB
	Scrub	FE	NCh	AA				URB
	Pastures	FE	SE	NCh	NC			URB
	Croplands	FE		AA	NCh			URB
	Rocky areas with scrub and trees	FE	SE			NCh		URB
	Reservoirs		SE				NCh	URB
	Urban							NCh

* This case is referred to forest cleared for livestock farming, a traditional use in Mediterranean areas.

Table 3. Territorial dynamics from 1975 to 2009. FE: Forest Encroachment; SE: Scrub Encroachment; AA: Agricultural Abandonment; NC: New Crops; URB: Urban Development; NCh: No Change.

		ha	% PNA	% change	% dynamics
DYNAMICS	FE	2,884.1	5.4	26.6	3.0
	SE	2,076.7	3.9	19.5	2.1
	AA	3,235.9	6.1	29.8	3.3
	NC	955.1	1.8	8.8	1.0
	URB	1,700.9	3.2	15.7	1.7
	NCh	42,081.8	79.5		43.1

Table 4. Total surface of each dynamic and percentage area over PNA, change and dynamics areas

These dynamics clearly showed the most dynamic sectors (mainly associated with urban development) that are located in the South of the PNA, close to the city of Madrid and to the main communications networks In contrast, the mountainous area located at North presented fewer changes which are associated with natural dynamics (forest and scrub encroachment).

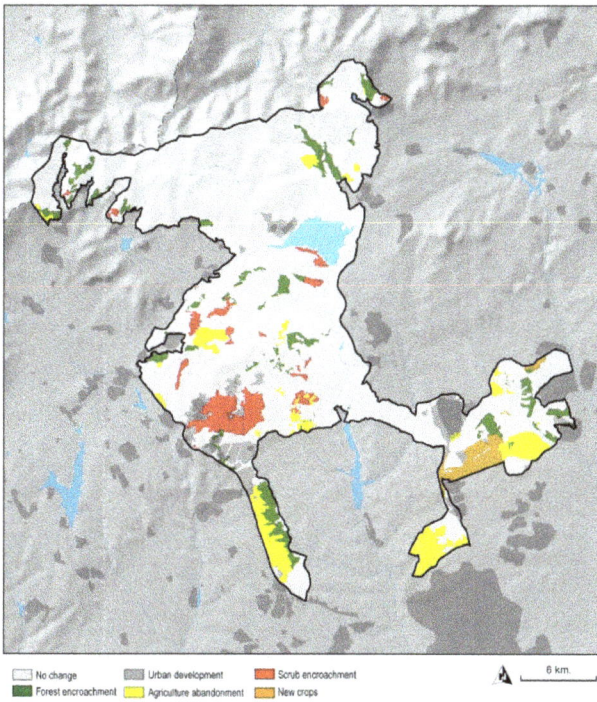

Figure 4. Change dynamics detected in the PNA

		ha	% PNA	% change	% dynamics
DYNAMICS	FE	2,884.1	5.4	26.6	3.0
	SE	2,076.7	3.9	19.5	2.1
	AA	3,235.9	6.1	29.8	3.3
	NC	955.1	1.8	8.8	1.0
	URB	1,700.9	3.2	15.7	1.7
	NCh	42,081.8	79.5		43.1

Table 5. Total surface of each dynamic and percentage area over PNA, change and dynamics areas

Percentage area of these types of dynamics in the PNA allowed us to establish four categories or degrees of change: very dynamic (ratio change > 60%); dynamic (ratio change 20-60%); stable (ratio change 12-20%); very stable areas (ratio change <12%) (Figure 5).

The more stables areas were located in the North part of the PNA. Dynamic areas increased from piedmont to Tagus Basin, especially in Madrid and its Metropolitan Area. This has facilitated to select three zones with different land use dynamism (1975-2009). First, a very stable area, named Cerceda that is located on rocky piedmont (over gneisses). The second, Torrelodones, is a transitional area between rocky piedmont and Tagus Basin where urban development is the most significant process (dynamic area). Finally, in the Southeast was located the third, very dynamic area (Fuencarral), over sedimentary materials, with significant changes in agricultural uses (Figure 1 and Figures 5 and 6).

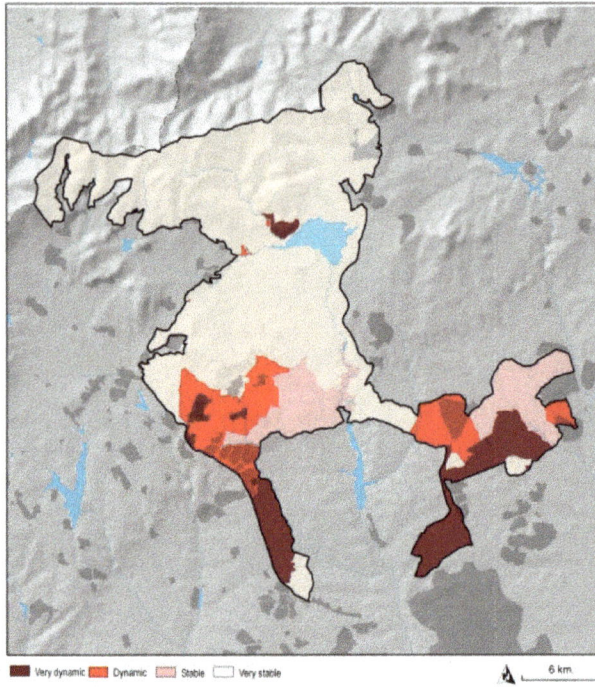

Figure 5. Categories or degrees of dynamism

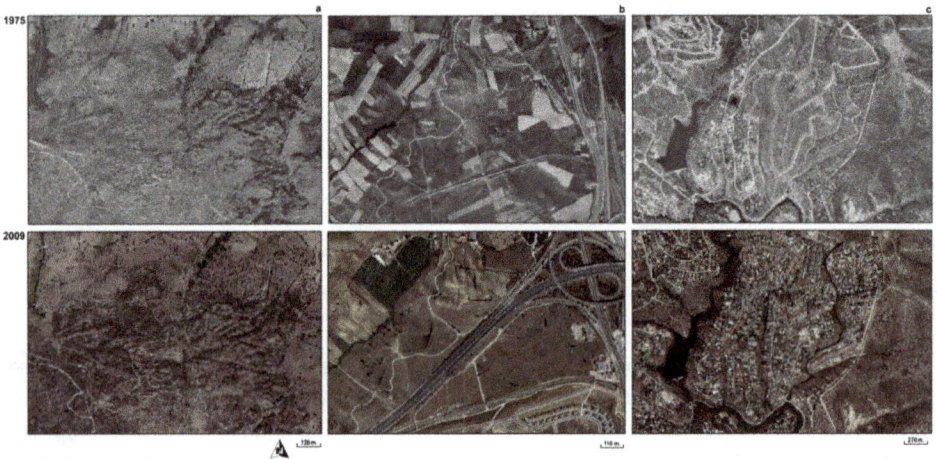

Figure 6. Orthophotographies. Cerceda, scrub encroachment (a); Fuencarral, agricultural abandonment (b); Torrelodones, urban development (c).

4.3. Changes at local scale in three study cases: Shrub encroachment, agricultural abandonment and urban development

As we had mentioned above three areas have been selected for a more detailed analyse (1:12,500 scale). The results obtained are showed below:

4.3.1. Cerceda: Open woodland and shrub encroachment

In Cerceda (Figure 7), there was no difference in R (u) so the number and type of land uses are the same along time. However, there had been some changes in the land use's relative abundance as showed diversity H′(u) and evenness E(u) values (Table 5). The changes in land uses over time have transformed the territory. One of them is dominant, the increase of understory density (Figure 8). These changes in surface are associated to pastures abandonment and the decrease of sheep and goat livestock. Woodland encroachment generated a mixed forest of *Juniperus*, *Quercus* and *Cistus*, with different densities: high density (1975: 15%; 2009: 30%) and medium density (1975: 25%; 2009: 40%). In particular the increase of open woodland areas is associated to intensification with beef cattle or bucking bulls. This intensification is determinant in the phytostructure of medium understory density in open woodland (30%, 1975 / 50%, 2009).

Figure 7. *Juniperus oxycedrus subsp. oxycedrus* and *Quercus ilex* subsp *ballota* open woodlands and farms in Cerceda.

	Cerceda		Fuencarral		Torrelodones	
	1975	2009	1975	2009	1975	2009
R(u)	10	10	13	13	12	10
H′(u)	2.38	2.25	2.05	2.97	2.81	2.49
E(u)	0.72	0.68	0.55	0.80	0.78	0.75

Table 6. Values of Richness R(u), Diversity H′(u) and Evenness E(u) of land uses in Cerceda, Fuencarral and Torrelodones

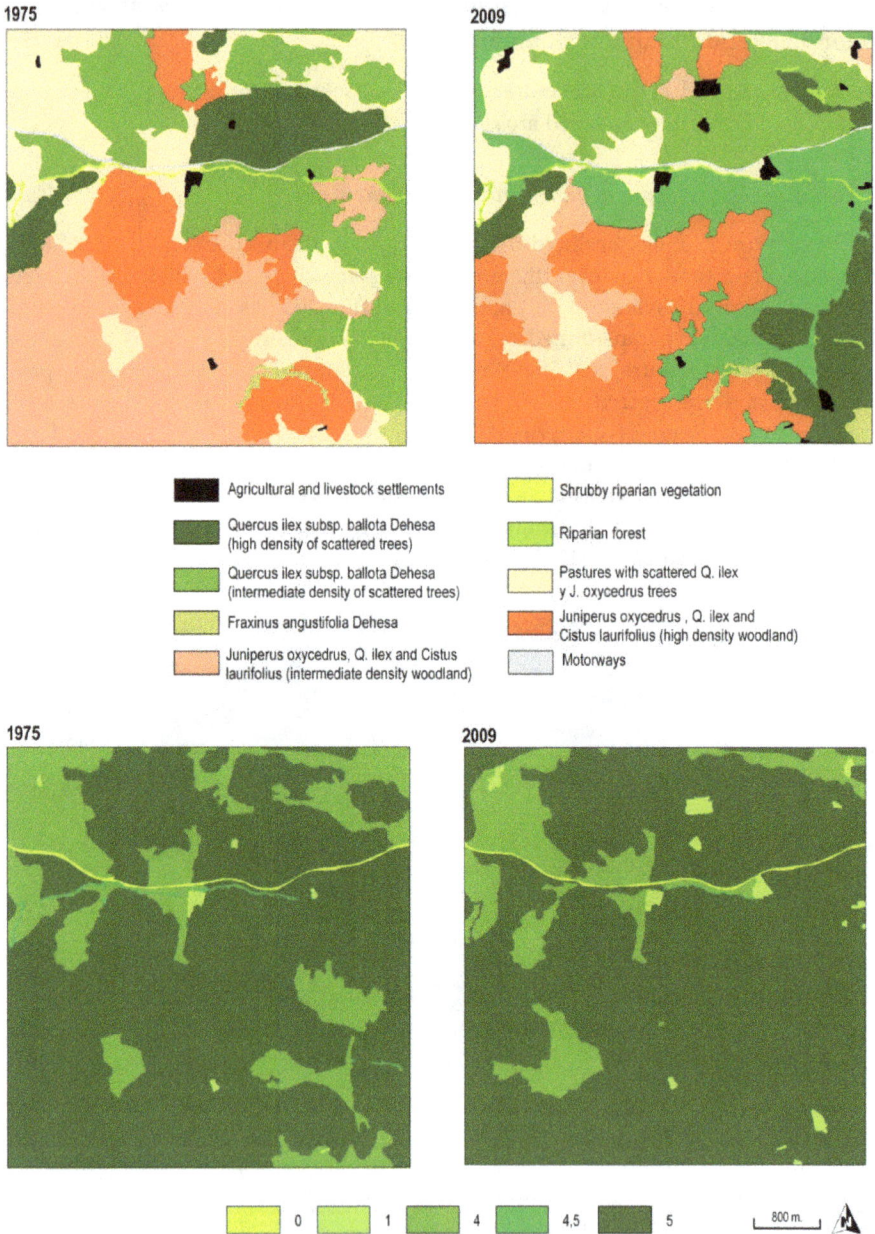

Figure 8. Land uses (1975-2009) and ecological connectivity in Cerceda

Fragmentation measured as total number of patches was significantly higher in 2009 than in 1975. The territory showed no changes and conserved the same permeability in 1975 and in

2009. So the exchange of matter and energy between patches it wasn't modify (Table 6). Finally, territorial connectivity presented the highest values of the three cases. This is an area with few transformations by anthropic uses and that maintains the same vegetation types in both dates (Figure 8).

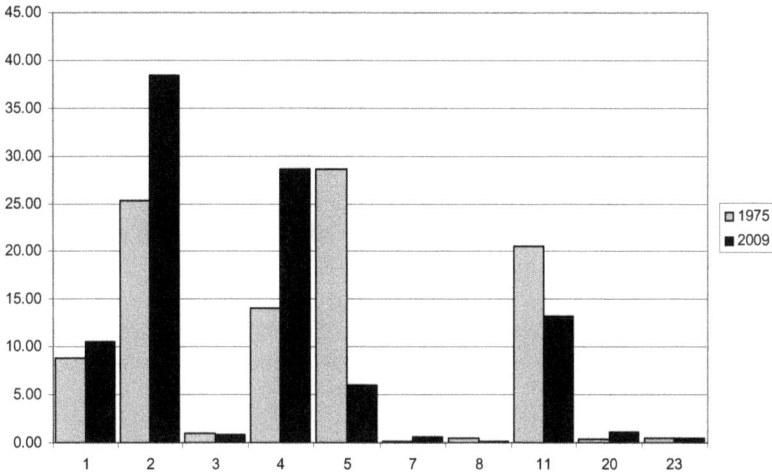

Figure 9. Relative frequency profile of land uses of each year in Cerceda. The codes and the land uses are the same for Table 6

Code	Land use	Number of patches	
		1975	2009
20	Agricultural and livestock settlements	6	13
1	Q. ilex subps. ballota Dehesa (high density of scattered trees)	4	5
2	Q. ilex subps. ballota Dehesa (intermediate density of scattered trees)	8	11
3	Fraxinus angustifolia Dehesa	2	2
23	Motorways	1	1
4	Juniperus oxycedrus subsp. oxycedrus, Q. ilex subps. ballota and Cistus laurifolius (high density woodland)	4	4
5	Juniperus oxycedrus subsp. oxycedrus, Q. ilex subps. ballota and Cistus laurifolius (intermediate density woodland)	3	3
11	Pastures with scattered Q. ilex subps. ballota y J. oxycedrus trees	13	11
7	Riparian forest	2	9
8	Shrubby riparian vegetation	7	1
	TOTAL	56	62

Table 7. Fragmentation value of each land use measured as number of patches in Cerceda. This parameter has been calculated in 1975 and 2009.

4.3.2. Fuencarral: Road infrastructures and urban pressure vs. agricultural abandonment

In the Fuencarral area (Figure 10) still stand the Richness R (u) at both dates (13). The values had certain variations in the H'(u) (1975: 2.05; 2009: 2.97) and E (u) (1975: 0.55; 2009: 0.80). These values showed an increase of landscape heterogeneity. It was also decisive the transformation of an agricultural environment with productive small and medium land parcels in 1975 (Figure 11) in an area characterized for high density of road infrastructures and the abandonment of traditional land uses in 2009 (Figure 12). Furthermore, percentage occupied for wastelands with fruit trees of *Ficus carica, Amygdalus communis* and *Vitis vitifera* and *Triticum/Hordeum* spp. is significant (1975: >50%; 2009: 10%). In 2009 traditional Mediterranean mosaic of land uses disappeared (fruit trees, dry farming crops, wastelands and scrublands) and was replaced by dry farming crops or wastelands with low-productivity. Shrub lands and wastelands had increased their surface area in 2009 especially in wastelands with *Retama sphaerocarpa* (20%) and shrub lands (5%). In many cases is difficult to distinguish between dry farming crops and wastelands on ortophotography. However wasteland's surface took a progressive reduced and transformed in non productive areas. In 2009 riparian forest area increased while in 1975 was fragmented (Table 7).

We analyzed that total patches number has been decreased between 1975 and 2009. Due to agricultural abandonment a homogeneous landscape was generated. Particularly this type of landscape maintains a great diversity of land uses, wastelands and habitats well preserved that have and interesting role such as ecological and territorial connectors [61]. In this area the connectivity at edges has decreased due to increase of road infrastructure and urban development.

Figure 10. Countryside with fruit trees and crops in Fuencarral

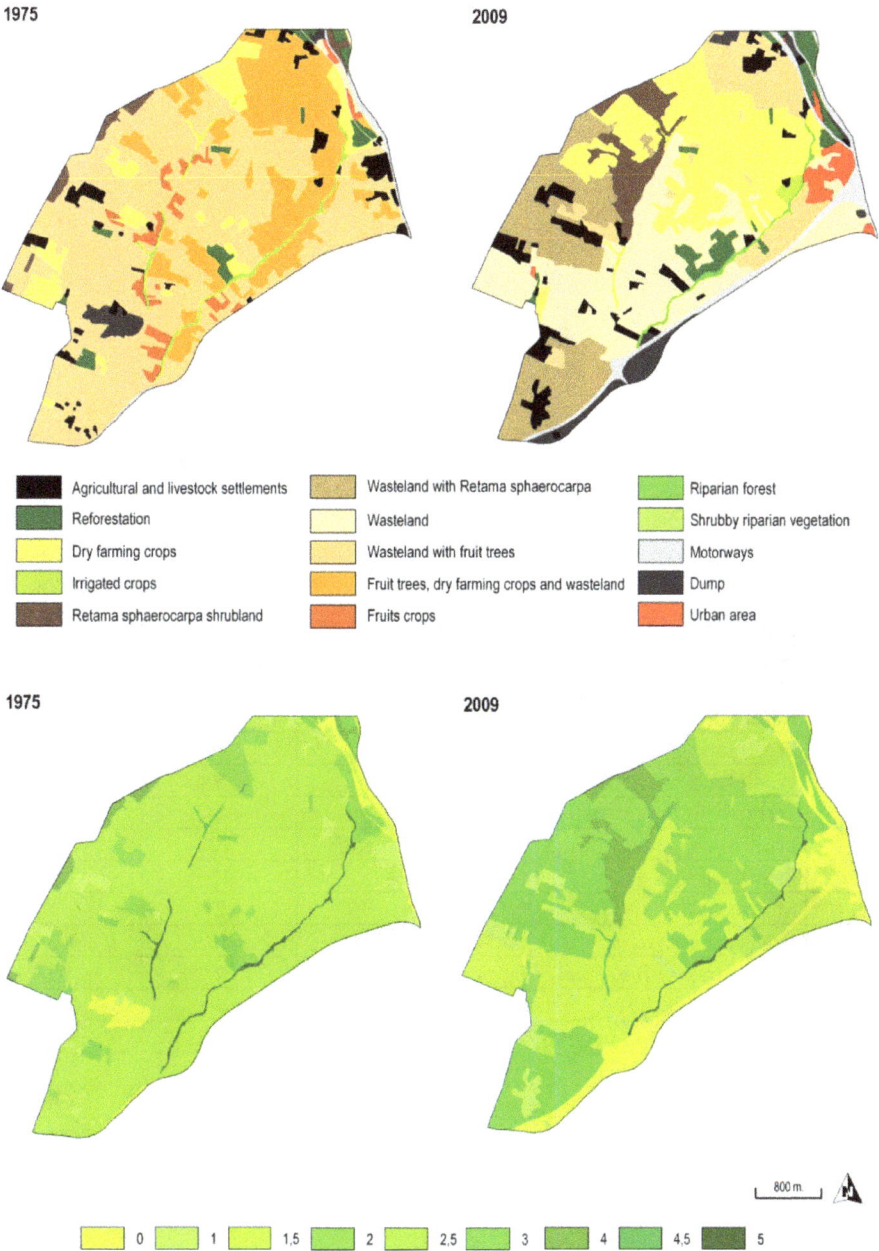

Figure 11. Land uses (1975-2009) and ecological connectivity in Fuencarral

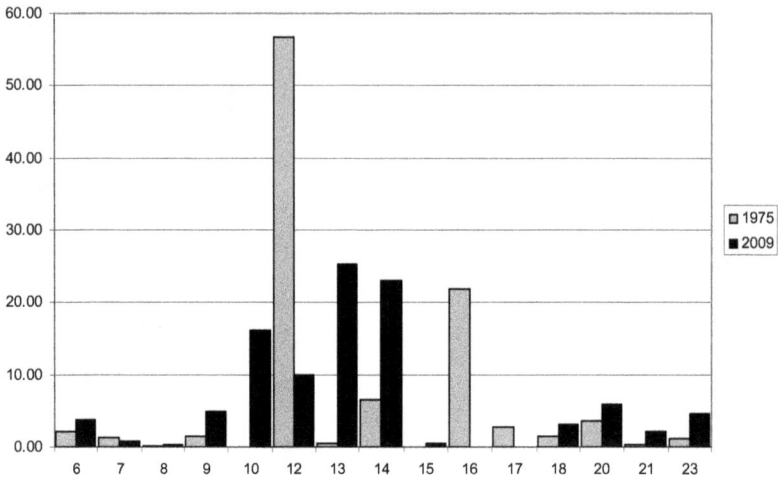

Figure 12. Relative frequency profile of land uses of each year in Fuencarral. The codes and the land uses are the same for Table 7.

Code	Land use	Number of patches	
		1975	2009
20	Agricultural and livestock settlements	36	29
15	Irrigated crops	0	1
14	Dry farming crops	13	6
6	Reforestation	14	13
13	Wasteland	3	10
12	Wasteland with fruit trees	5	9
18	Dump	2	8
12	Fruit trees, dry farming crops and wasteland	22	0
17	Fruits crops	15	0
23	Motorways	2	1
9	*Retama sphaerocarpa* shrubland	6	2
10	Wasteland with *Retama sphaerocarpa*	0	3
7	Riparian forest	4	1
8	Shrubby riparian vegetation	1	2
21	Urban area	2	8
	TOTAL	125	96

Table 8. Fragmentation value of each land use measured as number of patches in Fuencarral. This parameter has been calculated in 1975 and 2009.

4.3.3. Torrelodones: Urban development in a PNA

Urban development was the more significant process in Torrelodones area (Figure 3) The H′(u) values (1975: 2.81; 2009: 2.49) are remarkable and these, added homogeneity to landscape. The R (u) (1975: 12; 2009: 10) and E (u) (1975: 0.78; 2009: 0.75) values were almost unchanged (Table 5 and figure 15). The most important changes (Figure 14) were located in urban areas consolidated between both dates (2009: > 30%; 1975: 15%). The surface of mixture arborescens or shrubby woodlands with high density (*Juniperus, Cistus, Quercus*) decreased (1975: 20%; 2009: < 10%) and were replacement with urban areas. Into intermediate density woodlands the process is the contrary with a slight increase (2009: 30%). In this example values corresponded to disturbed areas near of urban areas or changes in livestock production (beef cattle into extensive o semi-extensive pastoral system).

As for fragmentation, although patches number decreased in 2009, the territory was more anthropic concentrating such growth in this sector analyzed in conjunction with the boundary of the PNA (Table 8 and Figure 14). This produced a boundary effect preventing the exchange of matter and energy in that direction and causing isolation of the territory whose consequences have been already indicated by several authors: loss of diversity and of species, less permeability, worst energy flow, etc. [62-64]. Such an example it was the large increase of road infrastructures in 2009 more greater than 1975. This is especially significant with N-VI, nowadays a highway that introduced fragmentation and negatives effects over territory (Figure 14). The connectivity of the territory is also affected overtime, showing a loss of permeability that results in a lower exchange of individuals between populations, lower the persistence of local and regional populations, increasing the rate of extinction and reducing the rate of colonization. Landscape connectivity favours not only movements of animal species, but also of plant and material and energy flows [65-67]. Therefore shows a loss of connectivity and increased fragmentation with the passage of time which is located predominantly in the edge of the ENP. This causes a barrier effect in that direction forcing the processes and natural movements to move into the space where connectivity is maintained higher and less fragmentation.

Figure 13. Single-family housing at the western edge of PNA (Torrelodones)

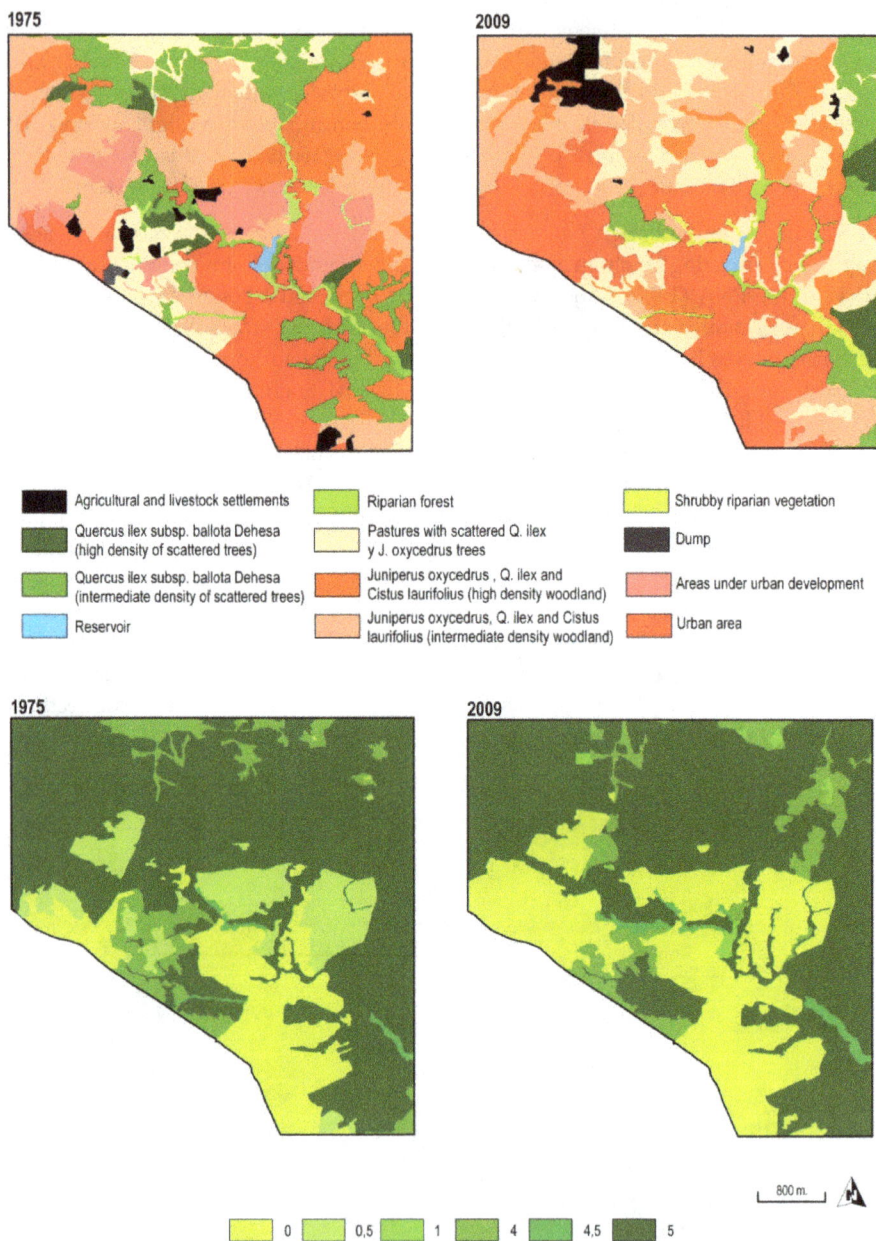

Figure 14. Land uses (1975-2009) and ecological connectivity in Torrelodones

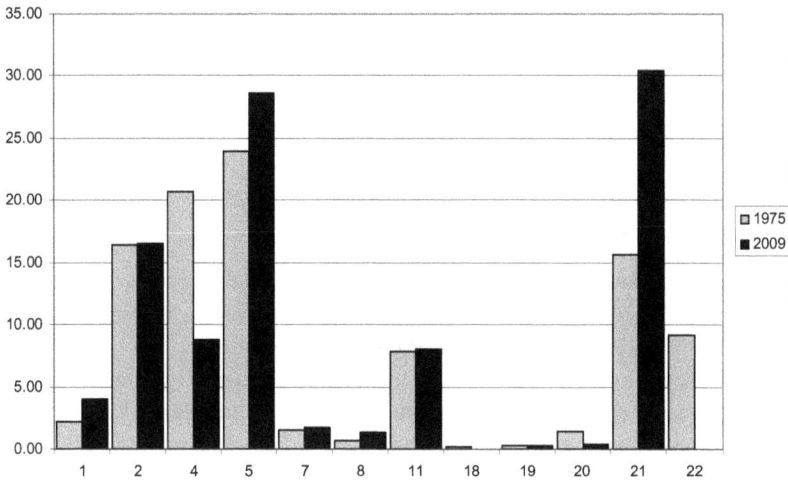

Figure 15. Relative frequency profile of land uses of each year in Torrelodones. The codes and the land uses are the same for Table 8.

Code	Land use	Number of patches	
		1975	2009
20	Agricultural and livestock settlements	15	5
1	*Quercus ilex* subsp. *ballota* Dehesa (high density of scattered trees)	8	4
2	*Quercus ilex* subsp. *ballota* Dehesa (intermediate density of scattered trees)	16	17
19	Reservoir	1	1
22	Areas under urban development	6	0
18	Dump	1	0
4	*Juniperus oxycedrus* , *Quercus ilex* subsp. *ballota* and *Cistus laurifolius* (high density woodland)	12	7
5	*Juniperus oxycedrus, Quercus ilex* subsp. *ballota* and *Cistus laurifolius* (intermadiate density woodland)	8	4
11	Pastures with scattered *Quercus ilex* subsp. *ballota* y *J. oxycedrus* subsp. *oxycedrus* trees	10	15
7	Riparian forest	6	6
8	Shrubby riparian vegetation	4	5
21	Urban area	9	12
	TOTAL	**148**	**104**

Table 9. Fragmentation value of each land use measured as number of patches in Torrelodones. This parameter has been calculated in 1975 and 2009.

5. Conclusions

This paper shows the multiscalar and multidisciplinary methods applied in landscape analysis using ecological, geographical and cartographical techniques. At PNA scale the comparative analysis of the vegetation and land use maps enabled us to identify five main change dynamics in the territory: urban development, scrub encroachment, forest encroachment, agricultural abandonment and new crops. Most of these dynamics were consequence of the abandonment of traditional activities or the increase in urbanised areas.

For a detailed scale three study cases were chosen. In Cerceda open woodland and shrub encroachment were more significant dynamics. Fragmentation measured was significantly higher in 2009 than in 1975 and conserved the same permeability. This is an area with few transformations by anthropic uses and that maintains the same vegetation types in both dates. The increase of road infrastructures, urban pressure and agricultural abandonment generated a homogeneous landscape in Fuencarral. This type maintains (2009) a great diversity of land uses, wastelands and habitats well preserved that have and interesting role such as ecological and territorial connectors. The most important changes in Torrelodones were located in urban areas consolidated between 1975 and 2009. At northwest edge N-VI highway introduced fragmentation and negatives effects with loss of permeability and connectivity.

Author details

N. López-Estébanez and F. Allende
Department of Geography, Autónoma University of Madrid,
Francisco Tomás y Valiente, Madrid, Spain

P. Fernández-Sañudo and M.J. Roldán Martín
Environmental Research Centre of Madrid Region, Madrid-Colmenar Viejo,
Soto de Viñuelas, Tres Cantos, Spain

P. De Las Heras
Department of Ecology, Complutense University of Madrid, Faculty of Biology,
José Antonio Novais s/n, Madrid, Spain

Acknowledgement

This research was partially funded by the Spanish Ministerio de Educación y Ciencia (projects CSO2009-12225-C05-02 and CSO2009-14116-C03-0102956-BOS).

6. References

[1] McGarigal K, Marks B, Holmes C, Ene E, (2002) Fragstats 3.3. Spatial Patter Analysis Program for Quantifying Landscape Structure, Department of Natural Resources Conservation, University of Massachusetts.

[2]   GIS (1997): FRAGSTATS*ARC User's Manual. Available: www.innovativegis.com/products/fragstatsarc

[3]   Berry J, Buckley D J, McGarigal K (1999) Integrating Landscape Structure Programs with GIS. Available: www.innovativegis.com

[4]   McGarical K, Marks B J, (1995) FRAGSTATS: *Spatial pattern analysis program for quantifying landscape structure*. USDA For. Serv. Gen. Tech. Rep.

[5]   Maldenoff DJ, Baker WL (2000) Spatial Modelling of Forest Landscape Change, Cambridge University Press.

[6]   Hong S, Barry H, De Zonia, Maldenoff DJ, (2000) An aggregation index (AI) to quantify spatial patterns of landscape, Landscape Ecology, 15 : 591-601.

[7]   Riitters KH, O'Neills RV, Hunsaker CT, Wickham JD, Yankee DH, Timmins SP,(1995) A factor analysis of landscape pattern and structure metrics, Landscape Ecology 10: 23-39.

[8]   Zhang J, Dai Z, Zhang Q, (2011) Two methods for detecting landscape associations based on a landscape map, Landscape and Ecological Engineering:145–151.

[9]   Bastian O (2001) Landscape ecology: towards a unified discipline? Landscape Ecology 16: 757–66.

[10] Boutin S, Hebert D (2002) Landscape ecology and forest management: developing an effective partnership, Ecological Applications12: 390–97.

[11] Naveh Z, Lieberman A S, (1984) Landscape Ecology. Theory and application. Springer. Series on Environmental Management. Springer Verlag. New York.

[12] González Bernáldez F (1991) Ecological consequences of the abandonment of traditional land use systems in central Spain, Options Méditerranéennes Serie Séminaire, 15: 23-29.

[13] González Bernáldez F (1981) Ecología y Paisaje, Madrid, Blume: pp. 250.

[14] Urban DL, O'Neil RV, Shugart HH, (1987) Landscape ecology: a hierarchical perspective can help scientist understand spatial patterns, BioScience 37: 119-127.

[15] Delcourt H, Delcourt P (1988) Quaternary landscape ecology: relevant scales in space and time, Landscape ecology 2: 23-44.

[16] Zonneveld IS, Forman RTT, (1990) Changing landscapes: an ecological perspective. Springer, New York.

[17] Forman RTT, Collinge SK, (1997) Nature conserved in changing landscape with and without spatial planning, Landscape and Urban Planning 37: 129-135.

[18] CORRELL D L (1991) Human impact on the functioning of landscape boundaries. In Holland M, Naiman RJ, Risser P G, Editors. The Role of Landscape Boundaries in the Management and Restoration of Changing Environments. Man and the Biosphere Program, UNESCO, Paris. pp. 90-109

[19] Jongman RHG (2001) Homogenisation and fragmentation of the European landscape: ecological consequences and solutions, Landscape and Urban Planning 58(2–4): 211–221.

[20] Phipps M, (1991) Diversity in anthropogenic ecological systems: the landscape level. In: Pidenad FD, Casado MA, De Miguel JM, Montalvo J, Editors: Diversidad biológica/Biological diversity. Fund. Areces, ADENA/WWF & SCOPE, Madrid: pp. 63-70.

[21] Risser PG (1987) Landscape ecology: state of the art. In: Turner MG (Editor). Landscape heterogeneity and disturbance. Springer-Verlag, New York.

[22] Palmer MW (1992) The coexistence of species in fractal landscapes. American Naturalist 139: 375-397.

[23] Rescia AJ, Schmitz MF, Martín de Agar P, de Pablo CL, Atauri JA, Pineda FD, (1994) Influence of landscape complexity and land management on woody plant diversity in Northern Spain, Journal Vegetation Science 5: 505-516.

[24] Ludwig JA, Tongway DJ, Bastin GN, Craig DJ, (2004) Monitoring ecological indicators of rangeland functional integrity and their relation to biodiversity at local to regional scales, Austral Ecology 29 (1): 108–120.

[25] Groome HJ 1990. Historia de la Política Forestal en el Estado Español. Agencia de Medio Ambiente, Madrid.

[26] OSE (2006) Cambios de ocupación del suelo en España. Implicaciones para la sostenibilidad. Ministerio de Medio Ambiente: pp. 485.

[27] Shannon CE, Weaver W, (1949) Mathematical Theory of Communication. University of Illinois Press, Urbana.

[28] De las Heras P, Fernández-Sañudo P, López Estébanez N, Roldán Martín MJ (2011) Territorial dynamics and boundary effects in a protected area of the Central Iberian Peninsula, Central European Journal of Geoscience 3: 1-11.

[29] BOCM (1985) Ley de la Comunidad de Madrid 1/1985 de 23 de enero del Parque Regional de la Cuenca Alta del Manzanares, BOCM 33 (8/02/1985).

[30] Riemann R; Tillman K. 1999. FIA photointerpretation in Southern New England: A tool to determine forest fragmentation and proximity to human development. USDA Forest Service Northeastern Reserch Station: pp. 709.

[31] Kelly M, Estes JE, Knight KA, (1999). Image interpretation keys for validation of global land-cover data sets, Photogrammetric engineering and remote sensing 95 (9): 1041-1050.

[32] Muchoney D, Strahler A, Hodges J, Locastro J, (1999) The IGBP DISCover confidence sites and the system for terrestrial ecosystem parameterization: Tools for validating global land-cover data , Photogrammetric Engineering and RemoteSensing 65 (9): 1061-1067.

[33] Madden M, Jones D, Vilchek L, (1999) Photointerpretation key for the Everglades Vegetation Classification System, Photogrammetric Engineering and Remote Sensing 65(2): 171-177.

[34] Riva-Murray Karen, Riemann Rachel, Murdoch Peter, Fischer JM, Brightbill R, (2010) Landscape characteristics affecting streams in urbanizing regions of the Delaware River Basin (New Jersey, New York, and Pennsylvania, U.S.) Landscape Ecology, 25 (10): 1489-1503.

[35] Gloagen JC, Roze F, Touffet J, et al. (1994) Study of successions in Brittany old fields, Acta Botanica Gallica, 141(6-7): 691-706.

[36] Aaviksoo K (1993) Changes of plant cover and land-use types (1950S to 1980S) in 3 Mire reserves and their neighbourhood in Estonia. Landscape Ecology 8(4): 287-301.

[37] Chamberlain K, (1992) Aerial surveys for Hertfordshire County Council, Photogrammetric record 14(80): 201-205.

[38] Buttner G, Biro M, Maucha G, et al. (2001) Land Cover mapping at scale

[39] 1: 50,000 in Hungary: Lessons learnt from the European CORINE programme In: Buchroithner MF, Editor. 20th Annual Symposium of the European Association of Remote Sensing Labs. pp. 25-31.

[40] Buttner G, Maucha G, Taracsak G, (2003) Inter-Change: A software support for interpreting land cover changes. In: Benes T, Editor. 22nd Symposium of the European Association of Remote Sensing, Prague, Czech Republic: pp. 93-98.

[41] Arnaez J, Lasanta T, Errea MP, (2011) Land abandonment, Landscape evolution, and soil erosion in a Spanish Mediterranean mountain region: the case of Camero Viejo. Land degradation & development 22(6): 537-550.

[42] Tavares A, Pato R L, Magalhaes M C, (2012) Spatial and temporal land use change and occupation over the last half century in a peri-urban area, Applied Geography, 34: 432–444.

[43] Poyatos R, Latron J, Llorens P, (2003). Land use and land cover change after agricultural abandonment - The case of a Mediterranean Mountain Area (Catalan Pre-Pyrenees), Mountain Research and Development 23 (4): 362-368.

[44] Chastain Robert A Jr, Struckhoff Matthew A, He Hong S, et al. (2008) Mapping vegetation communities using statistical data fusion in the Ozark National Scenic Riverways, Missouri, USA, Photogrammetric Engineering and Remote 74(2): 247-264.

[45] Ramos A, (Coord) (1985) Formaciones vegetales y usos actuales del suelo de Madrid. Memoria y Mapa, Consejería de Agricultura y Ganadería. Comunidad de Madrid. Escala 1:200,000: pp. 42.

[46] Muñoz Municio C, Gil Gil T, De las Heras Puñal P, González Bustamante N, (2004) Memoria del mapa de vegetación de la Sierra de Guadarrama (Vertiente Madrileña), Centro de Investigaciones Ambientales de la Comunidad de Madrid "Fernando González Bernáldez".

[47] SIOSE V.2, (2011) Sistema de Información de Ocupación del Suelo en España. Documento Técnico SIOSE 2005, DG Instituto Geográfico Nacional. Servicio de Ocupación del Suelo. S.G. de Cartografía. Madrid.

[48] De las Heras P, Fernández Sañudo P, Roldán Martín MJ, López Estébanez N (2006) Cambios territoriales y evolución de la gestión en el Parque Regional de la Cuenca Alta del Manzanares (Madrid, España), 2º Congresso Ibérico de Ecologia, SPECO-AEET, Lisboa. pp. 309.

[49] Forman RTT (1998) Landscape ecology, the growing foundation in land-use planning natural resource management. In: Proceedings of the CZ-IALE Conference on Present and Historical Nature—Culture Interaction in Landscapes (Experiences for the 3rd Millenium). Karolinum Press, Prague, pp. 13–21.

[50] Forman RTT (1998) Road ecology: a solution for a giant embracing us, Landscape Ecolology 13: iii–v.

[51] Forman RTT (1995) Land Mosaics: the Ecology of Landscapes and Regions. Cambridge University Press, Cambridge, England.

[52] Roldán Martín MJ, De Pablo CL, Martín de Agar P, (2006) Landscape changes over time: comparation of lands uses, boundaries and mosaics, Landscape Ecology 21: 1075-1088.

[53] Pielou EC (1977) Mathematical Ecology. Wiley, London, UK.

[54] O'Neill R, Krummel J, Gardner R, Sugihara G, Jackson D, Milne B, Turner M, Zygmunt B, Christensen S, Dale V, Graham R, (1988) Indices of landscape pattern, Lanscape Ecology 1: 153-162.

[55] Marull J, Mallarach JM, (2005) GIS methodology for assessing ecological connectivity: application to the Barcelona Metropolitan Area, Landscape and Urban Planning 71(2-4): 243–262

[56] Bennett G (2004) Integrating Biodiversity Conservation and Sustainable Use: Lessons learned from Ecological Networks. Gland, Switzerland and Cambridge UK. UICN: 55 p.

[57] Bodin Ö, Norberg J (2007) A network approach for analyzing spatially structured populations in fragmented landscape, Landscape Ecology 22: 31–44.

[58] IUCN (1994) Guidelines for protected areas management categories   CNPPA and WCMC, IUCN, GLAND, Switzerland and Camdbrige, UK.

[59] Gómez-Sal A, Belmontes JA, Nicolau JM, (2003) Assessing landscape values: a proposal for a multidimensional conceptual model, Ecological Modelling,168: 319-341.

[60] Gómez-Sal A, Álvarez J, Muñoz-Yanguas MA, Rebollo S, (1993) Patterns of change in the agrarian landscape in the area of the Cantabrian Mountains (Spain). Assessment by transition probabilities. In: Bunce RGH, Ryszkoewski L.

[61] Paoletti MG, Editors: Landscape ecology and agrosystems. Lewis Publishers, Boca raton, USA.

[62] Taylor DR, Fahrig L, Henein K, (1993) Connectivity is a vital element of landscape structure, Oikos 68:571-3

[63] Opdam P, Foppen R, Vos C, (2002) Bridging the gap between empirical knowledge and spatial planning in landscape ecology, Landscape Ecology 16: 767–779.

[64] Fahrig L (2003) Effects of Habitat Fragmentation on Biodiversity, Annual Review of Ecology, Evolution and Systematics 34: 487-515.

[65] Fry G (2005) Lectures on Landscape Ecology at the Norwegian University of Life Sciences. Ås, Norway.

[66] Beier P (1995) Dispersal of juvenile cougars in fragmented habitat, Journal of Wildlife Management 59(2):228-237.

[67] Berggren A, Birath B, Kindvall O (2002) Effect of corridors and habitat edges on dispersal behaviour, movement rates and movement angles in Roesel's Bush-Cricket (*Metrioptera roeseli*), Conservation Biology 16: 1562–1569

Open Source Tools, Landscape and Cartography: Studies on the Cultural Heritage at a Territorial Scale

Pilar Chias and Tomas Abad

Additional information is available at the end of the chapter

1. Introduction

The Task Group 2 "Open Source in use for the cultural heritage communication process" [1] is hosted by CIPA, the International Committee for Architectural Photogrammetry, one of the international committees of ICOMOS (International Council on Monuments and Sites). The main targets of the Task Group are related to the study and dissemination of the Cultural Heritage, considered from a wide perspective according to UNESCO's definition [2], and the current EU legal framework.

This scope includes "the entire corpus of material signs –either artistic or symbolic- handed on by the past to each culture and, therefore, to the whole of humankind" [3]. Thus, Cultural Heritage consists of more than *monuments* (architectural works, sculpture and painting elements, archaeological structures, ancient books, manuscripts, maps and charts), but of *groups of buildings* (historical towns, industrial heritage, historical infrastructures), and *sites* (cultural landscapes, natural values). Physical artifacts and intangible attributes inherited from past generations must be maintained for the future generations, and accordingly, their knowledge must be widely disseminated.

Due to the different particular types of cultural objects, different approaches, tools and techniques must be developed to get an accurate knowledge of them. Aspects and disciplines concomitant with the preservation and conservation of tangible and intangible culture include archival science, art and architectural conservation, audio recording, digital data collection, storage and management, and architectural, urban and landscape drawing, among others. From a multidisciplinary approach, the Task Group 2 aims to develop an intensive promotion of Open Source software as free alternate tools for all researches on Cultural Heritage, which are useful in every stage involved in the processes of knowledge, documentation, management, and dissemination of cultural heritage.

In a first epoch the Task Group 2 initiative was successfully chaired by Markus Jobst, of Research Group Cartography at Vienna University of Technology (Austria). Currently we are starting a new era chaired by Prof. Pilar Chias, at the University of Alcalá (Spain).

2. Main targets of the task group

Researches on Cultural Heritage have wide connotations related to the characteristics of the cultural objects that are considered, as well as to the particular targets and methodologies that can be applied in each case.

As a first stage, a deep knowledge of the cultural objects is needed, and adequate documentation processes become essential. But depending on the kind of cultural object that is being studied, it is necessary to collect and store a high amount of datasets in various formats, to draw accurate maps and sketches, to take pictures, and to write detailed descriptions (or transcriptions). Thus, useful software for writing, drawing, image manipulation and mapping is needed.

An easy and efficient data recording and archiving, as well as a quick access to all these digital materials become fundamental in the processes of documentation and storage of the cultural heritage. The other aim is to get high quality information from them. The later analysis processes must lay the foundations to propose the right conclusions, and to make the appropriate decisions. In this stage, GIS become a useful tool.

Finally, an efficient prospective plan must include management and preservation actions, as well as wide realist communication and dissemination programmes, including online platforms.

All these tasks and their related activities can be successfully approached and developed by using Open Source software. It provides the necessary office and prepress tools, multimedia and graphic editors (both vector and raster formats), friendly computer aided design (CAD) and 3D modelling software, as well as geographic information systems (GIS) and other publishing products.

According to it, the Task Group's strategies will correspond to two main objectives:

1. Related to the cultural heritage:
    *   Improved knowledge and understanding of the cultural heritage, covering the different 'cultural properties' [4].
    *   Improved methodologies in order to get a deep knowledge and an accurate documentation of each cultural object, encouraging an integrated interdisciplinary approach.
    *   Collaboration between involved research groups and other related initiatives.
    *   More effective preservation and management.
2. Related to the Open Source tools to be used to accomplish these targets:
    *   Dissemination of Open Source software capabilities in order to fulfill objective no. 1.
    *   Greater accessibility to the public via the Task Group web site, but also through courses, conferences and online materials, as well as other cultural actions.
    *   Wide offer of academic courses at the Universities involved.

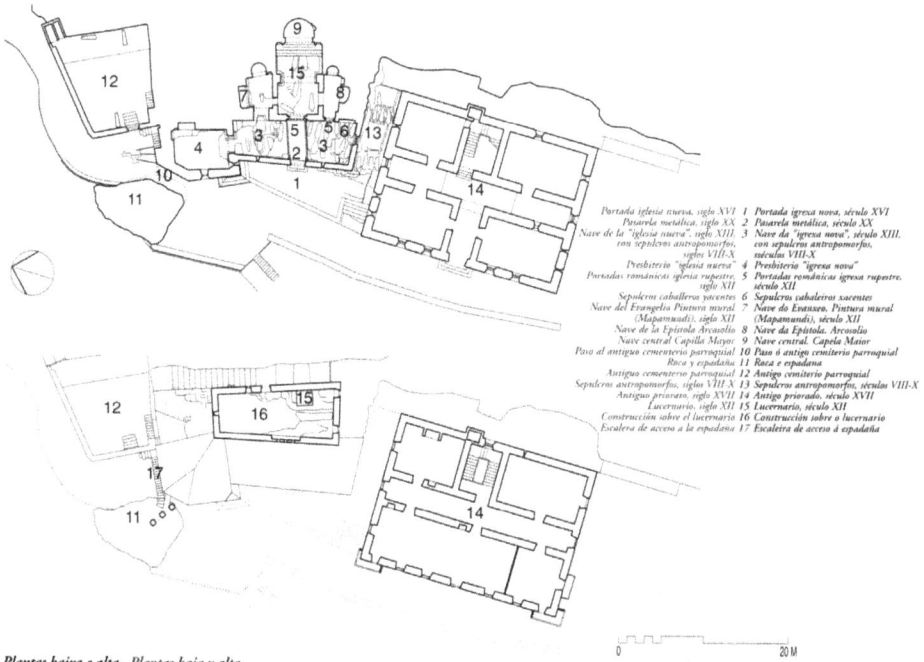

San Pedro de Rocas

Portada iglesia nueva, siglo XVI 1 Portada igrexa nova, século XVI
Pasarela metálica, siglo XX 2 Pasarela metálica, século XX
Nave de la "iglesia nueva", siglo XIII. 3 Nave da "igrexa nova", século XIII.
con sepulcros antropomorfos, con sepulcros antropomorfos,
siglos VIII-X séculos VIII-X
Presbiterio "iglesia nueva" 4 Presbiterio "igrexa nova"
Portadas románicas iglesia rupestre, 5 Portadas románicas igrexa rupestre,
siglo XII século XII
Sepulcros caballeros yacentes 6 Sepulcros cabaleiros xacentes
Nave del Evangelio Pintura mural 7 Nave do Evanxeo. Pintura mural
(Mapamundi), siglo XII (Mapamundi), século XII
Nave de la Epístola Arcosolio 8 Nave da Epístola. Arcosolio
Nave central Capilla Mayor 9 Nave central. Capela Maior
Paso al antiguo cementerio parroquial 10 Paso ó antigo cemiterio parroquial
Roca y espadaña 11 Roca e espadana
Antiguo cementerio parroquial 12 Antigo cemiterio parroquial
Sepulcros antropomorfos, siglos VIII-X 13 Sepulcros antropomorfos, séculos VIII-X
Antiguo priorato, siglo XVII 14 Antigo priorato. século XVII
Lucernario, siglo XII 15 Lucernario, século XII
Construcción sobre el lucernario 16 Construcción sobre o lucernario
Escalera de acceso a la espadaña 17 Escaleira de acceso á espadaña

0   20 M

Plantas baixa e alta · Plantas baja y alta
*A planta baixa recolle a igrexa rupestre do século XII escavada na roca, coa "igrexa nova" románica
do século XIII situada diante transversalmente. Á dereita, antigo priorato do século XVII.
La planta baja recoge la iglesia rupestre del siglo XII excavada en la roca, con la "iglesia nueva" románica
del siglo XIII situada delante transversalmente. A la derecha, antiguo priorato del siglo XVII.*

Ground and first floor plans: the ground floor plan depicts the 12th century rock church, with the Romanesque 13th century 'new church' which is placed in front and transversely. The old 17th century priory is on the right side of the plans. Legend:

1. New church's 16th century façade.
2. Metallic footbrige, 20th century.
3. Nave of the 'new church', 13th century; with anthropomorphic tombs dated to the 8th-10th centuries.
4. New church's presbytery.
5. Romanesque façades of the rock church, 12th century.
6. Tombs of the recumbent knights.
7. Wall map (world map) in the gospel aisle, 12th century.
8. Epistle aisle's arcosolium.
9. Main chancel in the nave.
10. Access to the old graveyard.
11. Rock and belfry.
12. Old graveyard.
13. Anthropomorphic tombs, 8th-10th centuries.
14. Old priory, 17th century.
15. Skylight, 12th century.
16. Structure over the skylight.
17. Stairs to the belfry.

(a)

San Pedro de Rocas

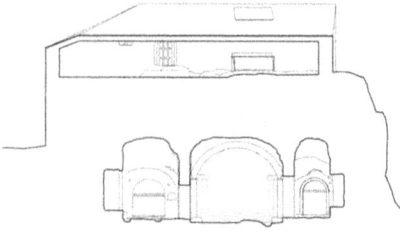

Sección transversal da igrexa rupestre · Sección transversal de la iglesia rupestre
As tres capelas hipoxeas, comunicadas entre sí, ventilaban a través dunha clarabóia excavada na roca (a trazos), hoxe protexida por unha construcción superior.
Las tres capillas hipogeas, comunicadas entre sí, ventilaban a través de un lucernario excavado en la roca (a trazos), hoy protegido por una construcción superior.

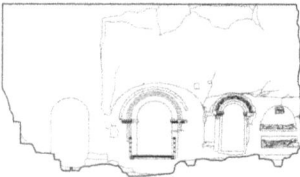

Sección-alzado da fachada románica · Sección-alzado de la fachada románica
Esta sección lonxitudinal sen a "igrexa nova" mostra a fachada da antiga igrexa rupestre coas dúas portadas románicas do século XII, a central e a da dereita, e o nicho cos sepulcros de dous cabaleiros xacentes.
Esta sección longitudinal sin la "iglesia nueva" muestra la fachada de la antigua iglesia rupestre con las dos portadas románicas del siglo XII, la central y la de la derecha, y el nicho con dos sepulcros de dos caballeros yacentes.

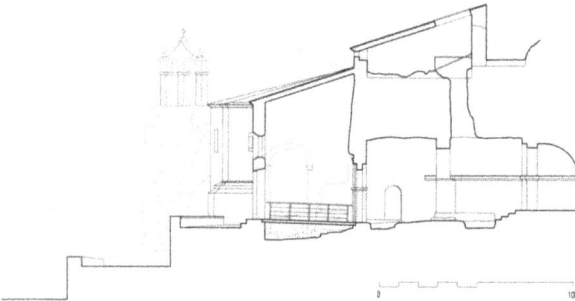

Sección lonxitudinal da igrexa · Sección longitudinal de la iglesia
Nesta sección pode apreciarse á esquerda a espadana sobre a roca, no centro a sección transversal da "igrexa nova" do século XIII coa pasarela metálica de acceso á igrexa rupestre. Encima pode observarse a clarabóia superior excavada na roca e a nova construcción superior que a protexe.
En esta sección puede apreciarse a la izquierda la espadaña sobre la roca, en el centro la sección transversal de la "iglesia nueva" del siglo XIII con la pasarela metálica de acceso a la iglesia rupestre. Encima puede observarse el lucernario superior excavado en la roca y la nueva construcción superior que la protege.

On top: Cross section of the rock church. The three hipogean chancels, which are connected, ventilated through a skylight dug in the rock (broken lines) that nowadays is protected by an upper building.
Centre: Cross-elevation of the Romanesque façade. This longitudinal section omitting the 'new church', shows the old rock church with its two 12th century Romanesque façades –the central and the right ones-, and the niche with the tombs of the two recumbent knights.
Below: Longitudinal section of the church. It depicts the bellfry on the rock at the right side; in the middle, the cross section of the 13th century 'new church' can be seen, with the metallic footbridge that leads to the rock church. The old skylight is placed over the rock church, currently protected with a modern upper structure.

(b)

Figure 1. a) Combined techniques in use for subterranean heritage mapping. San Pedro de Rocas, Esgos, Spain (Task Group's Partner: *GIRAP Research Team*, University of La Coruña, Spain).
b) Combined techniques in use for subterranean heritage mapping. San Pedro de Rocas, Esgos, Spain (Task Group's Partner: *GIRAP Research Team*, University of La Coruña, Spain).

3. Open Source tools in the methodologies applied to the study of territories and landscapes

Researches on cultural heritage at a territorial scale involve several stages, according to the different methodologies applied in each case. They can be structured into three main groups of tasks: documentation, analysis, and management and dissemination processes.

3.1. Documentation processes

A deep knowledge of the cultural objects requires previous and accurate measurements, reports and drawings.

Measurements of angles, areas, lenghts and distances must be managed to some level of quality (in terms of accuracy, availability, usability, and resilience). Apart from the traditional direct measuring methods, datasets and georeferencing can be achieved nowadays by means of several techniques and electronic devices. They can be lately transferred to a computer, and manipulated through CAD or GIS software, or by graphic or sound editors [5].

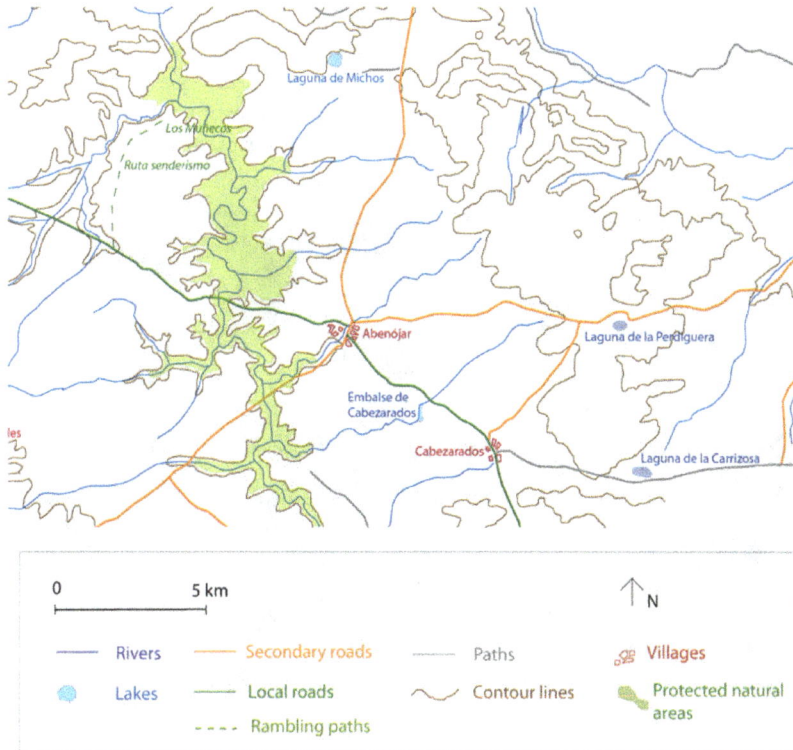

Figure 2. Abenójar, Ciudad Real. Natural resources as an essential part of the Cultural Heritage. Original scale 1:50.000 (Task Group's Partner: *Cultural Heritage and sustainable architecture Research Team*, University of Alcalá, Spain).

Thus, photogrammetry, angle and distance measuring with distance meters and GPS, or total stations, become useful tools with the adequate standalone or plugin software. Free software as Angle Meter, Distance Meter and GeoDistance provide easy to use tools to collect all these datasets [6].

Afterwards digital maps and plans can be drawn with free vector graphics editing programs as AutoCAD or Inkscape, which can easily integrate the measures and georeferences that were previously collected.

Depending on the type of heritage and the working scale, 2D and 3D designs as well as landscape drawings, can be produced in order to show their main characteristics and features with the required precision.

There are free specific tools for cartography and landscape as Cross GL Draw and Terragen. The first one is a general purpose 2D raster and vector graphics library. As a full featured vector graphics editor, it covers a wide range of utilization domains as business graphics, web graphics, DTP, cartography, printing, user interface graphics, serverside rendering, and many more. It is a useful tool to draw digital cartography and available for Windows platforms.

Terragen is a powerful solution for rendering and animating realistic landscapes and natural environments. It is useful in researches on landscape as a cultural heritage, and works both on Mac OS and Windows operating systems. Ancient maps are useful tools to study the historical features at a territorial scale. They provide interesting information about the history of the territory and the landscape, as they depict their main features, their topologic relationships, and the toponimy [7].

Figure 3. The *Via Augusta* near the Roman colony of *Hasta Regia*, Cádiz. Hypothetical reconstruction by means of aerial photographs (Task Group's Partner: *Cultural Heritage and sustainable architecture Research Team*, University of Alcalá, Spain).

Successful methodologies to compare and georeference ancient maps are being currently applied [8, 9], and provide interesting conclusions about their accuracy. Other projects aim to reconstruct the history of the territories through ancient cartography, aerial photographs and field works [10, 11]. Their targets can be easily reached with the capabilities of free open source tools as NorthGate's Kml Builder.

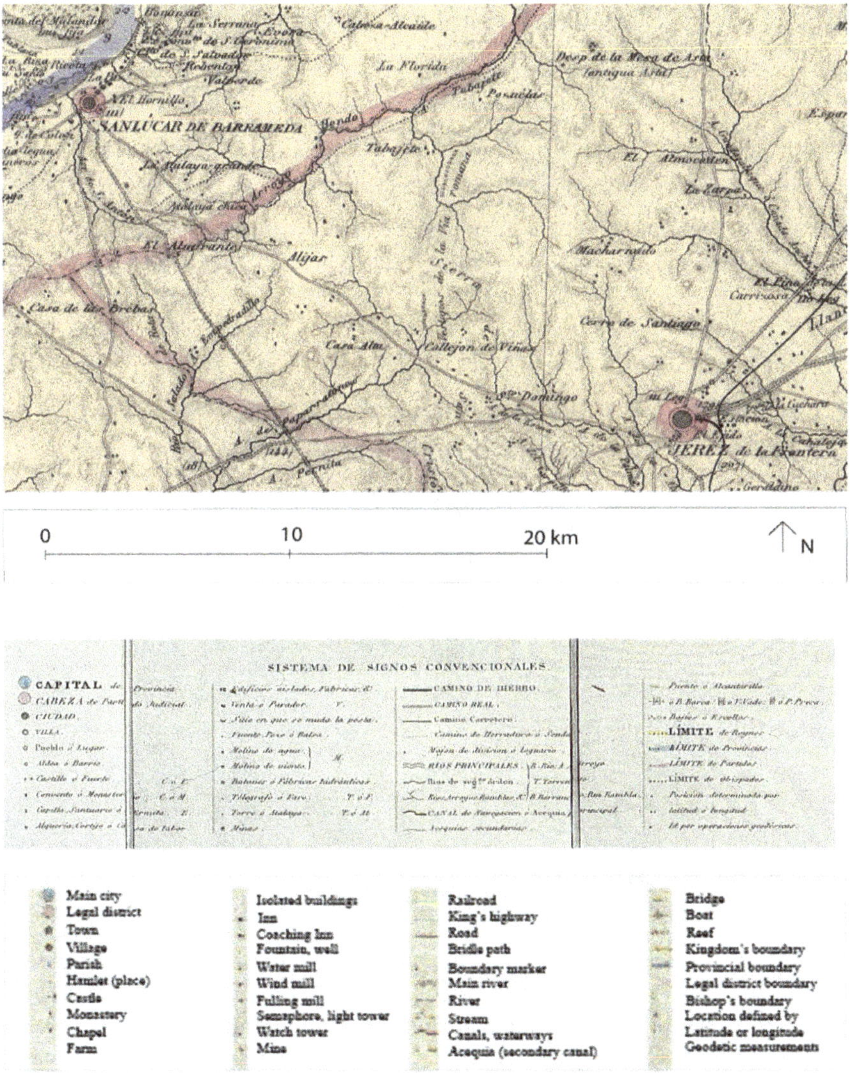

Figure 4. Francisco Coello 1868: *Map of the Province of Cadiz*, showing the *Via Augusta* between Jerez de la Frontera and Sanlúcar de Barrameda, lettered as "Vestigios de la Vía Romana " (remains of the Roman road) (Task Group's Partner: *Cultural Heritage and sustainable architecture Research Team*, University of Alcalá, Spain).

Terrestrial and aerial photographs can also be edited and manipulated with free programs as GIMP, Photoshop, or Image Analyzer, that run on Windows and Mac operating systems. They provide an array of useful features such as bitmap image viewing, management, comparison, resizing, cropping, warping, and colour adjustment.

In terrestrial photogrammetric works GIMP compensates easily perspectives, distortions caused by lens tilt, and eliminates len's *barrel distortion* and vignetting [12].

Figure 5. Accurate drawing of the 16th century façade of the Colegio de San Ildefonso (Alcalá de Henares, Spain), using photogrammetrical techniques (Task Group's Partner: *Cultural Heritage and sustainable architecture Research Team*, University of Alcalá, Spain).

Simultaneously, some researches require to collect other kind of data, which are related to perception. These relevant attributes of the cultural object must be also recorded, stored and drawn. Specific software and techniques are needed for studying some qualities as colour or textures [13], and there is also free image and photo open source software available to do it, as Photoshop and GIMPShop.

When tools for live video and sound recording are needed, Adapter free video and audio converter, and the audio editor Audacity, become useful free tools for studying soundscapes.

(a)

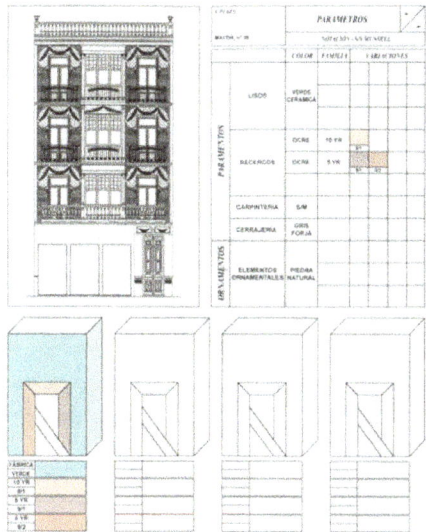

(b)

Figure 6. a) Plan of Cartagena, Murcia (Spain). Study of Cartagena's urban landscape colours (Task Group's Partner: *Cultural Heritage's Colour Research Team*, University of Valencia, Spain). In orange: Case studies in the city centre. b) Study of Cartagena urban landscape colours (Task Group's Partner: *Cultural Heritage's Colour Research Team*, University of Valencia, Spain).

Legend (right drawing):

Structuring features
1. Continuous abandonment of the academic traditional arrangement. Development of a free composition, particularly on ornamental tops.
Ornamental features

1. Wealth of details, mainly Modernist.
    1.1 ART NOUVEAU adornment: curves and organic shapes prevail.
    1.2 SEZESSION adornment: geometric shapes and brick bonding prevail.
2. Colour: polychromatic.
    2.1 Tendency to distinguish the features in the structure of the façade.
    2.2 Use of diverse materials.

Legend (left drawing):

Modernist building

Street / Square **Mayor, no. 35**	**Parameters**	
	S/S Munsell Notation	

		Colour	*Family*	*Variations*			
Faces	Plain	Ceramic Green					
	Frames	Ochre	10 YR	9/1			
		Ochre	5 YR	9/1	9/2		
	Carpentry	S/M					
	Locksmith's Craft	Grey Forge					
Adornments	Ornamental Elements	Natural Stone					

All these multiformat datasets previously collected are organized and stored in databases, that imply the use of general-purpose database management system.

As we have seen, databases related to Cultural Heritage involve multimedia contents, that must be processed, stored and disseminated. And there are some interesting and useful free multiformat databases (DBDesigner, Firebird, MySQL, OpenOffice.org BASE) that integrate database design, modeling, creation, and maintenance into a single, seamless environment. Some enable users to meet the database challenges of next generation web, cloud, and communications services with uncompromising scalability, uptime and agility. They are mainly developed for Windows, Unix and Linux.

Figure 7. Texture and colour data of the façade of San Juan de Letran; example of compensated distortions of a photograph (Task Group's Partner: *DAVAP Research Team*, University of Valladolid, Spain).

3.2. Analysis processes

Researches on cultural heritage use to create 3D models, renderings and animations in order to explain and analyze its formal features and qualities, and recreate virtual walks around and inside the object that is being studied.

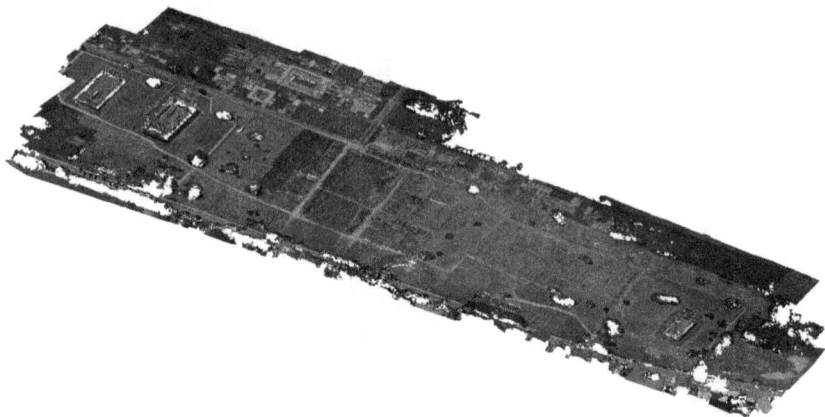

Figure 8. 3D Modell of Paestum, Italy (Task Group's Partner: *DAVAP Research Team*, University of Valladolid, Spain).

There are also free multiplatform tools available to achieve these targets as 3D Studio Max, Blender, Open FX and SketchUp, which provide powerful, integrated 3D modelling, animation, rendering, and compositing tools. Some include full renderer and raytracing engines, NURBS support, kinematics-based animation, morphing, and extensive plugin APIs. Plugin capabilities include image post processor effects such as lens flare, and depth of field.

For particular illumination effects, Lightsmark provides real time global illumination, colour bleeding, and penumbra shadows.

Useful tools to study the territory and the landscape are GIS, that are used for geospatial data management and analysis, image processing, graphics and maps production, spatial modeling, and visualization. They allow to work with detailed maps in raster, vector, or database format. Among them must be mentioned Quantum GIS, GRASS GIS, GDAL, OpenEV, Thuban, and JUMP.

Virtual Terrain Project osters the creation of tools for easily constructing any part of the real world in interactive, 3D digital form. For Digital Terrain Models (DTM) can be also used the above mentioned Terragen free software.

Finally, SAGA (System for Automated Geoscientific Analyses) is an open source GIS application with an emphasis on grid and raster functions: digital terrain analysis, geo-statistics, and image processing.

3.3. Management and dissemination processes

Project management can be successfully developed with OpenProj, which is a desktop project management application similar to Microsoft Project with a familiar user interface, and even opens existing MS Project files. It is interoperable with Project, Gantt Charts, and PERT charts. For Mac OS X.

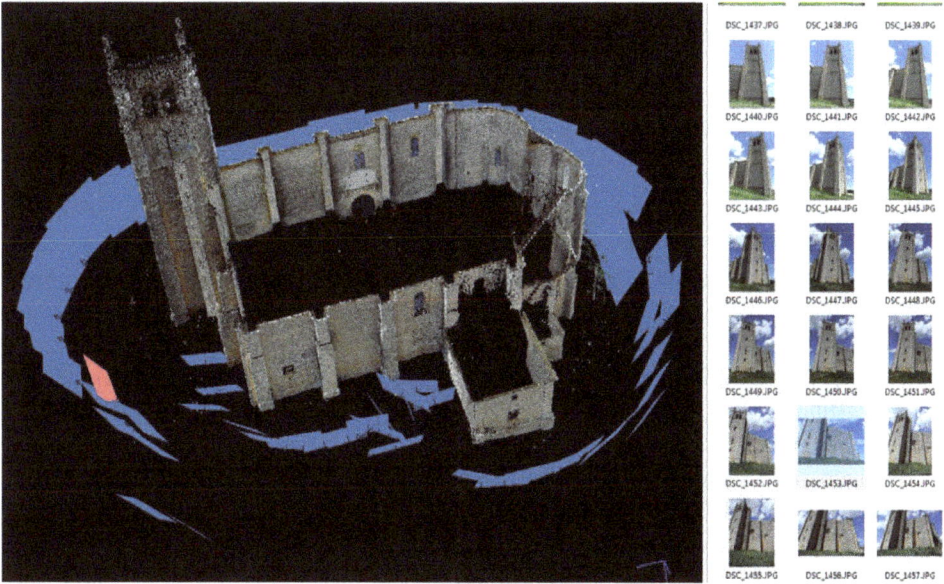

Figure 9. 3D Modell of the Church of the Natividad in Villasandino, Spain. It is the cartographic basis of an architectural GIS (Task Group's Partner: *DAVAP Research Team*, University of Valladolid, Spain).

On the other hand, dissemination processes of the Cultural Heritage involve presentation, publishing and prepress, including geoportal design and maintenance. But nowadays dissemination becomes supported by using the Internet and digital storage media [14].

According to Jessop [15], the presence of maps -and other cultural heritage informations- on the web, promotes "heritage material to new audiences, both specialists and general, across many social and cultural groups."

Books' editors as Scribus and SIGIL, as well as other office tools for document production and data processing (OpenOffice, AbiWord, LibreOffice, NeoOffice for Mac, among others) are suitable for a wide variety of word processing tasks, spreadsheets, presentations, graphics, databases and more. They store all the data in an international standard format and can also read and write files from other common office software packages.

There are also free multiplatform web editors, i.e. tools used to create and upgrade documents directly on the web. Browsing features are seamlessly integrated with the editing and remote access features in an uniform environment. Some combine web file management and easy-to-use WYSIWYG web page editing, and have been designed to be extremely easy to use, making them ideal for non-technical computer users who want to create an attractive, professional-looking web site without needing to know HTML or web coding. Thus, the web becomes a space for collaboration, and not just a one-way publishing medium.

The offer of free software is wide (Amaya, Aptana, Kompozer, and Nvu can be mentioned), but MapServer is a specific program for publishing spatial data and interactive mapping applications to the web, running on all major platforms (Windows, Linux, Mac OS X).

4. Members, partners, and initiatives

4.1. Members

The Task Group is currently hosted by the University of Alcalá (Spain), and is composed by the following international teams of researchers:

- University of Alcalá (Spain), Department of Architecture, Technical School of Architecture and Geodesy: Research Group 'Cultural Heritage and sustainable architecture'(http://www2.uah.es/cipa_opensource_taskgroup). Director: Prof. Dr. Arch. Pilar Chias (pilar.chias@uah.es). Researchers: Master in Civil Engineering Tomás Abad, Prof. Dr. Ing. León González Sotos, Dr. Arch. Ernesto Echeverría, Dr. Arch. Enrique Castaño, Dr. Arch. Fernando Da Casa, Ph. D. Ángeles Layuno, Dr. Ing. Raúl Fernández del Castillo, Dr. Ing. Francisco Maza, Arch. Manuel de Miguel, Arch. Paz Llorente.

  Their main research subjects are focused on cultural heritage documentation and dissemination, mainly at architectural, urban and territorial scales, including landscape studies. They lead and participate in many national and international research projects, and develop different methodologies to be applied to the documentation and dissemination of the cultural heritage at territorial, urban and architectural scales. They also collaborate with other Universities as the University of Grenoble (France), the CUJAE in Havanna and the University of Oriente (Santiago de Cuba).

  These methodologies use free open-source CAD software for cultural heritage documentation, as well as databases, digital cartography, terrestrial and aerial photogrammetry, and GIS for cultural heritage management.

  They are also leading an important effort in order to disseminate all data and informations related to cultural heritage via Internet. As an example can be cited the Ancient Spanish Cartography e-Library (http://www.ielat.es/).

- University of Valladolid (Spain), Technical School of Architecture: 'LFA-DAVAP Laboratory of Photogrammetry' (http://www.uav.es/davap/). Director: Dr. Arch. Juan José Fernández Martín (juanjo@ega.uva.es); Coordinator: Dr. Arch. Jesús San José Alonso (lfa@ega.uva.es); Research director: Dr. Math. Javier Finat (jfinat@agt.uva.es).

  Since 1995, this team is leading an important documentation and dissemination programme about the cultural heritage mainly in the Comunidad de Castilla y León (Spain), but also in Italy. They currently use both terrestrial and aerial photogrammetry.

- Politechnical University of Valencia, Technical School of Architecture: 'Colour' and 'Photogrammetry' research groups. Research directors: Prof. Ph. D. Ángela García Codoñer (angarcia@ega.upv.es) and Prof. Dr. Arch. Pablo Navarro (pnavarro@ega.upv.es). Researchers: Dr. Arch. Jorge Llopis, Dr. Arch. Ana Torres, Dr. Arch. Juan Serra.

  They all are members of the Assoziazione Internazionale dei Design.

  The team is developing some experiences on drawing and documenting the cultural heritage of the Comunidad Valenciana (Spain) by means of different techniques, including colour charts, architectural design, terrestrial photogrammetry, and CAD.

- Politecnico di Milano (Italy), II School of Architecture: 'Management and Urban Cultural Politics' research group. Director: Ph. D. Alessandro De Masi (alessandro.demasi@unina.it). As a member of UNESCO Forum, he is on the international research team for the 'Rural Vernacular Heritage 2007-2012' created by UNESCO World Heritage Centre (France).

  The team participates in international research projects with the Universities of Naples and Florence (Italy), the Simmons College of Boston (USA), and the West University of 'Neofit Rilisky' (Bulgaria).

- Università di Chieti-Pescara, Facoltà di Architettura: 'Theory and Practice in Conservation'. Director: Prof. Dr. Arch. Claudio Varagnoli (cvaragnoli@tiscali.it).

  They work mainly on architectural and urban restoration, as well as on indutrial heritage, focusing on the materials and their qualities.

- University of A Coruña (Spain), Technical School of Architecture: 'GIRAP Group for the Representation of the Architectural Heritage'. Director: Prof. Dr. Arch. Jose Antonio Franco Taboada (jafranco@udc.es). Researchers: Dr. Arch. Antonio Amado, Dr. Arch. Juan Manuel Franco, Dr. Arch. Antonia Pérez, Dr. Arch. Santiago Tarrio.

  The team focuses mainly on the documentation and analysis of the architecture and the archaeology using combined techniques (photogrammetry, direct measuring, radar). They are currently collaborating with the University of Puerto Rico.

- University of Málaga (Spain), Technical School of Architecture Research Group. Director: Dr. Arch. Guido Cimadomo (cimadomo@uma.es). Researchers: Arch. Javier Castellano, Arch. Antonio Álvarez, Arch. Jonathan Ruiz, and Arch. Dolores Goyanes.

  They usually collaborate with the School of Architecture of Rabat (Morocco) and the University of Bologna (Italy), leading projects about sustainable tourism, cultural landscapes, and databases of the architectural heritage in Andalusia (Spain) and the valley of Mgoun (Morocco), and about the european colonialism in Granada (Nicaragua).

- University Antonio de Nebrija (Madrid, Spain), Technical School of Architecture: 'Sustainable Architecture and Civil Engineering' research group. Director: Dr. Arch. Carlos González-Bravo (cgonzabr@nebrija.es). Team: Arch. Loreto Barrios, Arch. Emilio Mitre.

  Their main contributions are on the field of the sustainable rehabilitation of the cultural heritage, and the researches on green contruction solutions.

- IADE, Institución Artística de Enseñanza (Madrid, Spain). Director: Diego García de Castro (IADE@IADE.ES) (Website: http://www.iade.es/).

  The Institute is a centre with a long-established tradition in teaching interiors' design, fashion design, graphic and industrial design. According to their targets, they educate on the benefits of using open source software.

New members and collaborations are always wellcome.

4.2. Partners

There are some international teams of researchers that use to collaborate with the Task Group activities. Among them, the

- ICA Commission on Maps and the Internet (http://maps.unomaha.edu/ica/): chaired by Rex Cammack (rcammack@unomaha.edu), works on cartographic presentation, communication and processing methods.

4.3. Initiatives

Among the Open Source software dissemination initiatives that provide information and access to different tools and demos concernig to geospatial data are:

- Arc-Team Community (http://www.arc-team.com/): provides archaeological documentation and processing methods. Virtual Archaeology & 3D modelling.
- Cultural Heritage Computing (CHC) Salzburg, (https://www.sbg.ac.at/chc/chc_site_en/chc_ziele.html): located at the University of Salzburg, the team aims at the documentation and procurement of contents concerned with cultural heritage using modern computer technologies.
- FreeGIS Project (http://freegis.org/).
- The Open Source Geospatial Foundation (http://www.osgeo.org/).
- OpenGIS Consortium (http://www.opengeospatial.org/).

Other useful related initiatives and websites are:

- François Letellier's web site (http://www.flet.fr/) and blog.
- Open Source Initiative (http://www.opensource.org/).
- Opensource.com (http://opensource.com/).
- Opensource CMS (http://www.opensourcecms.com/).
- Free Open Source Software Mac User Group (http://www.freesmug.org/).
- The OSSwin Project: Open Source for Windows (http://osswin.sourceforge.net/).

5. Conclusions and prospective

The above mentioned Open Source software, though not and exhaustive list, is regularly checked by the Task Group, and tools are tested and used in different research projects and experiences. We aim to compare their results and to disseminate them worlwide.

We invite all communities and professionals, that are currently involved in researches or management of the cultural heritage, to share their experiences with Open Source software products.

We include not only professional works, but other initiatives as courses, conferences, congresses, publications, related blogs and websites, and other materials wich contribute to the dissemination of both the cultural heritage and the free Open Source software.

All of them will be included in the Task Group's web page, as Information and Communication Technologies provide powerful media for education, research and the dissemination to a wider audience than ever before.

Author details

Pilar Chias and Tomas Abad

Technical School of Architecture and Geodesy, University of Alcala, Madrid, Spain

Acknowledgement

Researches developed by the Team of Alcalá are supported by the Junta de Comunidades de Castilla-La Mancha, and the Benjamin Franklin Institute (University of Alcalá).

6. References

[1]   Open Source in use for the Cultural Heritage communication processes. [Internet] 2012 [updated 2012 March 28; cited 2012 April 7]. Available from: http://www2.uah.es/cipa_opensource_taskgroup/index.html and also http://cipa.icomos.org/

[2]   UNESCO Convention concerning the Protection of the World Cultural and natural Heritage [Internet] 1972 [updated 2012 Jan 20; cited 2012 April 7]. Available from: http://whc.unesco.org/archive/convention-en.pdf

[3]   ICCROM Working Group 'Heritage and Society' [Internet]. 2005 [updated 2012 Jan 20; cited 2012 April 7]. Available from: http://cif.icomos.org/pdf_docs/Documents%20on%20line/Heritage%20definitions.pdf

[4]   UNESCO (1954) Convention Protection of Cultural Property in the Event of Armed Conflict. The Hague, §1.

[5]   Chías P, Abad T (2008) A GIS in Ancient Cartography: A New Methodology for the online accessibility to the Cartographic Digital Libraries. In: Ioannides M, Addison A, Georgopoulos A, Kalisperis L, editors, VSMM 2008. Digital Heritage. Proceedings of the 14th International Conference on Virtual Systems and Multimedia. Limassol, Ciprus: VSMM / CIPA / ICOMOS (2008). P. 125-130.

[6]   Franco JA, Tarrio S (2008) Monasterios y Conventos de Galicia. La Coruña: Xunta de Galicia.

[7]   Bitelli G, Cremonini S, Gatta G (2009) Ancient map comparisons and georeferencing techniques: a case study from the Po River Delta (Italy). e-Perimetron 4 (4): 221-233. Available: http://www.e-perimetron.org/Vol_4_4/Vol4_4.htm. Accessed: 2012 Jan 20.

[8]   Chías P, Abad T (2011) The Bay of Cadiz: Fortified territory and landscape. In: Chías P, Abad T, editors, The Fortified Heritage. Cadiz and the Caribbean: A Transatlantic Relationship. Alcalá: University of Alcalá Press (2011). P. 26-171.

[9]   Chías P, Abad T (2010) The nautical charts of the Spanish Mediterranean coasts in the 18th and 19th centuries: digital methods to compare the cartographical techniques of the main European Navies. e-Perimetron 5 (2): 49-57. Available: http://www.e-perimetron.org/Vol_5_2/Vol5_2.htm. Accessed: 2012 Jan 20.

[10] Heuvel C van den (2006) Modelling historical evidence in digital maps: A preliminary sketch. e-Perimetron 1 (2): 113-126. Available: http://www.e-perimetron.org/Vol_1_2/Vol1_2.htm. Accessed: 2012 Jan 20.

[11] Benavides J, Koster E (2006) Identifying surviving landmarks on historical maps. e-Perimetron 1 (3): 194-208. Available: http://www.e-perimetron.org/Vol_1_3/Vol1_3.htm. Accessed: 2012 Jan 20.

[12] Fernández JJ, Morillo F, San José JI (2004) Las ruinas de Dios. Valladolid: Ed. Universidad de Valladolid.

[13] García A, Llopis J, Torres A, Villaplana R, Saiz B (2005) La arquitectura tradicional de Cartagena. El color del Mediterráneo. Valencia: Ayuntamiento de Cartagena, Universidad Politécnica de Valencia.

[14] Jobst M (2006) Hybrid considerations on the sustainability of cartographic heritage. e-Perimetron 1 (2): 127-137. Available: http://www.e-perimetron.org/Vol_1_2/Vol1_2.htm. Accessed: 2012 Jan 20.

[15] Jessop M (2006) Promoting cartographic heritage via digital resources on the web. e-Perimetron 1 (3): 246-252. Available: http://www.e-perimetron.org/Vol_1_3/Vol1_3.htm. Accessed: 2012 Jan 20.

Chapter 13

Imaging the Past:
Cartography and Multicultural Realities
of Croatian Borderlands

Borna Fuerst-Bjeliš

Additional information is available at the end of the chapter

1. Introduction

Maps have long been central to geographical inquiry. The most usual approach to maps and cartography until recently dealt with its role in presenting a factual statement about geographical reality within the frames of actual survey techniques and skills of a cartographer. Recent researches since the end of 20th century tend to subvert the traditional, positivist model in analyzing the maps, replacing it with one that is grounded in iconological and semiotic theory of the nature of maps. According to J.B. Harley, one can understand a map as a social construction of the world expressed through a medium of cartography, or as a socially constructed image of reality.

Maps always represent much more than merely physical nature and inventory of space. Maps understood and considered as social construction of reality have a number of layers, including the symbolic one. They are conveyors of meanings, messages and perceptions of the world – and not only of an individual cartographer, but also of common societal and cultural values. They reveal what may be called the spirit of time: philosophical, political, religious and general socio-cultural context.

As images, maps should be put and studied in the appropriate context, i.e. period and place. Moreover, maps as images are never neutral or value-free; they are all social, political and cultural. Understood as images, maps can be used on one hand as a tool of disseminating messages, and, on the other hand as a source in analyzing the perceptions of past places, territories and societies.

Researching past images through maps is of particular interest in multicultural spaces, where a variety of different cultures, religious systems, complex ethnic structures and imperial systems have met. Borderlands are typical spaces where a multiplicity of such contacts reflect and produce a multiplicity of perceptions and images.

Early modern period in Croatian history is burdened with frequent changes of borders between three imperial systems with different religious systems and cultural traditions that have intertwined on the Croatian territory, and consequently reflect different attitudes toward borderlands. Accordingly, a map could and often did represent an image with multiple layers of meaning and perceptions. What one can put into relation here is Habsburg and Venetian cartography. Through a number of examples of the Croatian borderlands, the main aim is to reveal the symbolic layer of the map that leads us into the process of imaging the past, i.e. opening the abundance of different perceptions in the multicultural realities of the Croatian borderlands.

Through an analysis of the symbolic layer through graphic elements, place-names and other inscriptions, maps of Croatian borderlands have revealed two distinct levels of meaning. The first one is related to the specific relation of the state authorities to the border region, their particular interests and understanding of its importance. Maps have been used as a tool for disseminating the political message of power and control primarily through methods and techniques of emphasizing (over-exaggerating) or ignoring and omitting. At this particular level of meaning, we are dealing with directly opposing images of the borderlands realities, depending on the political sides and their official cartographies.

At the second level of the meaning maps have revealed the most common socio-cultural images of the borderlands that are, unlike cartographic expressions of different state power interests, expressed equally in all European cartographic traditions. These images include: environmental perceptions of the borderlands as depopulated and devastated area; distinction of social groups, related systems of beliefs, territorialization and de-territorialization of borderland communities; perception and formation of regional identities; and comprehension of the temporality of the border and the continuity of Croatian territory.

2. Theoretical and methodological frame

2.1. Image-reality dualism

The approach to imaging the past through a medium of cartography links two key concepts, such as image and map. Imagery is a subject of enquiry in fields as diverse as cognitive science, literature (imagology), human geography or cartography. These concepts, on different sides in the image-reality dualistic model in most modern writings, are being rethought and are actually converging only in recent postmodern works. Image–reality dualism opposed subjective and objective spaces, unreal and real geographies, mental images and cartographic representations. Reality was thus related to objective geographic fact, represented by a map, while images were considered as "false understanding", or a "coherent, logical, rule-governed system of errors" [1, 2]. Phillips [2] is questioning image-reality dualism arguing that the "general characterization of images as unreality is contradicted by a tendency to privilege certain types of images as reality". Maps as geographic representations have been commonly accepted as realistic, although constructed according to the conventions of artificial perspective [3]. Geographic faith in maps has been made possible largely by the development of techniques of scientific cartography and the maps "conquered the world of representation under the banner of reason, science and objectivity" [1]. However, geographic "reality" is not a nonimage, as argued by Phillips [2]. "Reality" is humanly constructed and merely conventional, and the "truth" is constructed, theoretically and politically committed. At this point we start to question the "unquestionable scientific objectivity" of the cartographic representation of the world and to question the map as a "mirror of reality". Recent researches witness these developments as "epistemic break between a model of cartography as a communication system, and one in which it is seen in a field of power relations, between maps as presentation of stable, known information and mapping… in which knowledge is constructed" [4,5].

On the other hand, since the 70s, the subjectivity and "*naïveté*" of images have been questioned by iconographers and iconologists as well [2]. They have shown that images can be read as explicitly social and political texts and not just as mental representations. Iconography defines images as a sign system and locates them at the social level [6].

2.2. Deconstruction as a methodological strategy

Eventually the two concepts begun to merge particularly in Harley's understanding of maps as socially constructed images. Although some scholars anticipated main ideas earlier, for instance in well known Korzybski's statement that "the map is not the territory it represents" [4] or that "every map is… a reflection partly of objective realities and partly of subjective elements" [7]. Harley formulated a broad strategy for understanding how maps redescribe the world, like any other document, in terms of relations of power and of cultural practices, preferences and priorities [8]. "…Maps are at least as much an image of the social order as they are measurements of a phenomenal world of objects" [9]. He derived basic ideas from writings of Michel Foucault about the "omnipresence of power in all knowledge even it is invisible or implied", including the particular knowledge encoded in maps, as well as Jacques Derrida's work on the rhetoricity of all texts. The concept of "text" does not imply the presence of linguistic elements, but the act of construction, so that maps, as "construction employing a conventional sign system become texts. By accepting the textuality of maps we are able to embrace a number of different interpretative possibilities [9].

In his seminal work on deconstructing the map Harley [9] argues that deconstruction as discourse analysis, demands a closer and deeper reading of the cartographic text and may be regarded as a search for alternative meaning. It means reading between the lines of the map – "in the margins of the text" and a search for metaphor and rhetoric in the textuality of the map [9]. Deconstruction is, as Harley sees it, a broad strategy, more than a precise method or set of techniques.

However, there are some important presumptions, or contexts in the research agenda of map deconstruction. Harley articulated the importance of context around three issues [8]. The first one is the context of the cartographer, including the appreciation of personal views, attitudes and skills, including the local knowledge that is related to the internal power of map. This context is related to the general statement that maps are, like art, a particular human way of looking at the world. Second context is a context of other maps that ensure and emphasize the importance of multiple maps, perspectives and polysemy. Third context is the context of society and points out to the importance of positioning the map within societal-power relations, i.e. within specific historical, social and political conditions, from which it cannot be extracted or generalized [2,10]. The contexts of other maps and society (or societal environment) are directly connected and linked to the external power of the map. It can be seen through maps made by different actors that are reflecting different or even opposed approaches to the territory that were embedded in the society and culture of the particular period and place [11].

Based on the iconographic studies by E. Panofsky [6], Harley has defined a number of semantic layers of the map. The symbolic one often has ideological connotations. It refers to power relationships, distinction of social groups and system of beliefs, to worldviews and to what may be called as a spirit of time.

Scholars in Croatia have recently addressed these topics from various perspectives. The first writings embrace the topics of cartographic perceptions and the state power interests of the multiple borderlands of Croatia [11,12], different perceptions of Croatian lands in Croatian and other European cartographic traditions [13], toponymy and perceptions [14,15], the relation of cartography, place-names and regional identities [16], the political rhetoric of maps [17] and recently the cartographic visualization and the image of Other [18].

Imaging the past of the multicultural space of the early modern Croatian borderlands was based primarily on deconstructing the maps of the time, as the main research and methodological approach. Reading between the lines, in the margins of the textuality of map, searching for metaphors, evaluating the presence or absence (silencing) of information; in short - tracing the rhetoric of map and its symbolic meaning and/or political messages. Key elements of analysis were place-names and smaller cartographic transcriptions and objections as they are as much related to an invisible social world and to ideology as they are to the material world that can be seen and measured. All the contexts were appreciated, especially the importance of the multiplicity of perspectives.

The analysis is based on the cartographic originals of the time from the map collections of the Croatian State Archives, The National and University Library and the Museum of Croatian History, as well as on the numerous published facsimiles [19, 20, 21]. A selection of maps and a comparative approach enable an insight into the different cartographic representations and images of the borderlands within different traditions and even within the framework of a single, overarching tradition.

3. Spatial, temporal and cultural context

In the course of three centuries (16th – 19th), the territory of Early Modern Croatia was determined by the borderlands of three imperial systems of the time: Habsburg Monarchy,

Ottoman Empire and Venetian Republic (Figure1). Borders were, consequently, significantly influential in political, social, cultural, and demographic sense.

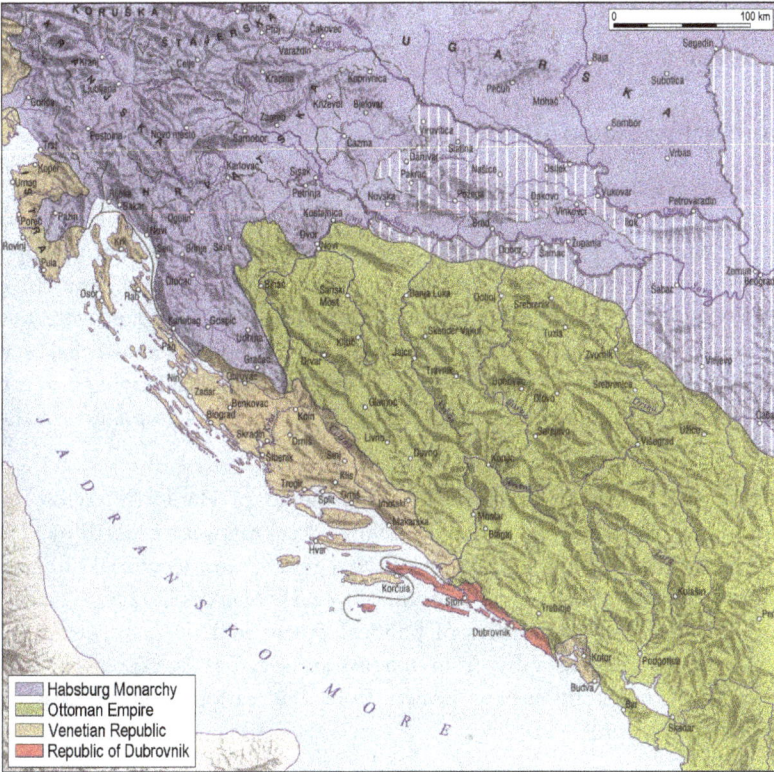

Figure 1. Croatia and triple border, 18th century, [22]

In the history of mapping the Croatian territory, the Early Modern Period was directly connected with military operations i.e. the process of Ottoman retreat. This is the period when cartography developed into so-called "military cartography", practiced in military institutions. Thus, military engineers were mainly the creators of new maps. However, the cartographers were rarely independent decision makers, free of financial, military or political constraints. The context of the cartographer, as Harley has pointed out [8], also included personal skills and the cartographer as a person living in a particular society and in particular political circumstances. Accordingly, map could and often did represent an image with multiple layers of meanings and perceptions and, but also, emphasized features of strategic importance of the state or empire, i.e. exercising the external and internal power of cartography [9].

Triple border conditioned a true multicultural surrounding. Croatian territory was a "meeting point" of Western and Eastern world, Christianity and Islam as well as maritime and continental traditions. Frequent changes of borderlines were followed by population shifts and migration, introduction of new (other) social and cultural groups, as well as leading to mixed cultural, religious, ethnic groups and lifestyles in borderlands. Appreciation of these differences, sense of uniqueness and perception of otherness, through the territorialization, conditioned the creation of spatial images and eventually resulted in regional identity.

4. Disseminating the political message of power and control

Maps are part of a general discourse of power [5]. Throughout history, as much as other weapons, maps have been an intellectual weapon of imperialism and of territorial pretensions of empires and states. In this imperial context, maps regularly supported the direct execution of territorial power. The specific functions of maps in the exercise of power range from global empire building to the preservation of the nation state and to the assertion of local property rights. Maps with their hidden agendas and texts beyond the text, speak political language. As George Orwell said that the "political language is…. designed to make lies sound truthful…" (See [5]), scholars consider maps as politicized documents with the ethical concerns [17,23].

In his most influential work on deconstructing the map, Harley [9] wrote about the power/knowledge matrix and stated that especially where maps are ordered by government it can be seen how they extend and reinforce the legal statutes, territorial imperatives and values stemming from the exercise of political power. He also distinguished *external* power when maps are linked to the centers of political power and when power is exerted *on* cartography and *with* cartography from *internal* power that is expressed through the political effects of maps in society drawn from the cartographic process (selection, generalization, abstraction).

Following the selection of Venetian and Habsburg maps as well as map by Croatian cartographer Vitezović, representing the borderlands is a very good example of these relations. Venetian cartographic policy was primarily subordinated to the Republic's political and administrative purposes. They have, generally, more information about political or administrative divisions and contain much less of geographical inventory. This is an example of direct dissemination of the political message of power and control over the territory. Many of the conventional tools in map making were used in doing this; such as deliberate or "unconscious" (but ideological) distortions and omissions (the "silence", see [9]) on the maps.

4.1. Opposing images as messages

The examples of Coronelli's map of Dalmatia of 1700 [19] and Alberghetti's map of Dalmatia of 1732 [20] enable us to distinguish two different stages for approaching the Venetian borderlands. Coronelli's map was still based mainly on the compilations, while Albeghetti's is already based on field survey. However, apart from technical differences these maps express the political message in the corresponding way. Coronelli, as the official Venetian cartographer, was the most prominent figure in promoting Venetian politics regarding the territorial pretensions on his maps. They were an important instrument for emphasizing the Venetian conquest over the Ottomans. Coronelli's map of Dalmatia is a general regional map on a rather small scale. The map charts Venetian Dalmatia, the territory of the Republic of Dubrovnik, parts of Croatia, Bosnia, Serbia and surrounding lands (Figure 2).

Figure 2. Coronelli's Map of the Kingdom of Dalmatia, **La Morlaquie**, Bosnia and Serbia ..., 1700, facsimile in [19]. Emphasized by the author.

The whole inland area between the river Sava and the Adriatic is compressed along its north-south axis. But, on the other hand, the territory of Venetian Dalmatia is unproportionally vast, especially the inland part. These "distortions" are a testimony to the expression of state power interests and an approach to the border area; emphasizing and over-exaggerating its possessions and importance, while ignoring the Ottoman side at the same time [11]. The rhetoric of the map is accompanied with the well known tool of expressing the Venetian possession of the Adriatic aquatorial space as *"Mer ou Golfe de Venise"* and in addition, introducing the *"Mer et Isles de la Dalmatie"*, that as a hydronym does not really exist apart from the political context of exercising power and control over Venetian Dalmatia.

The beginning of the 18th century was a time of relatively numerous changes of the border in Dalmatia and a time of intensive cartographic work. Demarcation maps from that time represent the very first topographic presentation of the Dalmatian hinterland. Alberghetti's supplemented map of Dalmatia from 1732 presents three borderlines with the Ottoman Empire; the old one from 1671 (Linea Nani), the one from 1700 (Linea Grimani) and the newest one from 1720 and 1721 (Linea Mocenigo). The map contains the administrative division of the territory that most Venetian maps have. The topography is very detailed except orography and communication. Beyond the border, there is no presentation whatsoever, except for some very general textual notifications of what may be found: "*Parte della Licca*", "*Parte della Bossina*", "*Ercegouina*". Thus, the central element of the map is the development of the Venetian – Ottoman border in terms of territorial extension of Venetian republic, disseminating the message of the Republic's power and control over the territory. An obvious ignorance of Ottoman presence (and even existence) is shown through the omission of recording their territory across the border [11]. In such a context the omissions are as important as emphasizing (the "silence").

These examples of decorative (Coronelli) and "scientific" cartography (Alberghetti), in spite of differences in a technical sense, have some unifying and constant elements that characterize the Venetian cartography of the borderlands. That is a strong rhetoric of power and control in promotion and giving legitimacy to the territorial occupation.

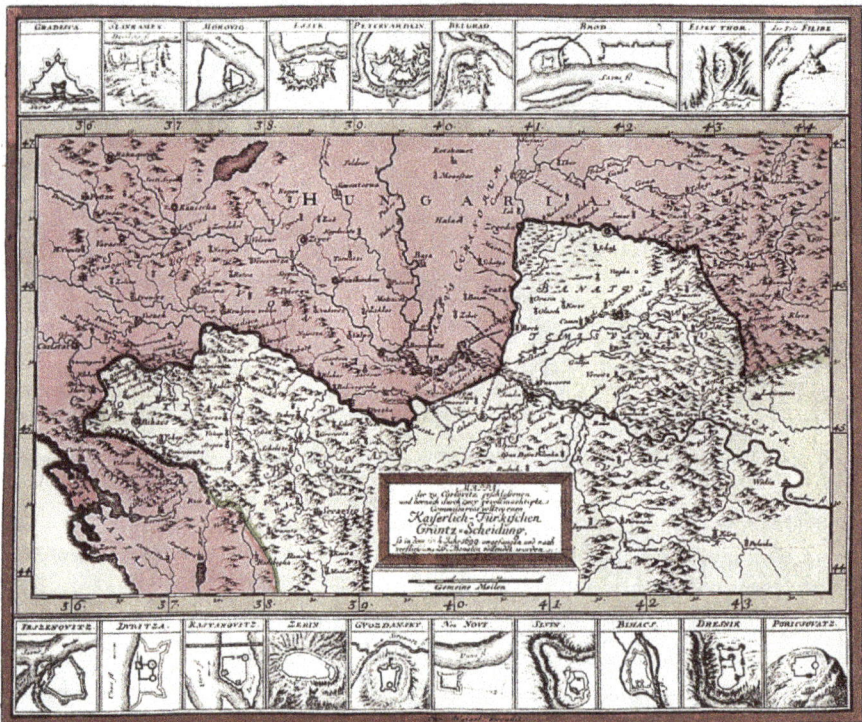

Figure 3. Weigl's Map of the Imperial – Turkish border, 1702; facsimile in [20]

Habsburg cartography was guided completely by military and strategic interests and needs. Highly aware of the extreme importance of knowing the border area in the strategy of warfare, Habsburg cartographers had already surveyed and mapped the territory long before the 18th century. The occasion of the peace treaty of 1699 and the need to fix the new border between the Habsburg Monarchy and the Ottoman Empire was the direct cause of the first topographical survey along the border.

The central interest in the new border is shown on Christoph Weigl's Map of the Imperial – Turkish border after the peace treaty, presumably around 1702 (Figure 3) [20,24]. The main theme of the map is the border, which is the most expressive element on the entire map as well as the territory of Habsburg Monarchy, which is the only colored element. The color has always been a very strong tool. Different colors send different messages to the audience. Strong, cardinal colors like red color for instance, as employed on Weigl's map, were always the imperial colors etc. On the other side, we are discovering silencing: there is not much content outside the borderline, but again a rich inventory of military fortifications on the margins of the map, communicating the undisputable power, security and the organization of the Monarchy.

4.2. Communicating the political program

There is an example of more specific approach to the territory and borders of Croatia, exercising the internal power of map aiming to formulate a political program. Pavao Ritter Vitezović, Croatian cartographer and representative in the demarcation commission in the occasion of Peace Treaty of Srijemski Karlovci (1699) between Habsburg Monarchy and Ottoman Empire, dissatisfied with the newly established borders, tried to present to the Austrian court his view of the "real historical" borders and territory of Croatia [25].The Map of the whole Kingdom of Croatia (Figure 4) represents the entire kingdom of Croatia in its ancient, historical limits as confirmed by the king Ludovic in the 16th century. Along with the 1699 demarcation, he drew in the former borders. The eastern border follows the line of the Vrbas river in contrast to the actual one on the river Una, lying more westerly and thus, compressing the Croatian territory. The map was followed by the document "Croatia Rediviva" (1700) [26] altogether aiming to produce and communicate the knowledge of considerably larger territory of "historical" Croatia. Vitezović's map is, among other contexts which will be discussed later, exercising the internal power of cartography in power/knowledge matrix and communicating political ideas and program to the targeted audience.

All these examples clearly show the external and internal power of map that is not necessarily separated. One particular map can express both – the external power which is imposed from above, especially when cartography became nationalized, but also the internal power, exercised by cartographers themselves, that is related to the context of cartographer.

Figure 4. Vitezović's Map of the whole Kingdom of Croatia, 1699, facsimile in [19] *"Terra deserta olim nunc a Valachis habitata"* and *"Terrae desertae"* along the borderlands. Emphasized by the author.

Triple border and "meeting point" of different political systems on the territory of Croatia conditioned different imaging of the borderland through a medium of cartography. These images are politically informed and valued and often directly opposed, giving legitimacy, importance and power to one side and ignoring and silencing the other. Thus, these examples show that the rhetoric of map include also the concept of otherness.

5. Socio-cultural images

Opposed to images discussed above that reflect different attitudes of different imperial forces and that can be easily recognized through corresponding official cartographic traditions, there is another level of meaning that reveal common socio-cultural images to all European cartographies, regardless of political affiliation. These images reflect social recognition and territorialization through the distinction of social otherness and, on the other hand, perceptions of territorial continuity in the circumstances of border fluctuation, through the distinction of territorial otherness. There is a number of related concepts that are embedded in maps and leading eventually to the creation of regional concept and identity. That is appreciation of differences, uniqueness and otherness that, through the territorialization, result in specific spatial images and regional identity [11,16].

5.1. Morlacchia: otherness, territorialization and regional concept

The first image of the triple border area is related to the recognition of Morlacchi, a distinct social group as Other. Their presence in the borderlands is a consequence of the population shifts due to the warfare and border fluctuations. Autochthonous sedentary population abandoned land and migrated towards more secure areas, while a large portion of the Croatian borderlands became a destination of new semi-nomadic pastoral communities from the Dinaric mountain hinterland. These borderlands communities are generally called *Vlachs* or *Morlacchi* in the Venetian tradition. Morlacchi communities partly immigrated to the borderlands area spontaneously, combining the pastoral economy with military service, while they were partly colonized and settled by the official politics of Venice and Habsburgs.

The toponyms *Morlaccha* or *Morlacchia* with a number of some other corresponding forms, such as *Morlacca, Morlacha, Murlacha and Morlakia* can be found on the maps as early as the 16th century (Figure 5).

Originating from the Venetian term for social community, the derived toponymic forms became a common name for the border region for more than three hundred years in circumstances where three imperial forces met. Throughout the course of centuries, the term Morlacchi has been related to the territory they have settled. The term gradually has got the spatial connotation [14,16].

Territorialization is seen as a reflection of perceived otherness of Morlacchi community; primarily through different social organization, lifestyle and customs in relation to the prevailing population. Perception of otherness and uniqueness is the basis of regionality and regional identity that is leading to the construction of the regional concept of Morlacchia. The image of otherness is very well expressed on Vitezović's map (Figure 4). What we can read "between the lines" of the notification along the border: "*Terra deserta olim nunc a Valachis habitata*" (deserted, depopulated and uninhabited land, yet inhabited by Vlachs!) is that Vlachs are considered as Others in terms of social and religious differentiation [11,16].

It has been clearly proved [14] that the regional concept of Morlacchia is found to be common to all European cartographies, even if the term for the social group is Vlach (Habsburg tradition). Morlacchia was an important regional concept if looking at the significance given by the typography (see Figure 2, 5). On Coronelli's map for instance, *La Morlaquie* is listed in the title of the map along with Bosnia, Serbia, Hungary and Croatia.

With the disappearance of triple border conditions by the end of the wars with Ottoman Empire and the fall of Venetian Republic in 19th century, the context of significance within which the Morlacchi community have been evaluated throughout the centuries, was dissolved.

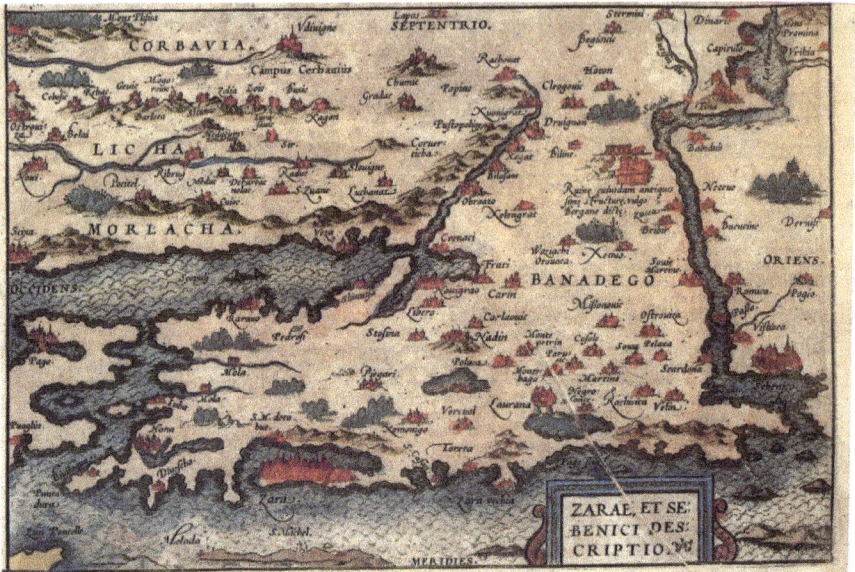

Figure 5. Bonifačić's Map of the surroundings of Zadar and Šibenik with the region of Morlacha, 1573, facsimile in [19]

This change is clearly recognizable in the disappearance of toponyms associated with Morlacchi. Constructed in the multicultural border circumstances of the 16th – 19th centuries, they disappear from the maps with the change of circumstances that created them. Following the change in the rhetoric of the maps, we can read about the territorialization as well as about the de-territorialization of borderland communities.

5.2. Environmental image of the borderlands

Coming back to the aforesaid statement *"Terra deserta olim nunc a Valachis habitata"* by Vitezović we are about to open a new question of environmental image of the borderlands. Now, the other part of the statement shall be emphasized: *"Terra deserta"*. Vitezović has repeatedly put the notification along the border, constructing the image of Croatian borderlands as deserted and devastated (see Figure 4). Apart from Vitezović's *terrae desertae,* one can find other examples of environmental rhetoric imaging the devastated borderlands on Coronelli's map of Istria and northern Dalmatia (1688) [21] through the records of destroyed and abandoned cities and fortresses: *"Starigrad Citta distructa"* or *"Carlobago distructa"* etc. Environmental image of borderlands as deserted, devastated area appears as common to all regardless their imperial or cartographic background.

5.3. Turkish Croatia: Territorial continuity, otherness and regional concept

The issue of old and new border, already discussed earlier in the context of communicating the political message, will be considered again, but in other contexts of a socio-cultural image, distinction of territorial otherness and regional identity.

Although Croatia regained a large parts of its territories by the peace treaty of Srijemski Karlovci (1699), it failed to get back some of its historical lands. That was, primarily, the area between the Una and Vrbas rivers – the area between the new and the old border. A number of cartographers, along with the new border drew in the old one as well. Some of the examples are already discussed Vitezović's Map of the whole Kingdom of Croatia (1699) (see Figure 4) and Müller's Map of Hungary (1709) [19]. There are cases where the inscription, either a general one like *"Croatia"*, or a more specific one like *"Turkish Croatia"* cover the interfluves territory that is beyond the new actual border, but integral part of the historical Croatian territory. On Weigl's Map (see Figure 3) for instance, the inscription *"Croatia"*, regardless of the actual borderline, is written more easterly across the river Una, over the Ottoman territory. Coronelli (1732) [19], although of different imperial and cartographic affiliation, used the color, the line and the text to differ the interfluves from the rest of Bosnia (Ottoman territory) as well as from the rest of Croatia, lying under the inscription of *"Croacie"*. The example of Schimek's map of 1788 (Figure 6), representing Viennese cartography, also shows the clear distinction of *"Turkisch Croatien"*.

There is quite a number of maps of different political backgrounds and cartographic traditions that are equally sharing the same image of Turkish Croatia, i.e. J. von Reilly's map (1790), map edited by Artaria and Comp. (1807), J. Szeman's map (1826), E. Zuchery's map (1848), Halavanja's map (1851) [19, 20].

"Reading between the lines" and searching for metaphors will lead us to the perception of temporality of border fluctuation in these centuries of their frequent changes. Consequently, the territory of Turkish Croatia reveals an image of the interfluves as integral Croatian territory in spite of the newly established border. This is an image of the new borderline as a temporary condition in relation to the "real historical" border. The image include the awareness of a temporality of the borders and understanding and appreciation of the continuity of Croatian territoriality (Turkish **CROATIA;** emphasized by the author). At the other hand, the image reveal the distinction of territorial otherness (**TURKISH** Croatia, emphasized by the author) that is grounded in the distinction of "Turkish"/Muslim as Other and the distinction of Christian Croatia versus Muslim Croatia. Thus, the image is pointing out to the awareness of different religious identities of the twofold region.

The example of Turkish Croatia opens two levels of reading: old and new border as real historical border versus temporary border; distinction of different religious and cultural identities, Christian versus Muslim Croatia. The consciousness of the otherness and uniqueness as related to the territoriality is leading to the creation of regional identity. These elements are formative elements of regional identity and the regional concept in both examples: in Morlacchia as well as in Turkish Croatia [11,16].

5.4. Reflections

Still, the development and reflection of these regional concepts are different. Turkish Croatia has undergone the process of conceptual translation. By the mid 19th century it has changed

Figure 6. Schimek's map of Turkish Croatia, 1788, facsimile in [20]

the name into *Bosanska Krajina*[1]. While the old name of Turkish Croatia emphasized the Croatian territoriality of different religious and cultural identity, the name of Bosanska Krajina is emphasizing the border character of the territory. Turkish Croatia / Bosanska Krajina retained its borderland character even later through the participation in the organization of Military Border that additionally sustained the image of otherness in terms of a particular military mentality, apart from multiculturalism. Still, Bosanska Krajina, as a regional concept, has preserved territorial coverage with an image of otherness and uniqueness in the multicultural and multiethnic sense. There has been a change in spatial image that conditioned the change, but also the preservation of regional identity and concept.

Morlacchia, on the other hand, has gone throughout its dissolution. The change of the multiethnic and multicultural triple border circumstances by the 19th century as a consequence of the end of the wars with Ottoman Empire and the fall of Venetian Republic, has led to the dissolution of context of political significance in warfare conditions, within which the Morlacchi communities have been evaluated throughout the centuries. Additionally, administrative measures brought by Habsburg government (the destruction of goat herding with the aim of forestation, confiscation of weapons, which exposed Morlacchi districts to devastation by wolves) and disorientation in peaceful conditions led to impoverishment and transformation of Morlacchi community [27]. Morlacchi descendants in the hinterland of Dalmatian cities, either Orthodox or Catholic faith, have gradually been merged with Croats and the Serbs, mostly by religious affiliation. The change in the multicultural architecture of the space and community as well as disappearance of triple border have led to the change of the spatial image and in this particular case conditioned the disappearance of the regional identity and concept.

All these examples of socio-cultural images and spatial and regional concepts are common to all relevant European cartographies, regardless of different (and often opposed) political affiliations, interests and attitudes towards the borderlands. They are not imposed from above, from the centers of political power, but reflect an internal and local knowledge and perceptions.

6. Conclusion

The paper discusses the role of cartography in imaging the past, particularly taking into consideration the multiculturalism of borderlands. The starting points are two concepts, image and map i. e. the understanding of map as a socially constructed image with a number of semantic layers which reflect power relationships, distinction of social groups and system of beliefs, worldviews and what may be called a spirit of time. Borderlands are typical spaces where a multiplicity of contacts reflect and produce a multiplicity of perceptions and images.

---

[1] *Krajina* has a meaning of borderlands.

Map deconstruction was employed as a basic research strategy in Harleian terms, signifying a search for alternative meaning, metaphor and rhetoric in the textuality of the map. Key elements of analysis were place-names and smaller cartographic transcriptions and objections as they are as much related to an invisible social world and to ideology as they are to the material world that can be seen and measured.

The selection of early modern maps of different European cartographic traditions has revealed two levels of meaning within the symbolic layer. The first one reflects different and opposed images of different cartographic traditions. These images are politically informed and valued giving legitimacy, importance and power to one side and ignoring and silencing the other, i.e. disseminating the political message of power and control and communicating the political program.

Contrasted to images that reflect different attitudes of different imperial forces and that can be easily recognized through corresponding official cartographic traditions, there is another level of meaning that reveals common socio-cultural images to all European cartographies, regardless of political affiliation. These images reflect social recognition and territorialization through the distinction of social "otherness" and, on the other hand, perceptions of territorial continuity in circumstances of border fluctuation, through the distinction of territorial "otherness".

The consciousness of the otherness and uniqueness as related to territoriality is leading to the creation of regional identity. These elements are formative elements of regional identity and the regional concept in examples discussed: in Morlacchia as well as in Turkish Croatia. These regional concepts, however, have undergone throughout different developments and have different reflections in present time. Morlacchia, as a regional concept, has dissoluted with the change of the multiethnic and multicultural triple border circumstances and the change in spatial image by the 19th century. On the contrary, Turkish Croatia, as a regional concept, has preserved territorial coverage with an image of multiculturalism till present time, but with the stronger accentuation of its borderlands character under the new name of Bosanska Krajina. The preservation of regional concept of Turkish Croatia / Bosanska Krajina is considerably due to the longer persistence of borderlands development even later through the Military Border and linking military and multicultural components of regional identity.

All examples clearly show the external and internal power of map that is not necessarily separated. One particular map can express both – the external power which is imposed from above, especially when cartography became nationalized, but also the internal power, exercised by cartographers themselves, reflecting internal and local knowledge and perceptions.

Author Details

Borna Fuerst-Bjeliš
University of Zagreb, Faculty of Science, Department of Geography, Zagreb, Croatia

7. References

[1] Mitchell W (1986) Iconology, image, text, ideology. Chicago: University of Chicago Press. 236 p.

[2] Phillips R (1993) The language of images in geography. Progress in Human Geography 17, 2: 180-194.

[3] Cosgrove D. E (1998) Social Formation and Symbolic Landscape. Madison, London: The University of Wisconsin Press. 293 p.

[4] Crampton J.W (1996) Bordering on Bosnia. GeoJournal 39: 353-361.

[5] Crampton J.W (2001) Maps as social constructions: power, communication and visualization. Progress in Human Geography 25, 2: 235-252.

[6] Panofsky E (1983) Meaning in the Visual Arts. Chicago: University of Chicago Press. 364 p.

[7] Wright J (1942) Map makers are human: comments on the subjective in maps. The Geographical Review 32: 527-544.

[8] Harley J. B (2001)The New Nature of the Maps, Essays in the History of Cartography. Laxton P editor. Baltimore: The Johns Hopkins University Press. 333 p.

[9] Harley J. B (1989) Deconstructing the Map. Cartographica 26, 2: 1-20.

[10] Black J (2000) Maps and History, Constructing Images of the Past. New Haven, London: Yale University Press. 267 p.

[11] Fuerst-Bjeliš B, Zupanc I (2007) Images of the Croatian Borderlands: Selected Examples of Early Modern Cartography. Hrvatski Geografski Glasnik 69, 1: 5-21.

[12] Fuerst-Bjeliš B (2000) Cartographic Perceptions of the Triplex Confinium and State Power Interests at the beginning of the 18th Century. In Roksandić D, Štefanec N, editors. Constructing Border Societies on the Triplex Confinium. Budapest. pp.215-220.

[13] Mlinarić D (2002) Različite percepcije ranonovovjekovnog prostora hrvatskih zemalja u domaćih i stranih kartografa. In Mežnarić S, editor. Etničnost i stabilnost Europe u 21. stoljeću. Zagreb: IMIN - Naklada Jesenski i Turk - Hrvatsko sociološko društvo. pp. 131-142.

[14] Fuerst-Bjeliš B (1999-2000) Toponimija i percepcija u prostoru Triplex Confiniuma: Morlakija. Radovi 32-33: 349-354.

[15] Faričić J (2007) Geographical Names on 16th and 17th Century Maps of Croatia. Cartography and Geoinformation 6, special issue: 148-179.

[16] Fuerst-Bjeliš B (2011) Slike i mijene regionalnoga identiteta - geografska imena na kartama ranoga novoga vijeka; odabrani primjeri. In Skračić V, Faričić J, editors. Zbornik radova s Prvoga nacionalnog znanstvenog savjetovanja o geografskim imenima. Zadar: 67-72.

[17] Slukan-Altić M (2005) Kartografski izvori između povijesti i politike ili kako lagati kartama. In Lipovčan S, Dobrovšak L, editors. Hrvatska historiografija XX. stoljeća: između znanstvenih paradigmi i ideoloških zahtjeva. Zagreb: Institut društvenih znanosti Ivo Pilar. pp. 313-334.

[18] Mlinarić D (2011) Kartografska vizualizacija i slika Drugog na primjeru višestruko graničnih prostora. Migracijske i etničke teme 27, 3: 345-373.

[19] Marković M (1993) Descriptio Croatiae. Zagreb: Naprijed. 372 p.

[20] Marković M (1998) Descriptio Bosnae et Hercegovinae. Zagreb: AGM. 496 p.

[21] Kozličić M (1995) Kartografski spomenici Hrvatskog Jadrana. Zagreb: AGM. 391p.

[22] Regan K, editor. (2003) Hrvatski povijesni atlas. Zagreb: Hrvatski leksikografski zavod "Miroslav Krleža". 386 p.

[23] Monmonier M (1996) How to Lie With Maps. Chicago, London: The University of Chicago Press. 207 p.

[24] Pandžić A (1988) Pet stoljeća zemljopisnih karata Hrvatske, Izložba Povijesnog muzeja Hrvatske. Zagreb: Povijesni muzej Hrvatske. 145 p.

[25] Marković M (1987) Prilog poznavanju djela objavljenih u zagrebačkoj tiskari Pavla Rittera Vitezovića. Starine 60: 71-79.

[26] Perković Z (1995) Croatia Rediviva Pavla Rittera Vitezovića. Senjski zbornik 22: 225-236.

[27] Modrich G (1892) La Dalmatia romana-veneta-moderna. Note e ricordi di viaggio. Torino. 506 p.

Permissions

The contributors of this book come from diverse backgrounds, making this book a truly international effort. This book will bring forth new frontiers with its revolutionizing research information and detailed analysis of the nascent developments around the world.

We would like to thank Prof. Carlos Bateira, for lending his expertise to make the book truly unique. He has played a crucial role in the development of this book. Without his invaluable contribution this book wouldn't have been possible. He has made vital efforts to compile up to date information on the varied aspects of this subject to make this book a valuable addition to the collection of many professionals and students.

This book was conceptualized with the vision of imparting up-to-date information and advanced data in this field. To ensure the same, a matchless editorial board was set up. Every individual on the board went through rigorous rounds of assessment to prove their worth. After which they invested a large part of their time researching and compiling the most relevant data for our readers. Conferences and sessions were held from time to time between the editorial board and the contributing authors to present the data in the most comprehensible form. The editorial team has worked tirelessly to provide valuable and valid information to help people across the globe.

Every chapter published in this book has been scrutinized by our experts. Their significance has been extensively debated. The topics covered herein carry significant findings which will fuel the growth of the discipline. They may even be implemented as practical applications or may be referred to as a beginning point for another development. Chapters in this book were first published by InTech; hereby published with permission under the Creative Commons Attribution License or equivalent.

The editorial board has been involved in producing this book since its inception. They have spent rigorous hours researching and exploring the diverse topics which have resulted in the successful publishing of this book. They have passed on their knowledge of decades through this book. To expedite this challenging task, the publisher supported the team at every step. A small team of assistant editors was also appointed to further simplify the editing procedure and attain best results for the readers.

Our editorial team has been hand-picked from every corner of the world. Their multi-ethnicity adds dynamic inputs to the discussions which result in innovative outcomes. These outcomes are then further discussed with the researchers and contributors who give their valuable feedback and opinion regarding the same. The feedback is then collaborated with the researches and they are edited in a comprehensive manner to aid the understanding of the subject.

Apart from the editorial board, the designing team has also invested a significant amount of their time in understanding the subject and creating the most relevant covers. They scrutinized every image to scout for the most suitable representation of the subject and create an appropriate cover for the book.

The publishing team has been involved in this book since its early stages. They were actively engaged in every process, be it collecting the data, connecting with the contributors or procuring relevant information. The team has been an ardent support to the editorial, designing and production team. Their endless efforts to recruit the best for this project, has resulted in the accomplishment of this book. They are a veteran in the field of academics and their pool of knowledge is as vast as their experience in printing. Their expertise and guidance has proved useful at every step. Their uncompromising quality standards have made this book an exceptional effort. Their encouragement from time to time has been an inspiration for everyone.

The publisher and the editorial board hope that this book will prove to be a valuable piece of knowledge for researchers, students, practitioners and scholars across the globe.

List of Contributors

Shao-Feng Bian
Department of Navigation, Naval University of Engineering, Wuhan, China

Hou-Pu Li
Department of Navigation, Naval University of Engineering, Wuhan, China
Key Laboratory of Surveying and Mapping Technology on Island and Reef, State Bureau of Surveying, Mapping and Geoinformation, Qingdao, China

Gabriele Bitelli and Giorgia Gatta
Department of Civil, Environmental and Materials Engineering (DICAM), Alma Mater Studiorum - University of Bologna, Bologna, Italy

Stefano Cremonini
Department of Earth and Geological-Environmental Sciences, Alma Mater Studiorum - University of Bologna, Bologna, Italy

Ricardo García, Juan Pablo de Castro, Elena Verdú, María Jesús Verdú and Luisa María Regueras
Department of Signal Theory, Communications and Telematics Engineering, Higher Technical School of Telecommunications Engineering, University of Valladolid, Campus Miguel Delibes, Paseo Belén 15, 47011 Valladolid, Spain

Romanescu Gheorghe
Department of Geography, "Alexandru Ioan Cuza" University of Iasi, Romania

Cotiugă Vasile
Departament of History, "Alexandru Ioan Cuza" University of Iasi, Romania

Asăndulesei Andrei
ARHEOINVEST Platform, "Alexandru Ioan Cuza" University of Iasi, Romania

Stanislav Popelka, Alzbeta Brychtova, Jan Brus and Vít Voženílek
Department of Geoinformatics, Palacký University in Olomouc, Czech Republic

Axente Stoica, Dan Savastru and Marina Tautan
National Institute of R&D for Optoelectronics – INOE 2000, Romania

Krystyna Szykuła
Wrocław University Library Department of the Cartographical Collection (retired), Wrocław, Poland

Carla Bernadete Madureira Cruz and Rafael Silva de Barros
Departamento de Geografia, Instituto de Geociências, Universidade Federal do Rio de Janeiro – UFRJ, Rua Athos da Silveira Ramos, CCMN, Cidade Universitária, CEP 21941-590 – Rio de Janeiro, Brazil

Janvier Fotsing
University of Buea, Faculty of Science/Department of Physics, Cameroon

Emmanuel Tonye
University of Yaounde I, National Advanced School of Engineering, Department of Electrical and Telecommunications Engineering, Cameroon

Bernard Essimbi Zobo
University of Yaounde I, Faculty of Science, Department of Physics, Cameroon

Narcisse Talla Tankam
University of Dschang, Fotso Victor Institute of Technology, Department of Computer Sciences, Cameroon

Jean-Paul Rudant
University of Marne-La-Vallée, Institut Francilien des Géosciences, France

Jorge M. G. P. Isidoro
IMAR – Marine and Environmental Research Centre, Department of Civil Engineering, University of Algarve, Campus da Penha, Faro, Portugal

Helena M. N. P. V. Fernandez and Fernando M. G. Martins
Department of Civil Engineering, University of Algarve, Campus da Penha, Faro, Portugal

João L. M. P. de Lima
IMAR – Marine and Environmental Research Centre, Department of Civil Engineering, University of Coimbra, Rua Luís Reis Santos, Campus II – University of Coimbra, Coimbra, Portugal

N. López-Estébanez and F. Allende
Department of Geography, Autónoma University of Madrid, Francisco Tomás y Valiente, Madrid, Spain

P. Fernández-Sañudo and M.J. Roldán Martín
Environmental Research Centre of Madrid Region, Madrid-Colmenar Viejo, Soto de Viñuelas, Tres Cantos, Spain

P. De Las Heras
Department of Ecology, Complutense University of Madrid, Faculty of Biology, José Antonio Novais s/n, Madrid, Spain

Pilar Chias and Tomas Abad
Technical School of Architecture and Geodesy, University of Alcala, Madrid, Spain

Borna Fuerst-Bjeliš
University of Zagreb, Faculty of Science, Department of Geography, Zagreb, Croatia